T0229695

Multidimensional Chromatography

CHROMATOGRAPHIC SCIENCE

A Series of Monographs

Editor: JACK CAZES
Sanki Laboratories, Inc.
Sharon Hill, Pennsylvania

Volume 52: Modern Thin-Layer Chromatography
    Edited by Nelu Grinberg

Additional Volumes in Preparation

Chromatographic Analysis of Alkaloids
Milan Popl, Jan Fahnrich, and Vlastimil Tatar

Multidimensional Chromatography

Techniques and Applications

edited by

Hernan J. Cortes

The Dow Chemical Company
Midland, Michigan

Marcel Dekker, Inc.   New York and Basel

ISBN 0-8247-8136-8

This book is printed on acid-free paper

MARCEL DEKKER, INC.
270 Madison Avenue, New York, New York 10016

Current printing (last digit):
10 9 8 7 6 5 4 3 2 1

PRINTED IN THE UNITED STATES OF AMERICA

Preface

In recent years, advances in the field of chromatography have provided a wealth of information on the principles, instrumentation, and applications in varied areas of chemistry, medicine, and the environment. The need to analyze samples of increasing complexity with ever lower detection limits has placed more stringent requirements on the separating power necessary for analysis. The potential of chromatographic systems to fulfill this need provides a strong incentive for further development.

Multidimensional chromatography (or coupled column chromatography) is a technique in which fractions from a separation system are selectively transferred to one or more secondary separating systems to enhance resolution and sensitivity, or to decrease analysis times. In the case of thin-layer chromatography, fractions are not transferred to another separation system, but rather, a secondary separation is developed orthogonally on the same plate.

Although multidimensional techniques are extremely powerful and have been under development and use in various forms for many years, in our opinion these techniques still remain underutilized. Therefore, it seemed timely to collect in book form the theoretical aspects and practical details as well as a wide number of applications of multidimensional technology in a variety of fields.

This volume summarizes all the important aspects of multidimensional separations. The information contained is expected to be of value to experienced chromatographers as well as those new to the field who wish to apply these principles. Information regarding the more established multidimensional techniques, such as gas, liquid and thin-layer chromatography, are covered in detail as well as more recent developments such as the coupling of liquid chromatography (LC) to gas chromatography (GC) and the techniques and applications of supercritical fluid chromatography (SFC) in the multidimensional mode. In addition, possible future directions of research in this field are stressed throughout the volume.

Although in the strictest sense supercritical fluid extractions (SFE) cannot be considered part of chromatography, information describing SFE coupled with GC, and SFC is included, as this is expected to be a major future direction of research.

It is our goal that by presenting the information in a collected form, the application of the principles described will see further growth and development in this important area of chemical analysis.

I am grateful to the contributors to this volume, without whose effort this work could not have been completed. I would also like to thank Drs. Mel Koch, Tom McGee, Gary Jewett, Curt Pfeiffer, and Warren Crummett, whose encouragement and support are appreciated. Finally, I am grateful for the understanding and patience of my family.

Hernan J. Cortes

Contents

Contributors

Keith D. Bartle, Ph.D. Senior Lecturer in Physical Chemistry, School of Chemistry, The University of Leeds, Leeds, England

Wolfgang Bertsch, Ph.D. Associate Professor, Department of Chemistry, The University of Alabama, Tuscaloosa, Alabama

Hernan J. Cortes, B.S. Research Leader, Analytical Sciences Laboratory, The Dow Chemical Company, Midland, Michigan

Frank David, Ph.D. Research Institute for Chromatography, Wevelgem, Belgium

Ilona L. Davies, Ph.D.* Chemistry Department, Brigham Young University, Provo, Utah

J. Calvin Giddings, Ph.D. Professor, Department of Chemistry, University of Utah, Salt Lake City, Utah

Ulrich K. Goekeler, Dipl. Ing.[†] Sales Manager, Department of Analytical Systems-Chromatography, Siemens AG, Karlsruhe, Federal Republic of Germany

---

Current affiliations:
*Senior Research Chemist, Dionex Corporation, Sunnyvale, California
[†]Manager, Department of Siemens Gas Chromatography, ES Industries, Voorhees, New Jersey

Toshinobu Hondo, M.Sc.   Applications Chemist, Liquid Chromatography Instrument Department, JASCO, Japan Spectroscopic Co., Ltd., Tokyo, Japan

Milton L. Lee, Ph.D.   Professor, Chemistry Department, Brigham Young University, Provo. Utah

Karin E. Markides, Ph.D.   Associate Professor, Chemistry Department, Brigham Young University, Provo, Utah

Friedhelm Mueller, Dipl. Ing.   General Manager Sales, Marketing, Research and Development, Department of Chromatography, Siemens AG, Karlsruhe, Federal Republic of Germany

Colin F. Poole, Ph.D.   Professor, Department of Chemistry, Wayne State University, Detroit, Michigan

Salwa K. Poole, Ph.D.   Research Associate, Department of Chemistry, Wayne State University, Detroit, Michigan

L. David Rothman, Ph.D.   Research Associate, Analytical Sciences Laboratory, The Dow Chemical Company, Midland, Michigan

Muneo Saito, Ph.D.   Engineering Manager, Liquid Chromatography Instrument Department, JASCO, Japan Spectroscopic Co., Ltd., Tokyo, Japan

Pat Sandra, Ph.D.*   Research Institute for Chromatography, Wevelgem, Belgium

Masaaki Senda, Ph.D.   Deputy General Manager, Liquid Chromatography Instrument Department, JASCO, Japan Spectroscopic Co., Ltd., Tokyo, Japan

Current affiliation:
*Professor, Laboratory of Organic Chemistry, University of Gent, Gent, Belgium

1

Use of Multiple Dimensions in Analytical Separations

J. CALVIN GIDDINGS   *University of Utah, Salt Lake City, Utah*

I. INTRODUCTION

In multidimensional methods of separation, different separation mechanisms or systems are coupled together. For reasons to be explained shortly, and to be amplified more specifically in subsequent chapters, multidimensional techniques are enormously powerful. They are also enormously varied, simply by virtue of the fact that the number of ways of combining different mechanisms is far greater than the number of mechanisms for combination (1). In theory (and increasingly exploited in practice), multidimensional methods can be developed by combining almost any of the different chromatographic mechanisms or phases, electrophoretic techniques, or field-flow fractionation subtechniques. One can go on to cross-couple a chromatographic mechanism with one selected from electrophoresis, field-flow fractionation, simple extraction, and so on. Clearly, the potential variety of resulting multidimensional combinations is almost beyond imagination. By one estimate, it might be possible to develop from 10^4 to 10^6 distinguishable binary combinations based on known separation techniques and variations (1). If the combination of more than two mechanisms is considered, the number of possible techniques becomes, for all practical purposes, unlimited.

This volume, fortunately, is largely limited to a more digestible menu, consisting primarily of combinations only of chromatographic mechanisms and conditions. The exception is this chapter, in which I deal generally with all the multidimensional separation methods, based on chromatography or not. However, I use chromatography as the main source of my examples here. Such is the unity of the field that most general concepts apply to chromatography anyway.

1

This chapter's expansion of the book's focus is not, as one might suspect, an attempt to interject nonchromatographic material into the content; it is instead an effort to show that the chromatographic case is part of a larger whole, and to explain how specific methods relate to one another across the broadest possible spectrum of separation methodology. However, along with this perspective, the material is intended to provide a considerable number of practical guidelines pointing the way to the optimization of multidimensional procedures.

I have indicated that multidimensional methods involve a combination of single mechanisms and systems. However, there are many ways of combining mechanisms; some are no more powerful than operation with a single mechanism (2). Therefore, it is necessary to be more specific on what constitutes multidimensional operation.

To clarify the scope of this chapter, I define a multidimensional separation by two conditions (2). First, it is one in which the components of a mixture are subjected to two or more separation steps (mechanisms) in which their displacements depend on different factors. The second criterion is that when two components are substantially separated in any single step, they always remain separated until the completion of the separative operation. This latter condition, as we shall show later, rules out simple tandem column arrangements in which components separated in the first column may re-merge in the second (2).

As a consequence of the above definition, two or more parameters (for example, successive retention times) are required to describe the final location of a component after a multidimensional separation. For any two components, a difference in only one of the parameters (that is, a separation based on that parameter), therefore, implies a difference in final position and the likelihood of successful separation. The fact that a difference in only one parameter is needed for success is largely responsible for the unique strengths of multidimensional systems.

The requirement for two or more (that is, multiple) retention parameters to describe the final position or emergence point of the component is consistent with the concept of multiple dimensions used throughout science and mathematics. As explained more fully in the last section, an n-dimensional coordinate system is one requiring the identification of n parameters to pinpoint "location." In chromatography, unfortunately, the word "multidimensional" has been used in a most varied and ill-defined way. By confining ourselves to the above definition, we circumscribe a family of powerful techniques whose extraordinary capabilities stem from the simultaneous utilization of the full range (or as large a part as possible) of two or more separation axes. The distinction between multidimensional combinations and those combinations falling outside the multidimensional class will be clarified as we proceed.

Two distinct multidimensional approaches can be carried out within the confines of the above definition (2). These will be compared more fully in a later section. The first is a two-dimensional (2D) separation using the two right-angle dimensions of a continuous (usually planar) surface or thin-film bed for carrying out the separative displacements (1,3,4). For simplicity we will refer to such methods as 2D planar separations whether actually planar or not. (The continuous bed concept can, in theory, be extended to three dimensions as well.) The other approach is coupled column separation (5-11), which for chromatography is often referred to as multidimensional chromatography, although as we see it is only one of the subclasses of the latter. Also, not all coupled column techniques satisfy the two criteria for multidimensional separation stated above.

In coupled column multidimensional separation, the effluent from one column is divided into sequential portions and shunted individually to one or more subsequent columns where they are subjected to new separative conditions. A given component may go through the first (primary) column and then, depending on its emergence time, be switched through a particular second (secondary) column, where a second elution time is observed. The sequence may continue into tertiary columns and beyond (2,12). This approach will be discussed at more length later.

The fact that multidimensional techniques occupy an expanding niche in separations/chromatographic methodology is strong evidence that one-dimensional (linear) separation systems (for example, single chromatographic columns), for all their resolving power and high plate counts, are by themselves inadequate for many problems in chemical analysis. In this chapter I will first describe the limitations of linear column systems for complex samples, then show how some major gaps in the analysis of these complex materials can be filled by a multidimensional strategy.

II. LIMITATIONS OF ONE-DIMENSIONAL SYSTEMS

Multidimensional separation procedures become especially important for complex samples consisting of numerous components (for example, > 100). For such samples, linear columns generally have inadequate power for cleanly resolving the components present; the failure becomes increasingly pronounced as the number of components increase (3).

The overall resolving power of linear columns can be formulated in terms of peak capacity n_c. The peak capacity is defined as the maximum number of conponent peaks that can be packed side-by-side into the available separation space with just enough resolution (usually unit resolution) between neighbors to satisfy analytical goals (13). It can be shown that the frequency with which component peaks overlap in occupying the separation space depends upon the peak capacity (14).

For nonprogrammed runs, peak capacity can be related to the number N of theoretical plates by (13)

$$n_c = \theta N^{\frac{1}{2}} \tag{1}$$

where θ, which depends on the retention time range, is approximately 0.5. For programmed and gradient systems, n_c retains its basic meaning while N loses its usual significance (15,16). Programmed or not, high-resolution linear separation systems generate an n_c in the range of several hundred (1,3).

A peak capacity somewhere in the hundreds would appear more than adequate to separate 100 components, or generally any number m of such components for which $m \leq n_c$, since by definition the peak capacity represents the largest attainable count of resolved peaks. However, to realize this maximum peak content, the peaks must be evenly spaced at their highest allowed density. Unfortunately, the components in most complex samples do not lie on a chromatogram with even spacing, as needed to actually separate n_c peaks. Rather, they tend to fall haphazardly across the chromatogram, frequently overlapping even in cases where the separation space is substantially larger than theoretically needed for full resolution (14).

The accidental overlap of peaks represents a serious problem for most chemically complex samples, no matter what separation system is used. For such complex samples, peak overlap can be ameliorated (but rarely eliminated) by greatly overdesigning the system, that is, by generating n_c values far in excess of m. Unfortunately, sufficiently high n_c values, if they are even achievable in linear systems, are costly in terms of increased analysis time and technical difficulty.

Multidimensional systems provide an alternate method to enhance n_c. The techniques, unfortunately, are complicated by the operational aspects of switching effectively from one separation step to another and by data acquisition and interpretation problems. However, the needed multiplication in n_c often cannot be achieved by any other means. These techniques, at their best, provide capabilities not even approached in linear systems.

In the following paragraphs we will describe the unexpectedly large peak capacities required to effectively handle various complex separations problems. While our basic results will apply to both one- and two-dimensional systems, we will orient our discussion in this section toward one-dimensional or linear systems in order to clarify the conditions under which systems suffice to solve a problem, and to identify conditions under which they fall short.

In accordance with the statistical model of overlap (SMO) developed earlier by the author and his former student Dr. Joe M. Davis, we will assume that component peaks fall randomly over the separation space (14,16—18). In one-dimensional systems the separation space consists of the column length axis or the elution time axis; in two-dimensional separation systems it is the two-dimensional space provided by the system. Theoretical reasons have been advanced for expecting the distribution of components to be random over such space (14). Comparison of the model with the experimental data has served to support the random hypothesis (17). Further support and development of the random mechanism has been provided by Guiochon and co-workers (19,20).

It is useful conceptually to represent the results of a one-dimensional separation by a line fractogram or a line chromatogram (17). In this representation (see Figure 1), each peak is represented by a line of effectively zero thickness. Because the lines have no real width, each component peak is separated from every other peak with which it is not identical. There is, with rare exceptions, no peak overlap. Therefore, this ideal chromatogram will reveal every existing component. (Strictly speaking, isotopic variants would also be revealed by a line chromatogram but we will not consider isotopic resolution to be involved here unless it is one of the goals of the separation.)

In most cases, a line chromatogram would provide all of the information desired from a real chromatogram. Each peak would be resolved and its characteristic migration parameter (for example, distribution coefficient) could be accurately measured for component identification. However, with a line chromatogram the information content of the higher statistical moments of peaks would be automatically lost. Nonetheless, because this information tends to be rather nondiscriminatory and easily perturbed by extracolumn effects, it is low in quality. At present one would greatly prefer to have the total resolution of the line chromatogram in place of the incomplete resolution of systems with peak broadening and peak overlap. There-

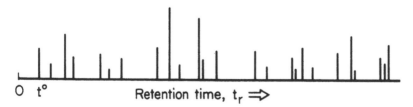

FIGURE 1   Representation of a one-dimensional separation by a line chromatogram/fractogram which, by omitting peak broadening, shows all the components of a mixture.

fore, the higher the resolution the better, with a line chromatogram representing the ultimate achievement.

Since entropy does not permit concentration pulses to exist as infinitely thin lines (15), we must consider now the consequences of line broadening. In that the gaps between lines are of random length because of the statistical distribution of peaks, there will be a number of gaps so narrow that they will be covered over and obscured in the initial stages of line broadening. The associated component peaks will merge and lose their identity as these gaps fill. When this happens, the chromatogram no longer "informs" us of the existence (or quantitative amounts) of the individual component peaks and the analytical information content begins to decline. If there are too many lines (components) or if the peaks broaden excessively, which is equivalent to reducing the peak capacity, most components will be obscured by the general diffusion of other components into their "space." The question we need to address is that concerning the degree of informational loss due to encroaching peak overlap.

The statistical model of overlap indicates that the frequency of peak overlap can be estimated if we have at our disposal only two parameters: peak capacity n_c and the number of components m. We should not, however, underestimate the difficulty of obtaining m for truly complex mixtures (for example, fuel liquids, biological fluids) that cannot be fully resolved. In most cases a direct count of component numbers is not possible. Fortunately our statistical model has provided a methodology for estimating m (providing m does not much exceed n_c) from a chromatogram despite overlap (17,18); we will henceforth assume that an m value (perhaps approximate) is available.

We also note that n_c is not a fixed number for any given chromatogram but varies with the choice made for the resolution level, $R_s^{\ddagger}$, needed to meet analytical (or theoretical) goals. While the peak capacity is clearly defined as the number of component peaks that can fit side by side in a chromatogram, we must specify the resolution required between neighboring peaks before a number can be unambiguously assigned for n_c. Very often a resolution of unity ($R_s^{\ddagger} = 1$) is used in defining peak capacity, that is, n_c is the number of components that can be crowded into the separation space when the gap between adjacent peaks is defined by unit resolution. However, an interesting case arises when we choose a critical resolution level of one-half ($R_s^{\ddagger} = 0.5$) because only above 0.5 do component peaks tend to emerge as distinguishable maxima with clear identities. Obviously, the peak capacity is greater (by a factor of 2) if we allow component peaks to crowd together with a resolution of 0.5 rather than 1.0. As long as we are consistent in our treatment of peak capacity, either choice of resolution can be made.

A few key equations from the SMO define the magnitude of the overlap problem. First, the number p of visible peaks (maxima) is given approximately by the equation (14)

$$p = me^{-m/n_c} = me^{-\alpha} \tag{2}$$

providing n_c is chosen as the value corresponding to a critical resolution $R_s^{\ddagger}$ of 0.5. Factor α is the saturation, defined as the number of components m over the peak capacity n_c. If peaks were uniformly spaced with $\alpha = 1$, the number of potentially observable (although very shallow) peaks would be $p = m$. However, lacking uniform spacing, a disappointingly large number of peaks are lost in peak coalescence. Theory shows that no more than 37% of the peak capacity can be used to generate peak resolution, that is, $p(max) \backsim 0.37$ n_c (14). Even worse, many of the peaks observed under these circumstances represent the grouping of two or more close-lying components. The number s of single component peaks is therefore even smaller than p. We have (14)

$$s = me^{-2m/n_c} = me^{-2\alpha} \tag{3}$$

for which s can never exceed 18% of n_c. This expression is not, of course, limited to $R_s^{\ddagger} = 0.5$; whatever $R_s^{\ddagger}$ value is judged necessary to resolve single components must be used in obtaining n_c.

Eq. 3 is the most important of the two preceding equations because it represents the number of single components isolated and thus subject to analytical measurement. The ratio s/m represents the fraction of all components isolated (to the defined critical resolution level $R_s^{\ddagger}$) as single components. We can think of this ratio as the probability P_1 that any given component is adequately isolated for analytical measurement. Thus

$$P_1 = \frac{s}{m} = e^{-2\alpha} \tag{4}$$

This equation shows that P_1 is directly linked to saturation α, a result that carries over to multidimensional separations. In some regards $\alpha = m/n_c$ is the most important single parameter governing analytical efficacy. From this point of view the primary value of large n_c values is the "thinning out" of component peaks as measured by reduced levels of α.

We note that when components are distributed at fairly constant density over the separation coordinate (time or distance), the resulting saturation is uniform throughout. However, if component peaks concentrate in some regions, the local saturation will correspondingly rise. Where such nonhomogeneous distributions are important, the local saturation α can be related to the mean value $\bar{\alpha}$ by

$$\alpha = \frac{\bar{\alpha}\rho}{\bar{\rho}} \tag{5}$$

where ρ is the local peak density (number of component peaks per segment defined as a fixed number of σ units) and $\bar{\rho}$ is the mean peak density. As α varies over the separation coordinate (e.g., of a chromatogram), P_1 will also vary in accordance with eq. 4.

Some examples help illustrate the impact of statistical overlap. Thus at unit saturation ($\alpha = 1$), under which circumstances all evenly spaced components would be adequately isolated, the probability that any desired component in a random sequence would reach the specified level of resolution with respect to both of its immediate neighbors is only 13.5%. This discouraging result can, of course, be counteracted by reducing saturation. At $\alpha = 0.5$, for example, the probability of isolation increases to 37%; at $\alpha = 0.25$, P_1 has only risen to 61%. One needs a tenfold reduction in saturation ($\alpha = 0.1$) to elevate P_1 to 82%. To work at the analytically desired level of 99% P_1, the saturation must be reduced to the phenomenally low level 0.005.

The cost of this slow improvement in P_1 is substantial. The tenfold reduction in saturation needed to get, for example, $P_2 = 0.82$ will normally be achieved by a tenfold increase in peak capacity. This increase requires a 100-fold gain in the number of theoretical plates as shown by eq. 1. Clearly, serious peak overlap problems cannot be effectively addressed by creeping gains in theoretical plates. The gains must be enormous to do much good. The costs in analysis time and refined instrumentation are expected to be correspondingly great, if possible at all.

The severity of the above trends can be stated in equation form. If we substitute n_c from eq. 1 into eq. 3 (or eq. 4) and solve for N, we get

$$N = \frac{4m^2}{\theta^2(\ln \frac{m}{s})^2} = \frac{4m^2}{\theta^2(\ln 1/p_1)^2} = \frac{4m^2}{\theta^2(\ln P_1)^2} \tag{6a}$$

For P_1 values close to unity, N acquires the limiting form

$$N = \frac{4m^2}{\theta^2(1 - P_1)^2} \tag{6b}$$

From these expressions we find N to increase rapidly with P_1, especially as the latter approaches unity. If we consider the case in which N' plates is observed to yield an isolation probability of P_1', then eq. 6a can be used to show that the elevation of P_1' to some desired level P_1 requires N plates where

$$\frac{N}{N'} = \frac{(\ln P_1')^2}{(\ln P_1)^2} \tag{7}$$

To further illustrate the case, we imagine a sample of 100 components ($m = 100$) run in a chromatographic column with $P_1' = 0.5$. Eq. 4 shows that the corresponding saturation α is equal to 0.3466 and thus $n_c = 288.5$. Such an n_c value could be obtained in a column of 10^5 plates and a relatively high $\theta = 0.912$, as shown by eq. 1. Since we have specified $P_1 = 0.5$, this means that only approximately 50 of the 100 components would be successfully isolated despite the large plate number, the large θ, and an m value of only 100. If we wish to increase P_1, then the required N goes up rapidly. Figure 2 shows on the lefthand ordinate scale the ratio of the re- quired N to the value, N (50%), needed for a 0.5 probability of com- ponent isolation. The righthand ordinate scale further shows the number N of theoretical plates required when the 50% probability of isolation, as in the $m = 100$ example just described, is achieved with 10^5 plates. The latter scale shows that we need over 4,000,000

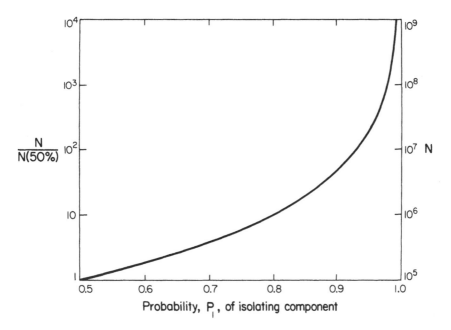

FIGURE 2   Curve illustrates dramatic rise in the number N of plates required to increase the probability P_1 of isolating a given compo- nent from 0.5 to various higher levels.

plates to increase the probability of isolation to a more satisfactory 0.9 and nearly 500,000,000 plates to increase that probability to 0.99. Clearly, such gains in resolving power are beyond the reach of present one-dimensional technology. We must seek more far reaching solutions.

III. TWO-DIMENSIONAL PLANAR SEPARATION

We next examine separation on a two-dimensional (2D) continuous bed. Although the bed need not lie on a flat plane, the planar configuration is most common. Separation on a planar bed is illustrated in Figure 3.

The planar 2D system is highly important in its own right, but it also serves as a simple reference system to which all other multidimensional systems can be compared. Furthermore, the separation process on a 2D plane provides the clearest example of multidimensional separation principles.

There are two distinct separative axes intersecting at right angles on a planar bed. These are shown as x_1 and x_2 in Figure 3. Although in some cases components can be displaced simultaneously along the two axes, many separations require a sequential operation (1): separation occurs first along an initial axis (which we label separative axis 1) and subsequently along the second axis (separative axis 2).

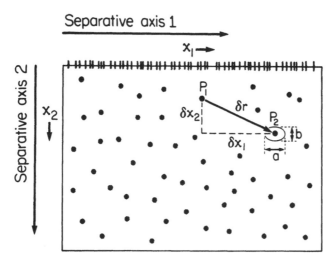

FIGURE 3 Line/dot representation of separation on a two-dimensional planar bed.

   Figure 3 illustrates a two-dimensional separation in an ideal line/ dot format analogous to the line chromatogram of Figure 1. Thus, along axis 1 we show the lines corresponding to the positions of every peak, although because of the reality of peak broadening many (generally most) peaks in complex samples will not be resolved from one another. After the subsequent development along separative axis 2, each component will form a two-dimensional spot. The center of that spot is represented by a dot. The dot representation shows all the components on the two-dimensional bed. The expansion of each dot into a two-dimensional spot will lead to component overlap. The situation is entirely analogous to the overlap of peaks in one dimension as they expand from infinitely thin lines to realistically broadened peaks.

   A 2D planar system clearly satisfies criterion 1 for multidimensional separation providing component displacement along the two axes is governed by different mechanisms. It also satisfies criterion 2 by virtue of the fact that the distance δr between two points (that is, spot centers) on a plane is given by the Pythagorean expression (see Figure 3)

$$\delta r = [(\delta x_1)^2 + (\delta x_2)^2]^{1/2} \tag{8}$$

Thus if _either_ axial separation, δx_1 or δx_2, is large enough to overcome zone broadening and yield good resolution, the separation cannot be nullified by any relative displacement, positive or negative, along the second axis (2). For example, if δx_1 is some distance S and δx_2 is $-S$, the opposite signs of the displacements do not lead to one displacement canceling another; instead, plus or minus, both δx_1 and δx_2, if not zero, contribute to δr as a consequence of the squaring of terms in the relationship of Pythagoras. Thus separation can fail only if components fail to resolve along _both_ axes, which is statistically far less likely than failure along a single axis.

   This profound advantage is not shared by all multistep separation systems. For example, when two elution columns are joined in tandem, the difference δt_r in elution times is the arithmetic sum of the δt_rs of the two parts

$$\delta t_r = \delta t_{r1} + \delta t_{r2} \tag{9}$$

In this case, a negative contribution to δt_r in one step can completely offset the positive contribution of another step, leading to coelution (2). If there are 100 or more components such offsets will, statistically, be a common occurrence and the fraction of components resolved will be no greater than that in a single homogeneous column of comparable efficiency. Thus the simple tandem arrangement does not satisfy criterion 2 and is not considered to be multidimensional.

It can be shown (1) that Gaussian-like zone broadening processes acting in two dimensions lead to the development of elliptically shaped spots centered around each dot as illustrated for dot (or point) P_2 in Figure 3. Clearly, these elliptical spots are subject to overlap, the degree of which will depend upon spot dimensions.

Resolution in two dimensions is thus determined by 2D zone broadening as well as by differential displacement. The length and breadth of a 2D spot, as shown in Figure 3, are a and b, respectively. These dimensions are given by (1)

$$a = 4\sigma_1 \left(\frac{1}{2} \ln \frac{n}{2\pi c_0 \sigma_1 \sigma_2} \right)^{\frac{1}{2}} \tag{10}$$

$$b = 4\sigma_2 \left(\frac{1}{2} \ln \frac{n}{2\pi c_0 \sigma_1 \sigma_2} \right)^{\frac{1}{2}} \tag{11}$$

where the σs are the standard deviations representing the Gaussian spreading processes along the individual axes, n is the spot content (in moles), and c_0 is the limit of detectability (moles per unit area) for the component contained in the spot. Although a and b vary slightly with c_0, and thus with the method of detection, this variation is not great because c_0 appears only in the logarithmic term. Thus, a and b are roughly comparable to the peak widths, $4\sigma_1$ and $4\sigma_2$, that would be observed in the development of a separation along axis 1 and axis 2, respectively.

The resolution of a component pair on a 2D plane is

$$R_s = \frac{\delta r}{4\sigma} = \left[\left(\frac{\delta x_1}{4\sigma} \right)^2 + \left(\frac{\delta x_2}{4\sigma} \right)^2 \right]^{\frac{1}{2}} \tag{12}$$

where σ is the mean standard deviation along the line joining the centers. The resolution for a separation evolving along axis x_1 is

$$R_{s1} = \frac{\delta x_1}{4\sigma_1} \tag{13}$$

while for an x_2 separation

$$R_{s2} = \frac{\delta x_2}{4\sigma_2} \tag{14}$$

If σ is used to approximate σ_1 and σ_2 in eqs. 13 and 14, and the resulting values are substituted back into eq. 12, we have

$$R_s \simeq [R_{s1}^2 + R_{s2}^2]^{\frac{1}{2}} \tag{15}$$

Thus, to an approximation, the components of R_s generated along the two axes contribute to the final resolution according to the geometrical law of Pythagoras. Most importantly, adequate resolution (usually $R_s > 1$) along either coordinate will generally assure adequate final resolution. Thus, the isolation of a component in a 2D planar system is far more probable than in a linear system because any given component will be much less likely to undergo two displacement steps coincident with another component than just one.

Another way of looking at the improved resolving power of 2D systems is through the peak capacity n_c and saturation α. This approach yields a simple numerical comparison of 2D and linear systems.

The peak capacity of a two-dimensional bed is given approximately by the area $L_1 L_2$ of the bed divided by the area ab taken up by the spot. (We use the area ab of the rectangle circumscribing the elliptical spot rather than the ellipse area because spots cannot fill the void area between their convexly curved surfaces.) Thus the peak capacity is

$$n_c \simeq \frac{L_1 L_2}{ab} = \frac{L_1}{a} \times \frac{L_2}{b} \simeq n_{c1} n_{c2} \tag{16}$$

In this equation we have used the approximation that the peak capacity n_{c1} along axis 1 can be represented by bed length L_1 over spot length a: $n_{c1} = L_1/a$. Likewise, n_{c2} is approximately L_2/b. With this we see that the overall peak capacity of the two-dimensional system is approximately equal to the product of the peak capacities that can be generated along the two individual axes (1). This general result has been developed and discussed in considerable detail by Guiochon and coworkers (4).

Associated with the gain in peak capacity is a corresponding reduction in saturation α. Since $\alpha = m/n_c$, we have

$$\alpha \simeq \frac{m}{n_{c1} n_{c2}} \tag{17}$$

The multiplicative form assumed by peak capacity in higher dimensional systems provides a mechanism for amplifying peak capacities far beyond that possible in one-dimensional systems. For example, a peak capacity of 100 along each of the two dimensions would yield a final peak capacity of approximately 10^4. It requires only about 40,000 plates to generate a peak capacity of 100. However, to achieve a peak capacity of 10^4 in a one-dimensional system would require approximately 400×10^6 theoretical plates. This enormous resolving power ($n_c = 10^4$) should prove capable of isolating 98 out of 100 total components present in a complex sample as discussed in connection with Figure 1. Even with 1000 components, approximately 800 would be isolated. Suh a system would clearly give us a greatly enhanced capability to deal successively with complex mixtures that are presently intractable.

One of the most powerful separations systems yet developed is of this two-dimensional form. The system is electrophoretic in nature, using isoelectric focusing along one axis and gel-gradient electrophoresis along another (21—24). Both of these separation processes are inherently powerful with the capability of generating peak capacities of 100 or more (15). They also fulfill the necessary condition that the two separative displacements must occur independently, that is, based on different properties. In the case of isoelectric focusing the displacement depends upon the isoelectric point, while for gel-gradient electrophoresis displacement depends upon the molecular weight. The combination of these two techniques in a planar two-dimensional format has led to the isolation of well over 1000 spots from complex protein mixtures (24).

Two-dimensional planar systems have also been used in the form of thin layer chromatography (25,26). Unfortunately, the resolving power of TLC is not extraordinarily high so one is not able to multiply together very large individual peak capacities ($n_c \sim 20$) in the attempt to reach high resolution. This matter has been discussed by Guiochon et al. (3). It is most unfortunate that a two-dimensional continuous column chromatographic system (assembled, for example, between flat plates), as proposed by Guiochon (3,4), has not been developed in order to take advantage of the high efficiency of column processes. The space between such plates might be packed with support, but an unpacked flat "capillary" configuration is also promising. The first displacement step would occur without elution, perhaps developed with gradients to yield good peak spacing on the 2D bed. The second step would likely involve elution from one edge into a spatially resolved detector system (e.g., a linear diode array).

Until such a system is realized, the closest thing presently to two-dimensional column chromatography is coupled column chromatography, which will be discussed in the following section.

One other advantage of two-dimensional planar separation relative to linear separation is that the measurement of the two coordinates of a spot provides two characteristic migration parameters (e.g., two distribution coefficients) that can be used to help identify the component in the spot. With two parameters characteristic of the component, one can greatly multiply the certainty of identifying the component relative to the case of a one-dimensional system in which only one parameter is available.

We note that the high peak capacity and low saturation theoretically available with 2D systems will be locally modified by nonhomogeneous distributions of spots in the plane of separation, much as noted for one dimensional systems (see eq. 5). On occasion there is a natural clustering of components in certain regions. More troublesome is the lack of independence between the two displacement processes. In this case the two displacement steps become correlated (1). For example, if components with a large displacement along axis 1 in Figure 3 also have an above average displacement along axis 2, and if the small displacements are also correlated, the components will crowd into a diagonal band extending from the upper lefthand corner of the plane to the lower righthand corner. If large displacements along one axis are associated with small displacements along the other, the band will run from the lower lefthand to the upper righthand corner. The stronger the correlation, the tighter the band of spots will be drawn together. This crowding of spots into a common region leads to high saturation levels and frequent overlap within the band and a surfeit of unoccupied space alongside. Thus most component spots will be found in a region of higher saturation than that implied by eq. 17. In this case the saturation down the center of the band will generally fall between the one dimensional value and the ideal 2D value of eq. 17, depending on the level of correlation.

IV. COUPLED COLUMN SEPARATION

A diagram of a simple coupled column system is shown in Figure 4. Here components are partially fractionated in the primary column and cuts from that column are fed into one or more secondary columns. In some cases only one secondary column is used with the fractions fed into this column one at a time. The varied means of transferring fractions will be detailed later in this volume.

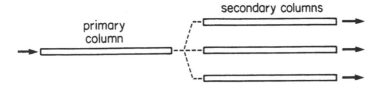

FIGURE 4 A simple coupled-column system in which cuts taken from the effluent of the primary column are routed to secondary columns.

   Migration in the secondary columns must be controlled by parameters that are relatively independent of the factors controlling migration in the primary column; otherwise the system has little effectiveness. With independent migration, we once again have two parameters characterizing each component peak emerging from a secondary column, the first identified by the position (or index number) in the sequence of cuts from the first column and the second measured by the retention time in the secondary column. However, if the cut from the primary column is fairly broad, as shown in Figure 5 by the interval Δt_r in the chromatogram, then the component peaks emerging from the secondary column will have considerable uncertainty (of magnitude Δt_r) in the retention time for their migration through the primary column. The greater Δt_r, the greater the degradation of the quality of information on each component's identity.

   Also, broader cuts (larger Δt_r) incorporate more components and increase the likelihood of interference in subsequent separation steps. Component separation will improve with decreasing Δt_r until the value of Δt_r approaches the width ($4\sigma_t$) of a single peak (see last section). However, with a small Δt_r, more fractions (i.e., more secondary runs) will be needed to cover a fixed elution time range.

   Clearly the coupled column system can be expanded to include tertiary and higher order columns. The use of a mixed secondary/ tertiary system is illustrated in Figure 6. As above, narrow cuts (small Δt_rs) will increase resolution but also increase the number of runs needed to examine a fixed portion of the sample.

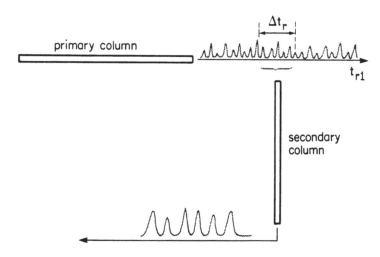

FIGURE 5 A coupled column system in which components emerging from the primary column in the time interval Δt_r are routed into a secondary column.

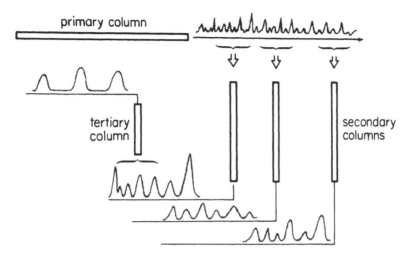

FIGURE 6   Illustration of mixed secondary/tertiary coupled column systems.

Various coupled column systems have proven useful in dealing with complex analytical samples (5—11). Huber has developed a powerful three-dimensional liquid chromatographic (LC) system capable of producing over 6000 peaks (12). Another example is provided by the coupling of LC columns with those of gas chromatography (GC) (8—10). In this system the LC (primary) column typically achieves separation by class and the GC column fractionates members within the class.

Whenever two such chromatographic operations are preceded by a solvent extraction step, the latter should be considered as part of the multidimensional system, adding its own dimension to yield a dimensionality of three.

While multidimensional coupled column systems have proven to be effective in solving a large number of practical problems, few general guidelines have been developed to predict the resolving power of such systems or the means to optimize them and push their performance to desired limits. The discussion below is intended to address these issues. The approach is very broad but hopefully the emergent guidelines will be useful in a variety of specific applications of coupled column systems.

We first need to establish the peak capacity of a coupled column system. As a first approximation we use the ceiling value theoretically attainable for the peak capacity: the sum of the peak capacities of the individual columns or stages of the system

$$n_c(\text{max}) = \sum_{i=i}^{k} n_{ci} \tag{18}$$

(If a single column is used repetitively on successive cuts, as in the Huber system (12), then n_c for each repetition must be counted.) Generally, n_c will be slightly lower than this summed value because the peaks partially resolved in the feeder columns should be counted only once and that is when they emerge as more fully resolved peaks (following the assumptions used to calculate peak capacity) in the final column stage, be it a secondary, tertiary, or higher order stage. However, we do not totally discount the feeder columns, such as the primary column in Figure 6, because not all segments of the chromatogram need necessarily be fed into subsequent columns; these segments are therefore available to separate additional peaks on their own.

An alternative way of writing eq. 17 is

$$n_c(\text{max}) = \bar{n} \times \text{number of columns} \tag{19}$$

where $\bar{n}_c$ is the average peak capacity of the columns used in the system.

While the estimation of peak capacity n_c in coupled column systems is fairly straightforward as shown above, saturation α is more difficult to assess because of inhomogeneous component distribution. For one thing, fractions containing many components are often discarded when they lack compounds of specific interest. In this way the effective m may be greatly reduced (and $\alpha = m/n_c$ correspondingly decreased) in a single step, often a binary separation stage. For example, a preliminary solvent extraction stage, which generates a peak capacity of only two (one for each phase), may generate a nonpolar fraction containing 90% of all components that is uninteresting in relationship to analytical goals. The remaining peak capacity may then be focused on only 10% of the starting components. Other unwanted compounds might be shunted aside in subsequent stages. The general advantage of coupled column systems is that there is enormous flexibility in the manner of coupling the columns (stages) together in order to finally reduce saturation to an acceptable level.

If we focus on a single compound in a mixture of m components we observe that its initial saturation (while still within the sample) is effectively m. (That is, m components occupy a volume that will accommodate only one pure component.) After implementing the first separation step in which the compound is collected as part of f_1m components, the effective saturation is reduced to f_1m. Here, f_1 is the fraction of all components with which the target compound is isolated. After a second step in which the compound is isolated with a fraction f_2 of the components left over from the first stage, approximately $f_1 f_2 m$ components remain. This process can be continued until the compound and its remaining contaminants are subjected to a terminal column process of peak capacity n_{ct}, leading to a final saturation

$$\alpha = \frac{(f_1 f_2 f_3 \cdots)m}{n_{ct}} \tag{20}$$

If components are not distributed homogeneously over the terminal column we have instead

$$\alpha = \frac{\rho_t}{\bar{\rho}_t} \frac{(f_1 f_2 f_3 \cdots)m}{n_{ct}} \tag{21}$$

where ρ_t is the local density of component peaks in the final column as defined earlier. If homogeneous component distributions are found in any of the column stages employed, then the corresponding fraction will be $f_j = \Delta t_r / (t_{rf} - t_{ri})$, where t_{rf} and t_{ri} are the upper and lower bounds, respectively, for the retention times of the final and initial emergent peaks. When Δt_r approaches $4\sigma_t$, then this f_j can be approximated by $1/n_{cj}$, where n_{cj} is the peak capacity of stage j. Generally, for $\Delta t_r \geq 4\sigma_t$, α can be approximated by the following product of terms, one for each stage

$$\alpha = \Pi \frac{\rho_j}{\bar{\rho}_j} \frac{\Delta t_{rj}}{4\sigma_{tj}} \frac{m}{n_{cj}} \tag{22}$$

For the final stage we can usually write $\Delta t_r = 4\sigma_t$. The term $4\sigma_t$, of course, can be replaced by another number of σ_t units if the critical resolution level $R_s^{\ddagger}$ is different from unity.

The last two equations help spell out the conditions necessary to most effectively reduce α. Foremost, as shown by eq. 21, the fraction f of contaminating components should be reduced to a minimum in each separation stage. This is done most successfully, as made clear by eq. 22, by taking narrow cuts (small Δt_r) from high n_c columns. With these conditions attended to, it is necessary to compound as many steps as needed to successfully disengage the m components, taking care to maximize the independence of steps so the local peak densities $\rho/\bar{\rho}$ do not much exceed unity.

As a final note on multidimensional coupled columns, it is instructive to consider the resolution between two given components after passing through n stages. Following eq. 15 we write a general Pythagorean resolution equation

$$R_s = (\Delta m_1^{\ 2} + \Delta m_2^{\ 2} + \cdots + R_{st}^{\ 2})^{1/2} \tag{23}$$

where $m_1 = 1, 2, 3, \cdots$ is a set of index numbers for the discrete fractions produced in stage 1 (and fed to stage 2) and m_2 is a similar set of index numbers for the products of the second separation step. Quantity R_{st} is the resolution realized in the final column or stage. Since the above equation is unconventional as resolution equations go, its use requires explanation.

Equation 23 has the unique ability to describe the level of separation realized when components are collected in different discrete fractions, represented by the Δm terms, as well as when they separate over a continuous path in the terminal column (R_{st}). Normally, Δm_1 and like terms will be 0, 1, 2, or 3, and so on. When $\Delta m_j = 0$, then stage j obviously contributes nothing to resolution, consistent with eq. 22. When $\Delta m_j = 1$, meaning the components end up in adjacent cuts after stage j, then the separation is "complete" as indicated by the resultant $R_2 = 1$, based on this stage alone. Thus when unit resolution is generated by $\Delta m_j = 1$, it signifies the occupancy of the two components in adjacent fractions rather than the 4σ separation conventionally implied. However, when $R_s = 1$ is generated by either mechanism, it has the common meaning of designating successful separation. When $R_s = 0$, zero fractionation is implied in both cases. The common meaning of $R_s = 1$ and $R_s = 0$ gives plausibility to the combination of terms in eq. 22. Other contributions to R_s by Δm_j (either Δm_j fractional or in excess of unity) require special consideration, as described below.

First, values of $R_s > 1$ deriving from $\Delta m_j > 1$ are meaningless because $\Delta m_j = 1$ and $\Delta m_j = 3$, for example, signify the same thing: the components have been shunted along different paths and are for all essential purposes completely separated. Fractional values, too, have no role if components are considered to be shunted cleanly along distinct paths. However, for component peaks divided in two parts by a cut boundary, a fractional resolution may be consistent with the overall treatment.

The most important point about eq. 23 is that it illustrates in mathematical form the essential multidimensional requirement that separation in only one of many stages is needed to resolve any two components.

V. EQUIVALENT TWO-DIMENSIONAL PLANAR AND COUPLED COLUMN SEPARATIONS

For perspective and ultimately for guidance on the choice of a multidimensional separation approach, it is useful to compare the resolving power of coupled column systems and two-dimensional planar

methods. It is most instructive to make this comparison by seeking
a system from the coupled column category (which is more flexible)
specifically molded to yield approximately the same result as a "standard" 2D planar system (2). For a two-dimensional rectangular bed
an equivalent coupled column system would be one, first, with a
primary column having a peak capacity equal to or greater than n_{c1},
the peak capacity of the first separative axis on the 2D bed. Equivalence further requires that we have a series of $\sim n_{c1}$ secondary columns, each taking a narrow cut from along the length (or elution
volume) of the primary column.

   By the above reasoning, a coupled column system with performance equivalent to that of a good 2D planar separation would need
to have many secondary columns, approximately equal in number to
the peak capacity of the primary 2D planar separation axis. In
other words, equivalence demands that the number of secondary columns in a coupled column system must equal approximately the peak
capacity generated along axis 1 of the 2D planar system (see Figure
7). This requirement is confirmed by comparing eqs. 16 and 19
subject to the condition that the overall peak capacity is the same.

   If we require equivalence to high resolution 2D systems in which
the one-dimensional peak capacities are 100 or more, the matching
coupled column system would become a very awkward device of 100
or more secondary columns (or 1 column subject to 100 sequential
runs). This is not expected to be a very practical system because
of the costs of construction and the difficulty of operation. None-

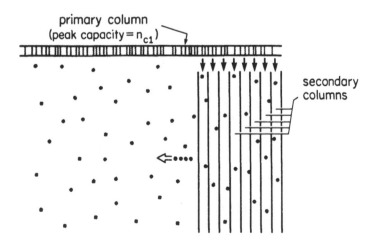

FIGURE 7   A schematic illustration comparing a two-dimensional planar separation system and an equivalent coupled column separation
system.

theless such a system provides a conceptual reference frame which
can be used to examine the performance of more specialized coupled
column systems.

We note that equal peak capacities, although not fully equivalent
performance, will result if fewer secondary columns are utilized than
the number calculated above, providing the peak capacity of the
secondary columns is higher than that achievable along the second
dimension of the 2D bed. For example, the coupling of LC and capillary GC can in theory provide very high peak capacities because
of the extraordinary efficiency of the secondary capillary GC systems.

Returning again to a coupled column system equivalent to a 2D
planar system, we note that a large number of options are possible
in modifying such a reference system to satisfy particular goals and
avoid costly overengineering mistakes that provide more resolution
for more components than absolutely necessary.

VI. MULTIDIMENSIONAL SEPARATION STRATEGIES

The above comparison of 2D planar operation and coupled column
separation brings into focus various strengths and weaknesses of
multidimensional methods that bear on the choice of effective practical systems.

The greatest strength of 2D planar methods, when properly implemented, is that they distribute components widely over a two-
dimensional space of high peak capacity. They thus provide a
broad but still relatively detailed overview of the distribution of
sample components.

The spreading out and mapping of component peaks in two dimensions is analogous to developing a map of the night sky. The
number of stars resolved depends on the resolution of the telescope
(and the scale of the map), but at best we are condensing an ex-
pansive distribution of higher dimensionality (three) onto a more limited 2D surface with an attendant loss of detail. Nonetheless the
2D representation is vastly richer than the projection of the distribution of stars along a single axis.

So, too, with 2D separation one is generally condensing a complex distribution of higher dimensionality onto a 2D plane. Overlap
is almost inevitable, but the amount depends on the complexity of
the sample and the resolution displayed on the map. Again, the detail tends to be much richer than that available in one dimension,
but we caution that a low resolution 2D distribution may contain no
more analytical information than a high resolution distribution represented in one dimension. Unfortunately, in chromatography (as op-
posed to electrophoresis) the existing continuous 2D methods, based

on thin layer technology, have peak capacities that are little higher
than found with high resolution linear columns. This problem was
discussed earlier.

Obviously, for complex samples it would be desirable to expand
the 2D planar separation methodology to three dimensions to gain
more resolution, but such a development will undoubtedly be difficult. Until such systems are developed, higher dimensionality is
better achieved with coupled columns.

As discussed in the last section, one can emulate the broad perspective of a continuous 2D system by the coupling of a primary column to a sufficient number of secondary columns, or by subjecting
the fractions obtained from the primary separation to an adequate
number of succeeding runs in one (or several) secondary columns.
These straightforward but laborious approaches can be expanded
rather directly to higher dimensionality. Obviously this requires
(if one is to benefit from the potential gain in overall resolving
power) an increased number of columns and/or runs. Huber has
demonstrated the power of such an approach using an LC system
operated in three dimensions (12). A single primary column applied
to a complex sample yielded 71 peaks. After subjecting succeeding
portions of column eluate to successive runs in a secondary column,
908 peaks were seen. Numerous fractions from the secondary runs
were then switched successively through a tertiary column yielding
a remarkable 6236 peaks. Although the procedure required 120.2
hours, it provided a wealth of detail not heretofore available for
complex samples.

Coupled column systems have another prominent strength over
and above their dimensional expandability and comprehensiveness.
This is their flexibility, which allows them to be used to examine in
great detail small regions of sample space. This is particularly valuable in seeking to isolate and quantitate active compounds from
complex natural materials.

Suppose we desire to isolate a single compound from a complex
mixture containing thousands of components. A well-designed planar 2D separation system would distribute these components over
the available planar space; our compound would then be located at
a fixed position in the plane, say at point P_2 in Figure 3. With so
many components the compound would likely still be overlapped with
other species.

This compound could be equally well separated in merely two of
the columns of the coupled column equivalent of the 2D planar system. Migration in the two columns, of course, would be based on
the same separation mechanisms as utilized in the 2D planar system.
First a narrow cut (a few σ_t in width) would be taken at a position
(measured in time units) equivalent to the x_1 coordinate of P_2 in
Figure 3. Then this fraction would be fed to the secondary column

and another fraction isolated from a position equivalent to the x_2 coordinate of P_2. All other fractions could be discarded. So far the procedure is simple and direct, and has the advantage over the 2D planar method of providing the capability to collect the desired compound in a small volume of liquid by the use of elution methods. With luck, the resolution of the applicable elution columns may be higher than that of the 2D bed and fewer contaminants would interfere with measurement. The main advantage, however, of a coupled column strategy is that a third separation step can now be taken to remove any remaining contaminants, or a fourth or fifth step if necessary. In this way a relatively simple procedure (considering the sample complexity), using a limited number of separation steps in sequence, can be used to focus on a specific component like that found at point P_2 in Figure 3. When resolution is inadequate, it is possible to further magnify the separation and resolve the component by applying additional stages or dimensions.

The strength of coupled column technology, then, is that columns can be coupled in flexible arrangements and used to focus on specific components or troublesome problems. If, for example, one had too high a peak density (saturation) in a particular region of a two-dimensional planar system for adequate resolution and characterization, a coupled column system could be brought to bear in this region and the resolution enhanced beyond that possible in the two-dimensional system by the use of tertiary and higher order columns. In this sense we can think of coupled column methodology as resembling a high-powered telescope, having enormous resolution in selected, but limited, regions. Clearly, considerable work is required to extend that resolving power over a broader range. Such a broader "mapping" capability is where the 2D planar approach excels.

VII. FURTHER CONSIDERATIONS

Separation of any kind requires the spatial disengagement of components. Therefore a separation axis can generally be defined for different kinds of separation steps or stages; after completion of the separative operation, components are distributed unequally along this separation axis. For most high resolution procedures the components are distributed along this axis as differently positioned Gaussian profiles, but other distributions are also observed. In isotachophoresis and solvent extraction, concentration profiles often resemble square pulses; for adsorption, components sometimes accumulate at the surface with an effective δ-function distribution.

Whatever the concentration profile, the center of gravity of a component's profile can be defined by a point or line. The character of the separation is established by the distribution of such points or lines along the separation axis and, of course, by the peak capacity which represents the ability to separate the points. The point or line distribution can be displayed as a line chromatogram/fractogram, as illustrated in Figure 1.

If a sample is subjected to another separation process, another distribution of points will be observed. This process can be repeated an arbitrary number of times, yielding a unique distribution on each occasion.

Effective multidimensional separation, as we have seen, requires not only a multiplicity of separation stages, but also that the integrity of separation achieved in one stage must be carried through to the others. Therefore the sequence and spacing of points in each distribution, although perhaps subject to scale changes, will be incorporated in the final distribution of components.

If we focus on a single compound, its position following each of the various separation steps can be designated by the coordinate positions x_1, x_2, x_3,...,x_n, respectively. If subjected to all of these steps in a multidimensional procedure, the designation of its final position requires that all n coordinate values be specified.

Mathematically, n-dimensional space is a space in which a single point is fixed uniquely by n coordinate positions (such as x_1, x_2, ...,x_n). Thus the final position of the compound after n separation stages is fully designated by a point in n-dimensional space. The final distribution of all m components is designated by m such points. Any two of these points are well separated from one another as long as they assume different positions along any one coordinate. This is consistent with our second criterion for multidimensional separation, namely that an adequate separation of two components along any axis is a sufficient condition for their ultimate resolution.

There is no requirement that the distribution of components along all axes have a similar form. For example, a solvent extraction axis with a binary (or at least bimodal) distribution can be combined with high resolution chromatographic and/or electrophoretic axes of rather uniformly distributed points. Separation into discrete fractions will appear as superimposed points at discrete intervals along the corresponding axis, indexed by the column number.

Not all components need be subject to all displacements represented by the n axes. For example, some of the components of Figure 6 are displaced through three separation steps (x_1, x_2, x_3) but most through only two (x_1, x_2). The latter can still be represented by points in the overall three dimensional space being mapped, but they will all lie on a plane at fixed x_3 (usually $x_3 = 0$).

Clearly, a point fixed by experimental measurement in n-dimensional space yields n parameters characteristic of the component measured. In general, the larger the n value, the more specific and unambiguous the information yielded by each measurement. A typical one-dimensional (1D) or linear chromatographic system yields only one parameter for each point, that is, a single distribution coefficient K. However, we note that a very precisely defined position along one axis can be less ambiguous than less certain positions indicated along multiple axes. As a rule of thumb, the certainty of identification increases with the peak capacity of the system.

On a somewhat different subject, it is important to bear in mind that when a complex sample is divided into discrete fractions, either by stream switching in coupled column systems or by a technique such as solvent extraction, some components will be present in both of the emergent fractions and will generally proceed in parallel through the remaining separation steps, unless the two fractions are treated differently (for example, one discarded and one further resolved). It is important to keep such component splitting in mind to avoid possible error in the final peak (component) count or in component quantitation.

Component splitting in coupled column systems will increase in frequency as the cut interval Δt_r is decreased. As Δt_r approaches peak width $4\sigma t$, nearly every component will be subjected to significant splitting. Yet a small Δt_r is desirable for high resolution as noted earlier. This point is explained in more detail below.

For maximum resolution in coupled column systems, the size of the cut must be only about 4σ ($4\sigma_t$ in time units) because otherwise some component peaks separated along the first axis would be jumbled back together (due to a reordering of migration rates) along the secondary axis; thus the initial separation would be nullified.

Another viewpoint on this comes from recognizing that the fractionation in the primary column, while not leading to complete resolution, is such that the composition along the primary separation axis will vary rapidly with a correlation distance of a few σ. To harvest the resolving power already developed along this first coordinate, one must make cuts every few σ and feed the fractions individually into secondary columns. If the fractions are collected from intervals much larger than a few σ, then we effectively remix components together that were initially separated, and we place a heavier demand on the secondary axis (column) to resolve these components a second time.

ACKNOWLEDGMENT

This work was supported by Grant GM10851-30 from the National Institutes of Health.

REFERENCES

1. J. C. Giddings, Anal. Chem. 56:1258A (1984).
2. J. C. Giddings, JHRC & CC 10:319 (1987).
3. G. Guiochon, M. F. Gonnord, M. Zakaria, L. A. Beaver, and A. M. Siouffi, Chromatographia 17:121 (1983).
4. G. Guiochon, L. A. Beaver, M. F. Gonnord, A. M. Siouffi, and M. Zakaria, J. Chromatogr. 255:415 (1983).
5. E. L. Johnson, R. Gloor, and R. E. Majors, J. Chromatogr. 149:571 (1978).
6. J. F. K. Huber, E. Kenndler, and G. Reich, J. Chromatogr. 172:15 (1979).
7. D. H. Freeman, Anal. Chem. 53:2 (1981).
8. J. A. Apffel and H. McNair, J. Chromatogr. 279:139 (1983).
9. K. Grob, Jr., D. Fröhlich, B. Schilling, H. P. Neukom, and P. Nägeli, J. Chromatogr. 295:55 (1984).
10. H. J. Cortes, C. D. Pfeiffer, and B. E. Richter, JHRC & CC 8:469 (1985).
11. G. Schomburg, LC·GC 5:304 (1987).
12. J. F. K. Huber, personal communication, May, 1987.
13. J. C. Giddings, Anal. Chem. 39:1927 (1967).
14. J. M. Davis and J. C. Giddings, Anal. Chem. 55:418 (1983).
15. J. C. Giddings, in Treatise on Analytical Chemistry, Part I, Vol. 5, (I. M. Kolthoff and P. J. Elving, eds.), Wiley, New York, 1981, Chapter 3.
16. J. C. Giddings, J. M. Davis, and M. R. Schure, in Ultrahigh Resolution Chromatography (S. Ahuja, ed.), American Chemical Society, Washington, D.C., 1984, ACS Symp. Ser. No. 250, p. 9.
17. J. M. Davis and J. C. Giddings, Anal. Chem. 57:2168 (1985).
18. J. M. Davis and J. C. Giddings, Anal. Chem. 57:2178 (1985).
19. D. P. Herman, M. F. Gonnord, and G. Guiochon, Anal. Chem. 56:995 (1984).
20. M. Martin and G. Guiochon, Anal. Chem. 57:289 (1985).
21. P. H. O'Farrell, J. Biol. Chem. 250:4007 (1975).
22. G. A. Scheele, J. Biol. Chem. 250:5375 (1975).
23. L. Anderson and N. G. Anderson, Proc. Natl. Acad, Sci. USA 74:5421 (1977).
24. N. L. Anderson, J. Taylor, A. E. Scandora, B. P. Coulter, and N. G. Anderson, Clin. Chem. 27:1807 (1981).
25. M. Zakaria, M.-F. Gonnord, and G. Guiochon, J. Chromatogr. 271:127 (1983).
26. B. R. Bochner and B. N. Ames, J. Biol. Chem. 257:9759 (1982).

2

Multidimensional Thin-Layer Chromatography

COLIN F. POOLE and SALWA K. POOLE *Wayne State University, Detroit, Michigan*

I. INTRODUCTION

Thin-layer chromatography (TLC) has changed rapidly during the past decade in response to theoretical and practical optimization of all aspects of the chromatographic process. To distinguish the modern practice of the technique from its conventional counterpart it has become known as high performance thin-layer chromatography (HPTLC) or instrumental thin-layer chromatography. In keeping with other chromatographic methods labeled "high performance" analysts have come to expect increased separating power and shorter analysis times for separations performed by modern TLC. A detailed account of the state-of-the-art in modern TLC is available in several books (1−7) and review articles (8−16) that have appeared recently.

This chapter is concerned with multidimensional thin-layer chromatography conveniently defined as separations involving the sequential application of more than one separation mechanism to resolve the mixture. For thin-layer chromatography three modes have most frequently been used for multidimensional separations. Unidimensional, multiple development involves repeated irrigation of the sample in the same direction alternated with segments in which the plate is dried prior to redevelopment (17). The migration length and solvent composition are easily changed at any segment in the development sequence to optimize the separation. Also, a natural refocusing of the spots occurs at each subsequent segment of the development sequence after the first and acts to reduce the controlling influence of diffusion on the efficiency of the chromatographic system. Since TLC separations are carried out on a thin layer of sorbent in the form of a rectangle, sequential development in two directions at right angles is possible (18,19). This method, commonly known as two dimensional thin-layer chromatography, has from a theoretical point of view a separating capacity of n^2 where n is equal to the unidimensional peak capacity. To reach this peak capacity the selectivity characteristics of the two unidimensional developments must be complementary. The technique has been widely applied for qualitative analysis but less frequently for quantitative analysis due to the difficulty of obtaining quantitative information by scanning densitometry. The third approach to multidimensional thin-layer chromatography is the automated coupling of two chromatographic techniques in which thin-layer chromatography is used as the second dimension and gas chromatography (GC) or liquid chromatography (HPLC) as the first (20). In this chapter we will discuss primarily those methods in which the two techniques are interfaced directly for continuous operation. Excluded from the discussion is the use of thin-layer chromatography as a sample preparation procedure for gas or liquid chromatography. In this instance thin-layer chromatography is employed for sample fractionation and specific zones of sorbent containing the components of interest are excised from the plate, the sample recovered by elution, and the analysis completed by gas or liquid chromatography in the normal manner. Thin-layer chromatography is still widely used as a sample preparation technique, but in recent years, its popularity has waned somewhat due to the introduction of micro extraction cartridges that are easier to manipulate and less demanding of the analyst's time.

II. DESCRIPTION OF MODERN THIN-
LAYER CHROMATOGRAPHY

One of the early milestones in the development of modern TLC was the commercial introduction in the mid 1970s of new plates prepared from specially sized sorbents having a smaller particle size (5—15 μm) with a much narrower particle size distribution than the sorbents used for preparing conventional TLC layers. These new plates provided higher efficiency and shorter analysis times compared with conventional TLC plates when sample sizes applied to the layer were reduced to the optimum level. Initially only silica gel layers were available (21) but in subsequent years a complete range of normal and bonded phase sorbents became available including alumina, cellulose (22), 3-aminopropylsilanized silica (23), reversed phase plates containing ethyl-, octyl-, octadecyl-, and diphenylsilanized silica (24—27), and most recently a chiral phase for the separation of enantiomers (28) and 3-cyanopropylsilanized silica (29). These new plates were not always well received; inferior results to conventional TLC were obtained occasionally due to a failure to appreciate the importance that sample loading had on efficiency.

Ideally, starting spots should have diameters of about 1.0 mm or
so, corresponding to sample amounts at or below the low-microgram
range and sample volumes on the order of 100—200 nL. Such low
volumes are difficult to apply manually in an accurate manner and
without damaging the layer. Instrumental methods capable of meeting all sample application problems are presently available (2,3,12,
30,31). Sample volumes of 100—200 nL can be applied to fine-particle layers using fixed volume platinum-iridium capillaries sealed into
a glass sleeve for ease of handling (dosimeters) or by using a micro-
meter-controlled syringe. In either case the dosimeter or the syringe is brought into contact with the layer using mechanical or
electromagnetic devices which ensure that all the sample is transferred to the layer without damaging it. These devices also ensure
that the coordinates of the starting spot are accurately defined to
facilitate scanning densitometry after development of the chromatogram. By evaporative concentration of solution volumes up to 100
μL on a PTFE tape and solid phase transfer to the sorbent layer it
is possible to quantitatively apply relatively large sample volumes
as spots of the required diameter to the sorbent layer (32,33).
Automated sample application devices are available that can remove
samples from a rack of vials and precisely deposit a preselected sample volume at a controlled rate to assigned positions on the TLC
plate (31,34).

Initially, because of the slow flow rate of solvent through the
new fine-particle-layer plates, solvent migration distances were kept
comparatively short (3—6 cm). Short migration distances were more
than adequate, however, to provide greater effective sample resolution than obtained by conventional TLC, and despite the lower sol-
vent-migration velocities, analysis times are substantially shorter.
For favorable situations, analysis times could be decreased by an
order of magnitude. Rapid solvent migration over short distances
renewed interest in the use of continuous and multiple development
techniques for the separation of complex mixtures (17). Further
stimulation was provided by the availability of equipment for short-
bed continuous development (3), automated multiple development
(31,35—38), and sequential development with variable positioning of
the solvent entry position (39). The short-bed continuous development chamber is shown in Figure 1 alongside a conventional twin-
trough chamber and a linear development chamber for HPTLC.
The apparatus for automated multiple development is shown in Figure 2. We will return to the properties of these chambers in subsequent sections of this chapter.

Both theoretical and practical considerations have indicated that
the restrictions imposed by solvent migration controlled by capillary
forces can be overcome if the mobile phase velocity is controlled by
external means (11,40—43). In forced-flow TLC (also called over-

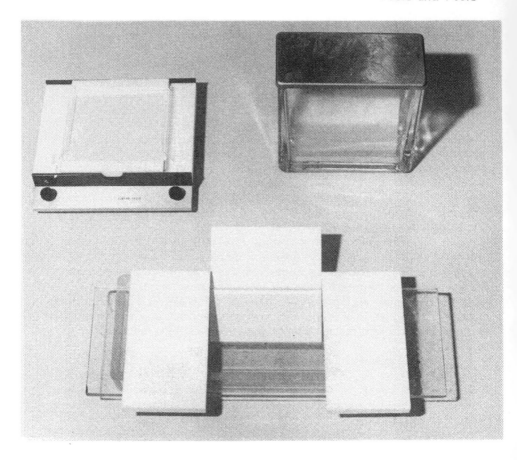

FIGURE 1 Development Chambers used for modern TLC. Short-bed
continuous development (front), linear development chamber (back
left), and twin-trough chamber (right). From Ref. 3 with permission.

pressured TLC), solvent is applied to the layer by a constant-vol-
ume pump permitting the plate length and solvent velocity for a separation to be optimized independently of each other. The new technique required construction of special developing chambers in which
the sorbent layer is sandwiched between the glass backing plate
and a flexible polymeric membrane forced into intimate contact with
the layer by application of hydraulic pressure. An overpressured
development chamber for forced flow linear development is shown in
Figure 3 (44). Versions of the overpressured development chamber
for circular and anticircular development have also been described
(45—47).

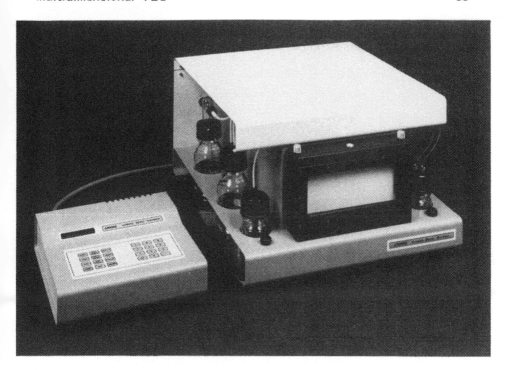

FIGURE 2   Apparatus for Automated Multiple Development.   Courtesy of Camag, Muttenz, Switzerland.

   The mobile phase velocity can also be controlled by centrifugal forces. Here, the mobile phase is pumped onto the layer at its center and the chromatogram developed in the circular mode while the plate is rapidly rotated by a variable, or fixed speed, motor. Commercial equipment is available for preparative centrifugal layer chromatography and some practical observations on how to optimize the performance of the system have been given (48). Superior resolving power has been demonstrated for sequential centrifugal layer chromatography in which the application of solvent to the layer can be varied both regionally and with respect to time (44,49,50). In this instance, the principal advantage is that a particular region of the chromatogram obtained after a preliminary separation can be redeveloped with a selective mobile phase while the other zones remain stationary. From an efficiency point of view, one disadvantage of this method, as currently practiced, is the wide particle size distribution of sorbents used to prepare the layers. Analytical applications of centrifugal layer chromatography are rare and therefore, we will not consider this technique further.

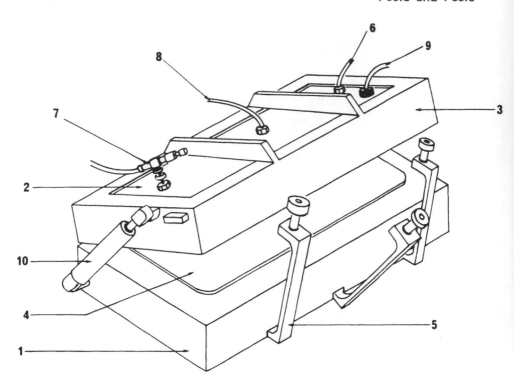

FIGURE 3   The Chompress 10 chamber for overpressured TLC.
1. bottom support block; 2. upper block with a polymethacrylate
support plate; 3. external frame holding 2; 4. position of TLC
plate; 5. clamps; 6. inlet for water providing hydraulic pressure
to one side of the membrane contacting the TLC layer; 7. position
of mobile phase inlet; 8. mobile phase outlet; 9. outlet valve for
relieving hydraulic pressure; 10. hydraulic lever. From Ref. 44
with permission.

     The method of development in TLC can be linear or radial and
the chromatogram produced described as linear, circular, or anticircular.   The linear development mode represents the simplest situation and is the most widely used of the three techniques.   The
samples are spotted along one edge of the plate and the solvent migrates to the opposite edge, effecting a separation.   Viewed in the
direction of development, the chromatogram consists of compact symmetrical spots of increasing diameter.
     In circular TLC, the samples are applied in a circle at the cen-
ter of the plate and then developed radially towards the edges of
the plate.   Spots near the origin remain symmetrical and compact

while those nearer the solvent front are compressed in the direction of solvent migration and elongated in the orthogonal direction (see Figure 4 (1)).

In anticircular TLC, the sample is applied along the circumference of an outer circle and developed toward the center of the plate (51). Spots near the origin remain compact, while those towards the solvent front are considerably elongated in the direction of migration, but changed very little in width, when viewed at right angles to the direction of development, as in Figure 4. For quantitation of circular or anticircular chromatograms a scanning densitometer capable of radial or peripheral scanning is required.

Instrumentation for in situ detection of TLC chromatograms first appeared around 1967 and is now considered essential for accurate measurement of both spot size and location, for a true measure of resolution, and for rapid, accurate quantitation (2,3,9,10,16,52–54). Slit-scanning densitomers are the most common, and provide a difference signal, based on the intensity of light reflected or transmitted from blank regions of the plate and from sample zones. Measurements are usually made by absorption or fluorescence depending on the properties of the sample zones.

The plate is scanned by moving the plate on a motorized stage through a fixed measuring beam. Each scan consists of a rectangular track of width defined by the dimensions of the measuring beam and of length determined by the migration of the sample zones. This method of scanning is suitable for unidimensional chromatograms, but as will be shown later, is unsuitable for scanning twodimensional chromatograms. A detailed review of commercial instru-

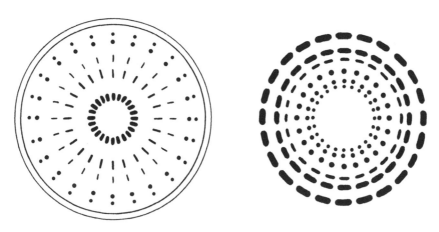

FIGURE 4 Circular development with the point of solvent entry at the plate center (right). Anticircular development from the outer circle towards the center (left). From Ref. 8 with permission.

ments for scanning densitometry, and their operating characteristics, has appeared recently (15).

A. Some Theoretical Considerations for Unidimensional Development

The fundamental parameter used to characterize the position of a spot in a TLC chromatogram is the retardation factor, or R_F value. It represents the ratio of the distance migrated by the sample compared to that traveled by the solvent front

$$R_F = \frac{Z \text{ substance}}{Z \text{ mobile phase}} = \frac{Z_S}{Z_F} \tag{1}$$

Z_S = distance migrated by the sample (spot) from its origin

Z_F = distance moved by the mobile phase from the sample origin to the solvent front

The R_F value measured in linear, circular, and anticircular chromatograms, for which the flow conditions are different, can be related to each other by equation (2) (1,51)

$$R_{F(L)} = \left[R_{F(C)} \right]^2 = 1 - \left[1 - R_{F(AC)} \right]^2 \tag{2}$$

$R_{F(L)}$ = R_F value for a substance determined by linear development

$R_{F(C)}$ = R_F value for the same component determined by circular development

$R_{F(AC)}$ = R_F value of the same component determined by anticircular development

For linear development with capillary-controlled flow, if fluctuations in the vapor phase, in contact with the sorbent layer, are negligible, the speed with which the solvent front ascends the TLC plate is given by equation (3)

$$(Z)^2 = \kappa t \tag{3}$$

Z = migration distance of the solvent front in cm from the mobile phase entry position in t seconds
κ = mobile phase velocity constant (cm^2/S)
t = time in seconds

A similar relationship holds for circular TLC, while a linear relation-ship is found in anticircular TLC, due to an equivalence between the quadratic decrease in mobile phase velocity and a similar reduc-tion in the plate area to be wetted by the mobile phase.

The velocity constant, κ of equation (3), can be related to the experimental conditions by equation (4) (43,55,56)

$$\kappa = 2K_0 \, d_p \, \frac{\gamma}{\eta} \, \cos\theta \qquad\qquad (4)$$

K_0 = permeability constant

d_p = mean particle diameter

γ = surface tension of the mobile phase

θ = contact angle

For capillary-controlled flow conditions, it can be seen from eqs. (3) and (4), that the rate of solvent advance will be greater for coarse-particle layers and will also, depend very strongly on the contact angle. As most organic solvents wet silica gel plates com-pletely, or very nearly so, the contact angle is generally favorable for rapid development. On the other hand, bonded-phase sorbents are generally poorly wet by aqueous organic solvent mixtures, re-stricting the range of mobile phases that can be sensibly used for reversed-phase TLC (57). Mobile phases should also be of low vis-cosity to promote rapid development.

Separations performed with fine-particle layers are characterized by a series of symmetrical, compact spots increasing uniformly in diameter with increasing migration distance (58—60). For fine-par-ticle layers, zone broadening is controlled by molecular diffusion, and mass transfer contributions are negligible. The distribution of sample within a spot is essentially Gaussian and the efficiency of the chromatographic system, determined as the number of theoretical plates, is adequately described by eq. (5):

$$n = 16 \left[\frac{Z_S}{W_b} \right]^2 = 16 \left[\frac{R_F Z_F}{W_b} \right]^2 \qquad\qquad (5)$$

n = number of theoretical plates

Z_S = migration distance of substance s

W_b = spot diameter of substance s

For coarse-particle layers, elongated and irregularly shaped spots are not uncommon, and in this instance, the contribution of mass transfer to zone broadening cannot be ignored.

There is also a striking difference between the efficiency as measured for a thin-layer and that for a packed column. In column chromatographic systems all substances travel the same migration distance (the length of the column) but have different diffusion times (retention time on the column)—the exact opposite conditions to those pertaining to TLC. Consequently, unlike equations used to describe the efficiency of column systems, equation (5) contains a term dependent on the migration distance, and is thus not a constant term for the layer. By convention, the efficiency of the thin-layer bed is measured, or calculated, for substances having an R_F value of 0.5 or 1.0, or an average value is used. Further assumptions are needed to derive an average value which we will not explore further here (58).

The resolution, R_S, between two closely migrating spots ($Z_{S1} > Z_{S2}$) is defined in the conventional way by equation (6).

$$R_S = 2 \left[\frac{Z_{S1} - Z_{S2}}{W_{b1} + W_{b2}} \right] \tag{6}$$

Z_{S1} = migration distance of component 1 with
 base width W_{b1}

Z_{S2} = migration distance of component 2 with
 base width W_{b2}

For two spots with similar R_F values, $W_{b1} = W_{b2} = W_b$, and substituting that $Z_S = R_F Z_F$ into equation (6), gives

$$R_S = Z_F \times \frac{R_{F1} - R_{F2}}{W_b} = \frac{Z_F}{W_b} \Delta R_F \tag{7}$$

$$\Delta R_F = R_{F1} - R_{F2}$$

The peak width at the base, W_b, can be expressed in terms of the number of theoretical plates, n, by equation (5), and substituted into eq. (7), to give

$$R_S = \frac{\sqrt{n}}{4} \times \frac{\Delta R_F}{R_{F1}} \tag{8}$$

The plate number, n, the equilibrium constants embodied in ΔR_F, and the denominator in equation (8) are all dependent on the R_F value. The relationship between resolution and R_F value is thus complex, with Figure 5 describing a bell-shaped curve (61). The resolution reaches a maximum at an R_F value of about 0.3 and is fairly constant for R_F values in the range 0.2 to 0.5 (within this range R_S is greater than 92% of the maximum value). Outside this range, the resolution declines rapidly, indicating the strong correlation between the separating power in TLC and the position of the zones in the chromatogram.

The spot capacity in TLC, is the number of spots resolved with a resolution of unity, that can be placed between the sample spot at the origin and the spot of an unretained compound. This is more difficult to calculate than the equivalent value for a column system, since the plate height in TLC is a complex function of the characteristics of the chromatographic system (61—62). Solving this problem for different sets of conditions indicates that it is very easy to achieve a spot capacity between 10 and 20, but it is extreme-

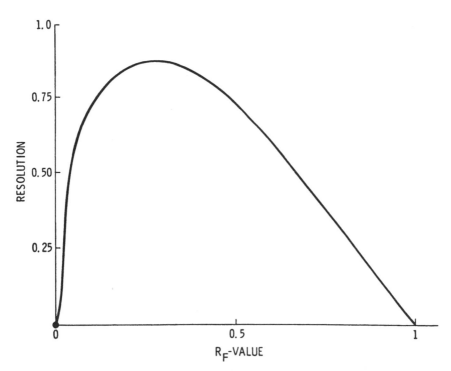

FIGURE 5   Change in resolution of two closely migrating spots as a function of the R_F value of the faster moving spot.   From Ref. 3 with permission.

ly difficult to reach 25, and practically impossible to exceed 30, by unidimensional development under capillary-controlled flow conditions.

In unidimensional, forced-flow development using a constant volume pump, the mobile phase velocity can be optimized independent of the development length, improving the performance of the chromatographic system, while simultaneously reducing analysis times. The principal advantages of forced-flow TLC are summarized in Table 1.

If the mobile phase is delivered to the sorbent layer at a constant volumetric flow rate, the position of the solvent front at any time t, is given by equation (9)

$$Z_F = u \cdot t \qquad\qquad (9)$$

u = mobile phase velocity

Under these conditions the average plate height is approximately constant, and the plate number increases linearly with migration distance (63). The relationship between the average plate height and migration distance for coarse- and fine-particle layers, using capillary-controlled and forced-flow development, is shown in Figure 6 (64). At longer migration distances, the efficiency of fine-particle layers is diffusion limited due to the low migration velocity under capillary-controlled flow conditions. This does not apply to forced-

TABLE 1   Advantages of Forced-Flow Thin-Layer Chromatography

1. The solvent velocity can be optimized independently of other experimental variables.

2. Higher efficiencies can be achieved using fine-particle sorbents and longer plates than is possible with capillary-controlled flow systems.

3. A linear increase in efficiency with increasing migration distance is obtained. The upper limit is set by the length of the sorbent layer and the pressure required to maintain the optimum linear mobile phase velocity for that length.

4. Analysis time can be reduced by a factor of 5 to 20.

5. Mobile phases with poor sorbent wetting characteristics may be used.

6. Solvent gradients, either step gradients or continuous gradients, can be employed to optimize the separation conditions.

Source: From Ref. 11 with permission.

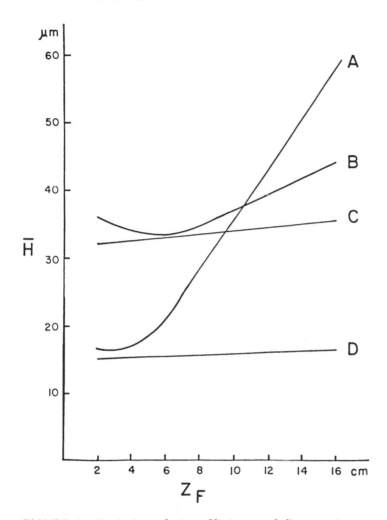

FIGURE 6   Variation of the efficiency of fine- and coarse-particle layers as a function of migration distance and development technique.   A.   Fine-particle layer with capillary-controlled flow; B. coarse-particle layer with capillary-controlled flow; C. coarse-particle layer with forced-flow development; and D. fine-particle layer with forced-flow development.   From Ref. 64 with permission.

flow conditions since the separation can be performed at the optimum
mobile phase velocity for relatively long migration distances. The
ultimate efficiency of the system is limited only by the particle size
and homogeneity of the layer, the length of the layer, and by the
pressure required to maintain the optimum mobile phase velocity.
Using commercially available equipment and plates, an upper limit
of about 50,000 theoretical plates seems reasonable. It should be
noted, that the above figure is a limit established by available technology, and is not a theoretical limit.

Using forced-flow development conditions, there is no maximum
for resolution as a function of migration distance, as was observed
for capillary-controlled flow conditions. Instead the resolution continues to increase as the migration distance increases (61,63). Similarly, the spot capacity of the system increases linearly as a function
of the square root of the plate length instead of going through a
maximum as it does for capillary-controlled flow conditions (65).
The spot capacity, therefore, becomes comparable to that achieved
in column chromatography; only the available pressure drop and
plate length limit the spot capacity. With commercially available
equipment a value of about 80 should be achievable, but theoretically, values two to three times higher are possible.

III. UNIDIMENSIONAL CONTINUOUS
DEVELOPMENT

For continuous development the mobile phase is allowed to traverse
the layer to some predetermined fixed position on the plate, at
which point, it is continuously evaporated. Evaporation of the mobile phase usually occurs at the plate atmospheric boundary using
either natural or forced evaporation. During the first part of the
development capillary forces are responsible for movement of the
solvent front. Once it reaches the boundary, additional forces are
applied by the evaporation of the solvent. Eventually a steady
state—constant velocity—is reached, at which, the mass of solvent
evaporating at the boundary is equivalent to the amount of new solvent entering the layer. Sandwich-type chambers for continuous
development have been reviewed by Soczewinski (66) and Perry has
outlined the use of the short-bed continuous development chamber
for optimized continuous development with variable selection of the
plate length (67).

The continuous development technique has recently been used
with forced-flow development for complex mixture analysis as illustrated in Figure 7 (68,69). Forced-flow, continuous development
with change of solvent at some time during the development, and
continuous development with linear or multistep gradients, has also

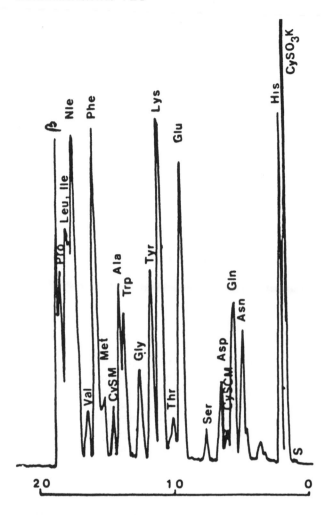

FIGURE 7 Separation of PTH-amino acids by forced-flow, multiple, continuous development TLC. First development was 20 min at 0.3 mL/min using the solvent chloroform-ethanol (95%)-acetic acid (90:10:2 V/V) and a development distance of 16 cm. The second development over 20 cm with dichloromethane-ethyl acetate (90:10) was carried out continuously for 34 min. From Ref. 68 with permission.

been discussed (70). These techniques are useful for the separation
of samples of wide polarity, not so easily handled by normal, continuous development, unless parallel development (71) is used (i.e.,
the separation is optimized by irrigating the sample, more than once,
in a series of solvents selected to separate individual groups of sample components with similar polarity).

Continuous development offers several advantages for capillary-
controlled, solvent-mediated systems. Instead of using long development lengths, several interstitial volumes can be passed through
a shorter bed length to yield the same effective path length for the
mobile phase, but in a much shorter analysis time, due to the rapid
flow of the solvent. A weaker solvent must be used to prevent the
faster spots from traveling beyond the end of the cover plate.
Under these conditions the resolution of the spots is enhanced by
the greater selectivity achieved by using weaker eluents. Perry
has summarized several studies indicating that for adsorption systems, the selectivity increases exponentially with a decrease in solvent strength, and thus, the number of theoretical plates (bed
length) required for resolution becomes very small (67). Spots are
then resolved by very short migrations, and remain compact and
are easier to detect. Detection limits may be enhanced by about
one order of magnitude compared to conventionally developed chromatograms.

A theoretical model for predicting separations by capillary-controlled, continuous development has been proposed, and verified,
by Nurok (72–75). One important observation from this model, is
that, for a given spot separation the analysis time will always be
shorter by continuous development than by conventional development, when the experimental parameters are correctly optimized
(72). The rate of solvent migration in continuous development TLC
is constant, when the steady state condition is reached, and can be
described by eq. (10)

$$U_c = \frac{\kappa}{2L} \tag{10}$$

U_c = mobile phase velocity in the continuous development
mode

L = Length of plate traversed by the solvent

κ = mobile phase velocity constant

The total analysis time is the sum of two components: the time during which the solvent front traverses the TLC plate, and the time
during which continuous development occurs, eq. (11)

$$t_L = t_1 + t_2 \tag{11}$$

t_L = total analysis time

t_1 = time during which the solvent front traverses the TLC plate

t_2 = time during which the continuous development occurs

Similarly, the total distance migrated by each solute is the sum of the distances that the solute migrates during t_1 and t_2, such that the following approximations are valid, eqs. (12) to (14)

$$M_D = d_1 + d_2 \tag{12}$$

$$d_1 = R_F (L - x) \tag{13}$$

$$d_2 = R_F \left(\frac{\kappa t_L}{2L} - \frac{L}{2} \right) \tag{14}$$

M_D = total distance migrated by each solute

d_1 = distance migrated during time t_1

d_2 = distance migrated during time t_2

x = distance between the sample origin and the height of the mobile phase in the tank

Combining the above equations leads to the generally useful expression for M_D, equation (15)

$$M_D = R_F \left[\frac{L^2 - 2Lx + \kappa t_L}{2L} \right] \tag{15}$$

The value of R_F can usually be predicted as a function of mobile phase composition when the latter consists of a binary mixture of a strong and weak solvent. For such a binary solvent system the relationship between the capacity factor, k, and the mole fraction, X_s, of the polar component can be expressed as

$$\ln k = a \ln X_s + b \tag{16}$$

The R_F value can be expressed as a function of the capacity factor and mole fraction, and substituted into eq. 15, to give

$$M_D = \frac{1}{1 + \exp{(a \ln X_s + b)}} \left[\frac{L^2 - 2Lx + \kappa t_L}{2L} \right] \qquad (17)$$

As the solvent velocity constant, κ, is also a function of the mole fraction of the stronger solvent, eq. 18, M_D can be evaluated as a function of X_s provided that L and t_L are specified.

$$\kappa = a_1 + a_2 (X_s) + a_3 (X_s)^2 \qquad (18)$$

a_1, a_2, a_3 = experimentally derived constants

X_s = mole fraction of the stronger solvent

Equations (17) and (18) can be used to predict the center-to-center spot separation of any two solutes by the difference in their M_D values. For multiple component mixtures overlapping resolution maps are very useful for identifying the optimal separation conditions based on eq. 18 (74).

IV. UNIDIMENSIONAL MULTIPLE DEVELOPMENT

In multiple development, the TLC plate is developed for some selected distance, the plate withdrawn from the developing chamber, and adsorbed solvent evaporated before returning the plate to the developing chamber and repeating the development process. This is a very versatile strategy for separating complex mixtures since the primary experimental variables of plate length, time of development if continuous development is used, and composition of the mobile phase, can be changed at each development step and the number of steps varied to obtain the desired separation. Quantitative measurements by scanning densitometry can be made at several steps in the sequence and, therefore, it is unnecessary for all components to be separated at one time provided that they can be resolved (chromatographically or spectroscopically) at different segments in the development sequence.

A further, unique feature of multiple development TLC, that leads to an increase in the efficiency of the chromatographic system, is the spot reconcentration mechanism. Every time the solvent front traverses the stationary sample it compresses the spot in the direction of development. This is illustrated in the upper part of Figure 8 (11). Compression in the direction of solvent migration occurs because the mobile phase first contacts the bottom edge of the spot

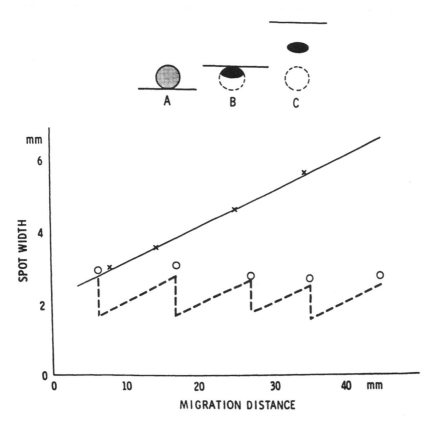

FIGURE 8   Schematic representation of the spot reconcentration
mechanism in multiple development.   At top: A.   advancing solvent
front contacts lower edge of spot;   B.   solvent front traverses spot
producing a compression in the bottom-to-top direction;   C.   spot is
developed normally after being reconcentrated.   At bottom, the
solid line represents the constant rate of spot broadening by normal
development and the broken line illustrates the effect of the spot reconcentration mechanism as a means of counteracting the natural
spot broadening mechanism during development.   From Ref. 11 with
permission.

where the sample molecules start to move forward, before those
molecules still ahead of the solvent front. Once the solvent front
has reached beyond the spot, the reconcentrated spot migrates and
is broadened by diffusion in the normal way. As illustrated in the
lower part of Figure 8, under optimized conditions, it is possible to
move a spot a considerable distance without significant broadening,
a balance or equivalence being struck between the spot reconcentration and band broadening mechanisms.

An example of the use of multiple development TLC for the separation of the 20 common protein amino acid PTH derivatives will be
presented next to illustrate the versatility of the approach, using
Figure 9 as a guide (76). Five developments with four changes in
mobile phase composition in a short-bed continuous development
chamber were used for the separation. The first development, Figure 9A, was made with methylene chloride for five minutes using a
plate length of 3.5 cm. This step was performed to order the PTH-
amino acid derivatives at the origin to enhance their resolution in
later segments. Although PTH-proline is baseline resolved from
other PTH-amino acids in this segment, there is no need to make
any determinations at this point since it remains baseline resolved
in the next four development steps. After the methylene chloride
had been evaporated from the plate, the plate was redeveloped for
10 min in methylene chloride-isopropanol (99:1) with a plate length
of 7.5 cm. In all subsequent segments the plate length was unchanged. At this development step, Figure 9B, the PTH derivatives
of proline, leucine, isoleucine, valine, and phenylalanine are separated. The third development segment was a repeat of the second,
Figure 9C, and provides a better separation of the peaks resolved
in step 2, as well as enabling the PTH-derivatives of methionine,
alanine/tryptophan, glycine, lysine, tyrosine, and threonine to be
identified. In this segment of the development sequence, tryptophan
frequently appears as a shoulder on the alanine peak, but is not
adequately resolved for identification purposes. The PTH-derivatives of alanine and tryptophan can be separated almost to baseline
by a 10 min development in hexane-tetrahydrofuran (9:1) with a
plate length of 3.5 cm (see Figure 9D) although not as part of the
development sequence currently being considered. The mobile phase
was changed to methylene chloride-isopropanol (97:3) for the fourth,
10 min development, segment. The separation between the PTH-
derivatives of lysine, tyrosine, and threonine is improved over the
third segment and the derivatives of serine and glutamine are now
baseline resolved also Figure 9E. A much more polar solvent was
required for the final 10 min development segment, ethyl acetate-
acetonitrile-glacial acetic acid (74.3:25:0.7), to separate the polar
PTH-derivatives of arginine, histidine, S-(carboxymethyl)cysteine,
and aspargine (see Figure 9F). The whole development sequence

requires less than 1 hr, only segments 3, 4, and 5 need be scanned
to indicate individual amino acid derivatives, and standards can be
run simultaneously with samples (up to 16 samples can be applied
to a 10 × 10 cm plate and separated simultaneously) to improve sample identification from peak position locations.

It has proven difficult to develop a theoretical model for multiple development under capillary-flow controlled conditions. Some
useful observations have been made for the simplest of cases, that
of n equal developments with a constant mobile phase composition
(77—81). The spot width and the position of the spot in the chromatogram are given, to a first approximation, by eqs. (19) and
(20), respectively.

$$W_{F_n} = W_i (1 - R_F)^n \tag{19}$$

$$R_{F_n} = 1 - (1 - R_F)^n \tag{20}$$

W_{F_n} = final spot width after n developments

R_{F_n} = apparent R_F value after n developments

W_i = initial spot width

n = number of identical developments

In this model, the extent of the separation of two components first
increases as the number of solvent developments increases, reaches
a maximum, and finally decreases as the solutes approach the end
of the layer. Thus, there is a certain optimum number of multiple
developments that will produce a maximum separation of two similar
solutes; this corresponds to the number of developments required
to move the spots an average distance of 0.632 Z_F. Since the maximum extent of separation increases as R_F decreases, difficult to resolve components should be repeatedly irrigated with solvents that
provide low R_F values (equivalent to using the most selective solvent system for the separation).

A. Unidimensional Multiple Development with Variable Solvent Entry Position

One disadvantage of the multiple development technique concerns
the separation of samples of similar polarity. These tend to migrate
together from the origin and become separated higher up the plate.
For difficult separations fairly long plate lengths will be needed,

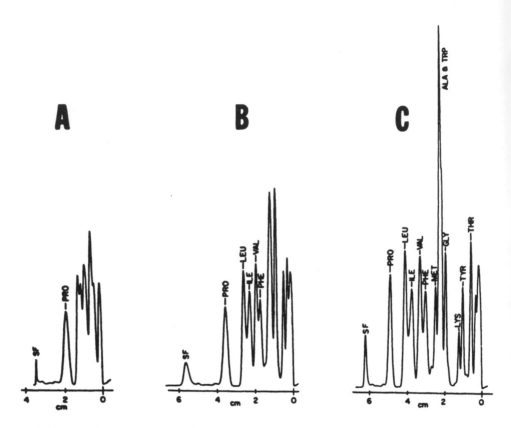

FIGURE 9 Separation of protein amino acid PTH-derivatives using
multiple development with changes in the mobile phase composition.
The stationary phase was a silica gel 60 HPTLC plate. From Ref.
76 with permission.

and at each subsequent development sequence, a substantial amount
of time is wasted while the solvent level reaches the level of the lowest spot on the plate. One possible solution to this problem would
be to decrease the solvent strength of the mobile phase and increase
the number of developments. This approach is not always effective
since at low solvent strengths streaking of the sample may be observed. An alternative approach is to use stronger solvents with
longer plate lengths, and to move the position of solvent introduction to higher positions on the plate at each successive development
(39,82), or to remove a portion of the lower edge of the plate at
some, or each, subsequent development segment (13,26). Provided
that the correct mobile phase velocity, number of development seg-

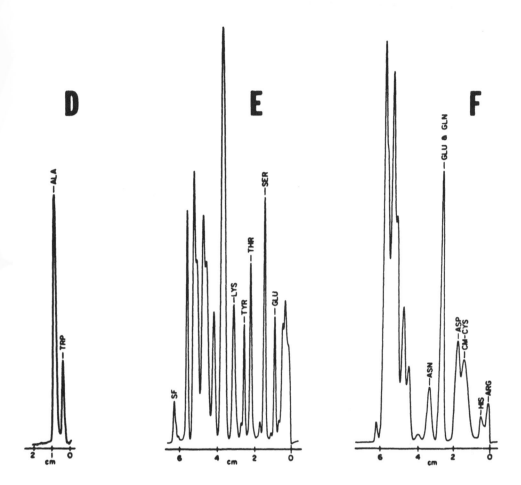

FIGURE 9   (Continued)

ments, and time for each development are selected then the plate cutting technique allows a spot to traverse virtually the whole length of the plate without experiencing significant band broadening beyond that introduced in the first few developments. This is illustrated in Figure 10, for the separation of a mixture of polycyclic aromatic hydrocarbons scanned after five developments in (A) and eleven developments in (B) (26). From the point of view of efficiency, by normal development an average value of approximately 2,000–3,000 theoretical plates was obtained; for normal multiple development this increased to 5,000–10,000; while, for multiple development with plate cutting at each segment a value of 15,000–25,000

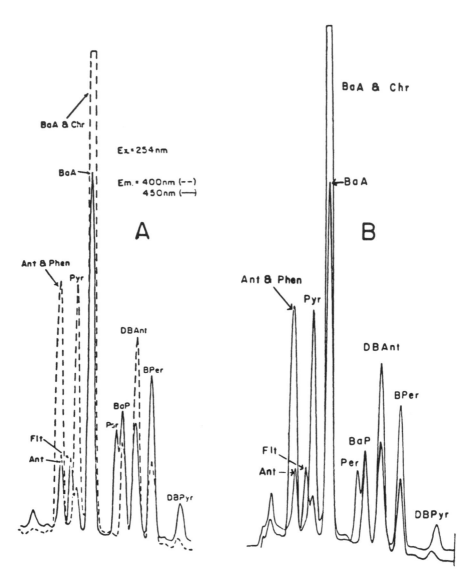

FIGURE 10 Separation of polycyclic aromatic hydrocarbons on octa-
decylsilanized silica gel HPTLC plates using the plate cutting technique.   The mobile phase was $CH_3OH-CH_3CN-H_2O$ (5:1:1).   The
plate was cut 0.5 cm from the lower edge before each successive development.   The multiple development sequence was five 10-min,
three 13-min, and two 15-min developments.   The separation obtained at the end of the fifth (A) and eleventh (B) development is
shown.   From Ref. 26 with permission.

was obtained.   In these experiments the mobile phase composition,
stationary phase, and total development time were identical and ten
segments were used in the multiple development experiments.

B.   Unidimensional Programmed Multiple Development

Programmed multiple development is a distinct version of unidimensional multiple development which can be fully automated (83—88).
It consists of a succession of initially brief but increasingly extended solvent advances separated by a controlled solvent removal step
while the plate remains in contact with the solvent reservoir.   The
plate is dried between developments by radiant heat, or by a flow
of inert gas, or both.   As the plate is in contact with the solvent
reservoir at all times, two spot reconcentration effects are observed
at each development segment.   Reconcentration during the solvent
advance was discussed earlier.   Reconcentration during solvent removal occurs because the solvent still flows in an upward direction
while the solvent front is receding.   Depending on the number of
developments and the R_F value of the substance in the solvent system, spots remaining close to the origin will be round, while those
close to the solvent front, sill nearly resemble lines at right angles
to the direction of solvent flow.   Programmed multiple development
provides improved resolution of complex mixtures and increased sample detectability compared to normal unidimensional development,
but generally at the expense of long analysis times.   It was difficult
to provide a predictive model for this technique, and except for a
few general theoretical observations (83), no complete model has
been developed.
     The instrumentation for programmed multiple development con-
sists of two units, a programmer and a developer.   The sequence
of operations is entered into the programmer which then controls
the developer, or developers, allowing unattended operation.   As
well as the commercially available units a simplified, inexpensive
version that could be easily assembled in any analytical laboratory
has been described (59).

C.   Unidimensional Automated Multiple Development

Automated multiple development has most of the characteristics of
programmed multiple development with some important differences
(31,34—38).   The separation sequence generally involves 10 to 25
individual developments with each development being longer than
the previous one by about 3—5 mm.   The mobile phase is removed
from the developing chamber and the plate dried under vacuum between each development.   A provision is also made for conditioning
the layer with an atmosphere of controlled composition prior to each

development. As fresh solvent is introduced into the developing chamber after each intermediate drying step, the use of solvent gradients with multiple developments, is easily accommodated. The use of eluents of gradually decreasing polarity for the individual chromatographic runs leads to a step gradient whose effect corresponds to a linear elution gradient, provided that a sufficient number of runs are performed. A separation of a multiple component test mixture carried out in this way is shown in Figure 11. The good resolution and sharp peak shapes observed indicate the general usefulness of this approach.

V. TWO-DIMENSIONAL DEVELOPMENT

So far the discussion of multidimensional TLC has dealt exclusively with unidimensional development. This method has the advantage of preserving the ability of the TLC plate to handle multiple samples (simultaneous separations) and of maintaining compatability with instrumental methods of recording chromatograms (scanning densitometry). In two-dimensional chromatography the sample is spotted at

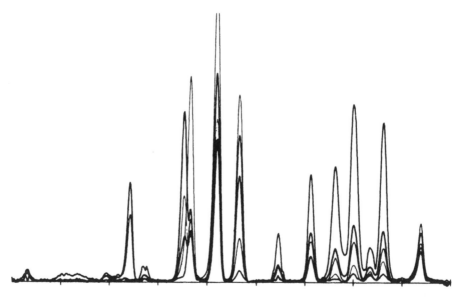

FIGURE 11 A separation of a test mixture using automated multiple development with a universal gradient from acetonitrile through dichloromethane to carbon disulfide on a silica gel HPTLC plate. The chromatogram was scanned at different wavelengths to enhance the chromatographic information. From Ref. 35 with permission.

the corner of the flat bed and developed along one edge of the plate.
The solvent is then generally evaporated, the plate rotated through
90°, and redeveloped in the orthogonal direction. If the same solvent is used for both developments then the sample will be redis-
tributed along a line from the corner at which the plate was spotted
to the corner diagonally opposite. In this case only a very slight
increase in resolution can be anticipated (a factor of $\sqrt{2}$) corresponding to the increased migration distance for the sample. The realiza-
tion of a more efficient separation system implies that the resolved
sample should be distributed over the entire plate surface. This
can be achieved only if the selectivity of the separation mechanism
is different in the orthogonal directions. Under these conditions
the spot capacity for two-dimensional TLC can be very large, on
the order of several thousand at the upper limit, and easily exceeds
that of column systems such as used in modern HPLC. The upper
limit of separation performance, however, will depend on the mode
of separation, whether by development or elution, and whether by
capillary or forced-flow dominated conditions. For our discussion,
it will be convenient to treat the different possibilities separately
as well as to look at the problems of optimization and detection
which remain the most critical obstacles to the wider use of two-
dimensional TLC. A comprehensive review of separations by two
dimensional TLC was published recently (18).

A. Separation by Development Under Capillary-Controlled Flow Conditions in Both Dimensions

For two-dimensional TLC under capillary-controlled flow conditions
it has been shown, in theory, that it should be possible to achieve
a spot capacity between 100 and 250, but difficult to reach 400 and
nearly impossible to exceed 500 except in very favorable circumstances (90). Such high spot capacities can only be achieved if
the separation mechanisms in the orthogonal direction are complementary to one another. In practice several methods have been
used to reach this goal and can be briefly summarized as follows
(18):
    1. A single sorbent layer is used and alternately developed in
orthogonal directions with two solvents exhibiting different selectivities for the sample components.
    2. A bilayer plate with two different sorbent layers abutting
each other is used, each layer effecting a separation according to
a different retention mechanism. Commercially available plates have
silica gel and reversed-phase layers on adjacent zones of the plate.
    3. A mixed sorbent layer can be prepared and developing sol-
vents selected such that the retention mechanism is dictated by the
properties of one of the sorbents in the mixture for each development.

4. Between the first and second developments the properties of the layer can be modified by impregnation with a reagent or solvent prior to the second development.

5. Between the first and second developments the properties of the sample are modified by a chemical reaction or derivatization prior to the second development.

The first approach is the easiest to implement in practice but requires that the two solvent systems provide substantial differences in selectivity for the sample to be successful in all but the simplest of problems. This may be very difficult to achieve in practice. Also, the solvent used in the first development conditions the sorbent layer that may modify the R_F values of the components in the second direction. The use of bilayer plates has become popular in recent years. The larger area of the plate used in the second development may be either normal or reversed-phase. To avoid distortion of the chromatogram it may be necessary to transfer the sample from the narrow to the larger layer using an intermediate development with a strong solvent. A useful alternative to bilayer plates is normal phase TLC on silica gel with simultaneous paraffin impregnation for subsequent reversed-phase separation in the second development (91). Liquid paraffin (7.5% V/V) is incorporated in the normal phase solvent used to obtain the initial separation. During chromatography in the first dimension there is a concomitant coating of the silica with paraffin that, following the evaporation of the solvent, results in a TLC plate that may be used for subsequent reversed-phase TLC in the second direction. Impregnating the sorbent layer with a selective reagent between developments is also relatively easy to carry out. For example, Grinberg and Weinstein (92) separated dansyl amino acids into their enantiomeric resolved forms by carrying out an initial separation of the amino acid derivatives on a reversed phase plate employing a convex gradient of aqueous sodium acetate buffer with varying concentrations of acetonitrile. Prior to development in the orthogonal direction the sorbent was impregnated with the chiral copper complex of N,N-di-n-propyl-L-alanine and developed in an aqueous sodium acetate-acetonitrile buffer in the second direction to provide the necessary selectivity for resolution of the enantiomers. Similarly, silver nitrate impregnation of silica gel has been used to effect a separation of saturated and unsaturated lipids after reversed-phase separation on a bilayer plate (93). As an example of the use of derivatization, Bakavolia et al. (94) separated pharmaceutically important benzodiazepines by normal phase chromatography followed by in situ acid hydrolysis to the benzophenones. Unresolved benzodiazepines in the first development were identified by separation of the benzophenones in the orthogonal direction.

B. Separation by Development in the First Dimension and Elution in the Second Under Forced-Flow Conditions

A further improvement in the resolving power of two-dimensional TLC can be achieved by using forced-flow development. In this case the plate length and particle size can be selected without any prejudicial influence on the solvent velocity which can be maintained at its optimum value. Theoretical calculations indicate that under forced flow development it should be relatively easy to generate spot capacities well in excess of 500 with an upper bound of several thousand depending on the choice of operating conditions (95).

Instrumentation for two-dimensional TLC with forced-flow development, also known as two-dimensional column chromatography or bidirectional column chromatography, is at an early state of development (95—97). The separation conditions are similar to those used in overpressured TLC. To facilitate detection elution is used in the second direction. The separated spots are eluted through a pair of quartz plates running along the complete length of the plate opposite to the solvent entry position. The quartz plates form the detection cell which is evenly illuminated with monochromatic UV light and focused onto a photodiode array. The separated spots can be reconstructed from the signal from the array as a three-dimensional plot of concentration against time with the plate length as abscissa. The detection and instrumental problems with this technique remain formidable and commercial instrumentation is unavailable at present.

C. Computer-Aided Optimization of Two-Dimensional Thin-Layer Chromatography

Several methods of computer simulation of two-dimensional chromatograms derived from unidimensional chromatographic experimental data have been devised using either visual inspection or a mathematical separation function to judge the quality of the simulated chromatograms (98—101). Since any spot in a two-dimensional chromatogram can be defined by a pair of x and y coordinates, the quality of a separation can be established by comparing the separation distance between all pairs of components in the chromatogram for any given combination of solvent systems and/or stationary phase layers. A straightforward computational approach is simply to seek the lowest correlation between unidimensional R_F values in each of two sequential solvent systems as was done by De Spiegeleer and co-workers (102). In this way twelve of fourteen local anesthetics were separated based on the optimum solvent system selected from those evaluated. Gonnord and co-workers (103) defined two mathematical functions, termed D_A and D_B, to judge the quality of simulated two-dimensional TLC separations using the R_F values for each solute in the two sequential solvent systems as the x and y coordinates.

$$D_A = \sum_{i=1}^{k-1} \sum_{j=i+1}^{k} (x_i - x_j)^2 + (y_i - y_j)^2 \tag{21}$$

$$D_B = \sum_{i=1}^{k-1} \sum_{j=i+1}^{k} \frac{1}{(x_i - x_j)^2 + (y_i - y_j)^2} \tag{22}$$

The function D_A represents the sum of the square of the distance between pairs of spots for all possible pairs. Optimization is achieved by maximizing D_A. The function D_B represents the sum of the inverse of the square of the distance between pairs of spots for all possible pairs. Optimization is achieved by eliminating unresolved pairs (or by defining an arbitrary minimum separation distance for all unresolved pairs) and then minimizing D_B. The function D_A gives equal weight to all spot pairs whereas D_B gives more weight to poorly separated pairs, and consequently, D_B tends to select fewer optimum separations with overlapping spots than D_A. Steinbrunner et al. (104) have used the above separation functions to predict the separation of a mixture of steroids by continuous development two-dimensional TLC using eq. (17) for each set of unidimensional data. The optimization strategy was further simplified by restricting the analysis to the maximum available plate length for the chamber used in the development and setting the analysis time in each dimension equal to the time required for the fastest moving spot to reach the end of the TLC plate for all mobile phase compositions. This reduces the number of variables from six to two, namely, the solvent composition in each dimension. Good agreement between the experimental and computer simulated chromatograms (see Figure 12) was found.

Nurok and coworkers have since modified the above approach and introduced four new separation functions (100,105,106). The distance function, D_F, and the inverse distance function, IDF, represent simple modifications to the functions D_A and D_B in which the sum of the distances between pairs of spots is taken in place of the sum of the squares as indicated in eqs. 21 and 22 as the basis for the separation function. For the IDF function, spots that are less than 1 mm apart are assigned an arbitrary separation distance of 1 mm for the same reason as was indicated for D_B. The critical pair separation function uses the distance between the most poorly separated pair of spots in a mixture as a measure of the quality of the separation. The planar response function, PRF, is defined by eq. 23

$$PRF = \sum_{i=1}^{k-1} \sum_{j=i+1}^{k} \ln \frac{S_D^{ij}}{S_D^*} \tag{23}$$

FIGURE 12 Computer-simulated (left) and experimental (right) chromatograms for the two-dimensional TLC separations of a steroid mixture performed on a bilayer plate. The vertical axis is a butyl ace-
tate-toluene development (mole fraction 0.1, development time 1430 s)
on silica gel and the horizontal axis is an acetonitrile-aqueous 0.5 M
sodium chloride development (mole fraction 0.4, development time
900 s) on a reversed-phase layer. The solvent path length was
fixed at 8.3 cm in both dimensions. From Ref. 104 with permission.

where k is the number of solutes in the mixture, S_D^{ij} the actual spot
separation and S_D^* the desired spot separation. All solute pairs with
$S_D^{ij} > S_D^*$ are assigned the value of S_D^* and made a zero contribution
to the PRF. Thus, the PRF considers only poorly separated pairs
of spots whereas the IDF considers all pairs.

No single criterion is ideal for evaluating all two-dimensional
chromatograms. The critical pair criterion is useful when the only
requirement is that all solute pairs be separated by a defined minimum distance. It provides information about only the most poorly
separated solute pair in a mixture and provides no information about
other unseparated pairs that may be present. In such cases the
PRF and IDF will provide a better indication of which two-dimensional system will provide the best overall spot separation. DF is in-
sensitive to the presence of poorly separated spots and, therefore,
is a poor criterion of how well a specific mixture is separated. It
is, however, a good criterion of how well dispersed the spot centers
are across the plate surface and may be of value for evaluating the
separation of very complex mixtures where spot overlap is inevitable
and where utilizing all the area of the TLC plate is important.

There can be no guarantee that there will be good agreement between the experimental and the corresponding computer-simulated

chromatogram. Plate to plate variations in the solvent velocity constant, changes in experimental conditions (for example, humidity, age of TLC plates), and solvent demixing can all adversely effect the agreement obtained. Published examples show good agreement in overall spot pattern but some errors in actual spot positions. Different separation functions may predict different optimum experimental conditions for separation leading to confusion in making a final selection. Also, the values for the individual separation functions gives no indication of whether an optimum value is narrow or broad. This latter problem may be solved by representing the data in the form of contour diagrams (105). However, since computer simulation requires only a fraction of the time and materials required for performing actual separations this approach is very promising for aiding methods development by two-dimensional TLC.

D. Limitations of Detection Systems for Two-Dimensional Thin-Layer Chromatography

After development, the sample components are distributed evenly (ideally) over the whole plate surface. To record this information, the detector must locate the exact x and y coordinates of all spots for qualitative analysis (identification) and the whole spot profile, in both directions, must be integrated for quantitative analysis. Mechanical slit-scanning densitometers designed for recording unidimensional chromatograms are unsuited to the above task. To define the spot profile a minimum of ten data points are required. These would have to be obtained by scanning the spot with a beam of small dimensions as a series of parallel scans of small displacement. Extending this type of scanning to the whole plate would require from several hours to days. Also, software would have to be developed to align these scans, detect the spots, and segment and integrate them. Software has not been developed for this purpose but, presumably, would be similar to software developed for scanning two-dimensional electrophoretic gels. High resolution scanners developed for two-dimensional electrophoretic gels cannot be successfully applied to two-dimensional TLC because of excessive noise developed at very small beam dimensions arising from the granular nature of the plate surface. A lack of a suitable, commercially available, and reliable detector for two-dimensional TLC chromatograms remains one of the principal reasons for the limited use of two-dimensional TLC.

An alternative to slit-scanning densitometry will be needed for progress to be made in quantitative two-dimensional TLC. One potential solution, the use of a photodiode array to detect spots eluted from one edge of the plate was discussed in Section V.B. An alternative approach is electronic scanning of the plate using advanced image analysis techniques (15,16,107—109).

The equipment requirements for image analysis are a computer
with video digitizer, light source and appropriate optics such as
lenses, filters, and monochromators, and an imaging detector such
as a vidicon tube or charged-coupled video camera (107—119). Measurements can be made by transmission (114,115) or reflectance (111,
117,118) in the UV-visible region for absorption or by reflectance
for fluorescence (110,115). Suitable arrangements for image analysis in the transmission and reflectance modes are shown in Figure
13 (107). The plate is evenly illuminated with monochromatic light
and the reflected or transmitted light focused as a scaled image of
the plate directly onto the active element of the vidicon. To obtain
wavelength discrimination for spectroscopic identification or spectroscopic resolution of overlapping spots a second monochromator between the plate and the camera is needed. Similarly, for fluorescence analysis either a filter or monochromator is required to shield
the vidicon from light originating from the source. The vidicon
functions as a two-dimensional array of unit detectors continuously
scanned by an electron beam. These unit detectors are periodically
discharged by the electron beam and the signal digitized for analysis by computer. Image resolution is limited by the number of detectors, known as picture elements or pixels, forming the detector
array. This is typically 512 × 512 pixels for commercial vidicon cam-
eras and falls somewhat short of the resolution needed for quantitive
two-dimensional TLC (90). Sensitivity and linearity of response are
problems for the simultaneous recording of trace and major components from the same plate.

For analysis the captured images are collected, stored, and
transformed by computer into chromatographic data. To a large extent the number of images that can be stored and the type of calculation algorithms used is a function of the capacity of the computer.
Image analysis is a data intensive technique, and although many of
the simpler systems described employ personal computers for data
analysis, their facilities for image storage, image subtraction, filtering, thresholding, and the use of false color for image enhancement,
is greatly restricted. Common to all techniques of data handling is
background subtraction in which the accumulated images of a blank
plate are subtracted from the analytical plate, preferably, or a shut-
ter is placed in front of the camera and background "noise" from
the detector array subtracted on a pixel by pixel basis. Thresholding is also routinely used to ensure that negative values for the
plate luminance do not occur. In most cases details of the methods
used to convert images into x and y coordinates and density measurements from which concentrations are obtained by calibration are
not given. An exception is the matrix annihilation approach used
by Burns et al. (117). Clearly, the future of image analysis as a
detection technique for two-dimensional TLC requires the develop-

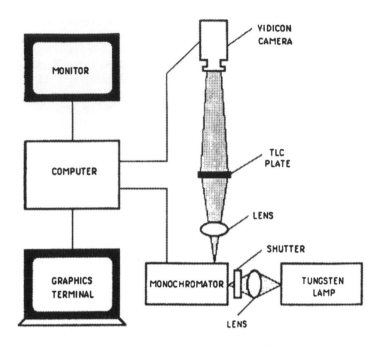

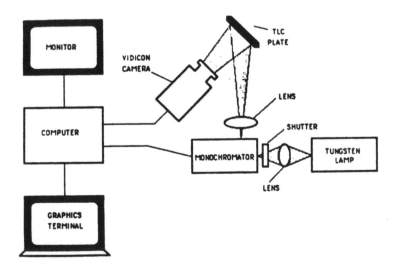

FIGURE 13 Optical layout for an image analyzer used to record two-dimensional TLC chromatograms in the transmission (top) and reflectance (bottom) mode. From Ref. 107 with permission.

ment of both new hardware and software to make it as routine as slit-scanning densitometry. With technological advances that are occurring quite rapidly in the general area of image analysis, it seems reasonable to anticipate that the current equipment limitations will be rolled back and image analysis will one day be a viable detection method for TLC without compromising chromatographic resolution, sample detectability, or the calibration response range.

VI.   BIMODAL CHROMATOGRAPHY INCORPO-
RATING THIN-LAYER CHROMATOGRAPHY
AS ONE DIMENSION

Interfacing gas or liquid chromatography to thin-layer chromatography is not particularly difficult but is not widely practiced. Gas chromatography/thin-layer chromatography (GC-TLC) enjoyed some popularity in the late 1960s when several working instruments were developed (20). The gas chromatographic effluent was applied directly to the TLC plate using a heated applicator as shown in Figure 14 (120). In practice the capillary orifice must have a diameter

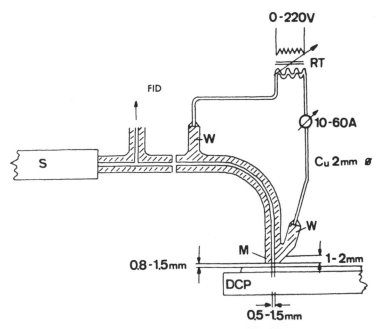

FIGURE 14   Heated applicator for applying a gas chromatographic effluent to a thin-layer plate. A splitter to a flame ionization detector (FID) is included to enable one-dimensional GC data to be recorded. From Ref. 120 with permission.

of 1 to 2 mm for normal, packed column flow rates. The position
of the jet with respect to the plate is important and should be maintained within 1 to 2 mm of the plate surface. This should ensure
that the recovery of sample will exceed 80% without damaging the
layer while, at the same time, focusing the sample as a narrow band
on the plate. During sample deposition the plate is moved at a fixed
speed in a continuous or discontinuous manner. Discontinuous mo-
tion is used to collect fractions that are subsequently resolved by
TLC. One critical feature of sample deposition is that the sample
components must be either overlayed or spaced adequately to prevent lane merging on development. As TLC spots develop they
spread in all directions simultaneously, and consequently, their initial spot centers must be separated by a distance greater than one
spot diameter if they have similar R_F values in the solvent system
used for the thin-layer development. An example of a GC-TLC separation is shown in Figure 15 (121). A summary of applications,
largely qualitative in nature, is given in reference 20. In most instances TLC was used to either confirm the identity of a GC peak
or to test for peak homogeneity. This problem is generally solved
by gas chromatography/mass spectrometry (GC/MS) today.

    In most organic synthesis and natural products laboratories TLC
is used routinely to characterize fractions obtained from liquid chromatographic columns (122). Here the attraction of TLC is its ability
to separate multiple samples simultaneously and thus allow fractions
of similar composition to be identified and grouped together for further analysis, if needed. This process is usually performed off-
line perhaps aided by an automatic fraction collector. van Dijk has
described the use of an effluent splitter to simultaneously apply a
minor fraction of the column eluent to a TLC plate with the remainder going to a fraction collector (123). The column eluent is applied to the plate by a narrow tube held in close proximity to the
TLC plate that was moved continuously beneath the applicator.
Forced convection was used to evaporate the solvent and maintain
the sample applied to the plate as a focused narrow band. Boshoff
et al. (124) have used TLC as a transport detector for high performance liquid chromatography (HPLC). The interface, which includes
a splitter to match the column flow to the adsorption capacity of the
thin-layer plate (0.6 mL/min.), is shown in Figure 16. The eluent
was transferred by a fine steel capillary tube (0.25 mm i.d.) to the
surface of the TLC plate. The gap between the steel capillary and
the plate was adjustable, but for the best results, a solvent bridge
between the capillary and plate surface was maintained at all times.
To aid rapid solvent evaporation the TLC plate was mounted on a
temperature-controlled brass block and the capillary applicator
shrouded by a vacuum manifold. Samples transferred to the plate
were then detected by fluorescence induction. Thus, the TLC plate

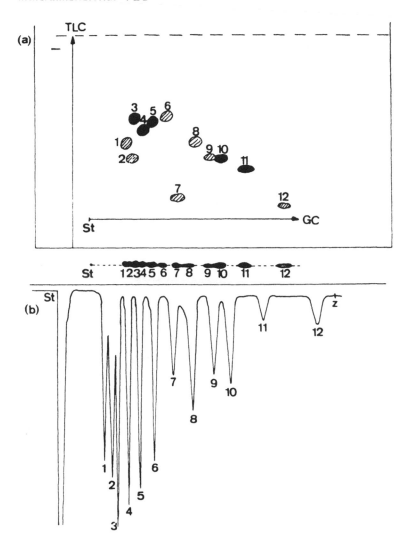

FIGURE 15 GC-TLC separation of a mixture of twelve trimethylsilyl ether steroid derivatives. The lower portion (b) is the one dimensional GC separation and the upper portion (a) the bimodal GC-TLC separation. From Ref. 121 with permission.

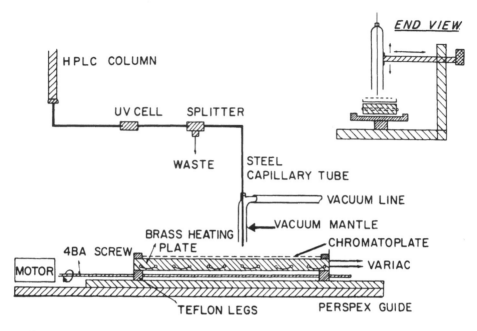

FIGURE 16 Interface for depositing the effluent from an HPLC
column to a TLC plate moved mechanically below the applicator.
From Ref. 124 with permission.

was used as a storage detector, and no chromatography was involved.
Still, the interface described, would function well for bimodal (HPLC-
TLC) separations. Similarly, Karmen et al. (125,126) have described
the use of a contact spotter as a microfraction collector for determining the concentration of low-activity radiolabeled compounds in an
HPLC eluent. In this case the eluent was fractionated in a continuous manner into selected aliquots of 100 to 350 µl, the solvent evaporated in a controlled manner, and all the spots transferred simultaneously to a paper strip impregnated with a scintillator for recording by autoradiography. Again no chromatography was involved,
but the approach should be applicable for HPLC-TLC applications.
Hofstraat et al. (127–129) described an interface for depositing the
effluent from a narrow bore HPLC column (0.7–1.0 mm) onto a TLC
plate with partial removal of the solvent. The interface consisted
of a 30 cm × 100 µm i.d. fused silica capillary and a Linomat III
(Camag, Muttenz, Switzerland) spray-jet assembly. The deposited
samples were suitable for spectroscopic identification, the TLC plate
acting as a storage device for spectroscopic techniques requiring immobilized analytes.

Supercritical fluid chromatography/thin-layer chromatography (SFC-TLC) remains an unexplored technique. Stahl has described an apparatus for extraction by supercritical fluids with direct application of the fluid to a thin-layer plate (130,131). A very fine capillary (25 to 50 μm) was required to accomodate the large pressure drop between the high pressure fluid used for extraction and the low pressure gas phase applied to the layer. The TLC plate was moved backwards and forwards below the capillary and the extracted substances were precipitated as a 2 cm band. Component identification was then performed by thin-layer chromatography in the usual way.

VII. CONCLUSIONS

The revolution in modern TLC that contributed to its present status as a powerful separation tool postdated developments in high performance gas and column liquid chromatography (132). Thus, experience with modern TLC lags behind that of the column techniques and our understanding of its optimization is less crystalline. However, within this framework certain trends now seem well established. For difficult separations unidimensional multiple development employing several segments incorporating continuous development or step changes in solvent composition is the method of choice. Multidimensional development has the advantages of requiring simple equipment, is compatible with scanning densitometry for solute identification and quantitation, and enables the spot reconcentration mechanism to be exploited for minimizing zone broadening. When more sophisticated equipment is available, forced-flow development enables the chromatographic conditions to be optimized independently of the plate length and particle size, and long plates to be used for difficult separations. The principal limitation of the technique at the moment is that equipment for forced-flow development is in use in very few laboratories and it seems that further efforts are needed to popularize the technique. Theoretical considerations indicate that for the most difficult separations two-dimensional forced flow TLC should be used. Reaching the very high spot capacity available in this case will require further advances in instrumentation, in particular, sample detection. The detection problem is not satisfactorily solved at present so the technique is used primarily for qualitative analysis.

True bimodal systems incorporating TLC have been rarely used. Interfacing TLC to GC, HPLC, or SFC is not particularly difficult and can be achieved within established technology. The driving force for such systems does not seem to have been established largely because of the perception that column systems provide superior separation capacity as well as the difficulty of quantifying the TLC

68 Poole and Poole

chromatograms obtained. Here, image analysis could provide a breakthrough not only in in situ quantitation, but also on account of its ability to simultaneously provide spectroscopic data for all spots in the chromatogram. The thin-layer plate could serve as an efficient storage detector for those techniques, such as HPLC and SFC, where on-line detection lacks adequate sensitivity or selectivity for the analysis of some samples.

REFERENCES

1. A. Zlatkis and R. E. Kaiser, eds. HPTLC: High Performance Thin-Layer Chromatography, Elsevier, Amsterdam, 177.
2. W. Bertsch, S. Hara. R. E. Kaiser, and A. Zlatkis, eds., Instrumental HPTLC, Huethig, Heidelberg, 1980.
3. C. F. Poole and S. A Schuette, Contemporary Practice of Chromatography, Elsevier, Amsterdam, p. 619, 1984.
4. R. E. Kaiser, ed., Planar Chromatography, Vol. 1, Huethig, Heidelberg, 1986.
5. B. Fried and J. Sherma, Thin-Layer Chromatography. Techniques and Applications, Marcel Dekker, Inc., New York, 1986.
6. F. Geiss, Fundamentals of Thin-Layer Chromatography, Huethig, Heidelberg, 1987.
7. F. A. A. Dallas, H. Read, R. J. Ruane, and I. D. Wilson, eds., Recent Advances in Thin-Layer Chromatography, Plenum, London, 1988.
8. D. C. Fenimore and C. M. Davis, Anal. Chem., 53:252A (1981).
9. M. E. Coddens, H. T. Butler, S. A. Schuette, and C. F. Poole, LC-GC Magzn. 1:282 (1983).
10. C. F. Poole, M. E. Coddens, H. T. Butler, S. A. Schuette, S. S. J. Ho, S. Khatib, L. Piet, and K. K. Brown, J. Liq. Chromatogr. 8:2875 (1985).
11. C. F. Poole, Trends Anal. Chem., 4:209 (1985).
12. C. F. Poole, H. T. Butler, M. E. Coddens, and S. A. Schuette, in Analytical and Chromatographic Techniques in Radiopharmaceutical Chemistry (D. M. Wieland, M. C. Tobes, and T. J. Manger, eds.), Springer-Verlag, New York, p. 3, 1986.
13. S. Khatib and C. F. Poole, LC-GC Magzn. 4:228 (1986).
14. C. F. Poole, S. Khatib, and T. A. Dean, Chromatogr. Forum, 1:27 (1986).
15. C. F. Poole and S. Khatib, in Quantitative Analysis Using Chromatographic Techniques (E. D. Katz, ed.), Wiley, Chichester, p. 193 (1987).
16. C. F. Poole and S. K. Poole, J. Chromatogr., 492:539 (1989).

17. K. Y. Lee, C. F. Poole, and A. Zlatkis, in Instrumental Thin-Layer Chromatography (W. Bertsch, S. Hara, R. E. Kaiser, and A. Zlatkis, eds.), Huethig, Heidelberg, p. 245 (1980).
18. M. Zakaria, M.-F. Gonnord, and G. Guiochon, J. Chromatogr., 271:127 (1983).
19. D. Rogers, Amer. Lab. 16:65 (1984).
20. J. Janak, in Progress in Thin-Layer Chromatography and Related Methods Volume II (A. Niederwieser and G. Pataki, eds.), Ann Arbor Science Publishers, Ann Arbor, MI, p. 63, 1971.
21. H. Halpaap and J. Ripphahn, Chromatographia, 10:613 (1977).
22. H. E. Hauck and H. Halpaap, Chromatographia, 13:538 (1980).
23. W. Jost and H. E. Hauck, J. Chromatogr., 261:235 (1983).
24. H. Halpaap, K.-F. Krebs, and H. E. Hauck, HRC & CC, 3:215 (1980).
25. U. A. Th. Brinkman and G. de Vries, HRC & CC, 5:476 (1982).
26. C. F. Poole, H. T. Butler, M. E. Coddens, S. Khatib, and R. Vandervennet, J. Chromatogr., 302:149 (1984).
27. H. E. Heilweil and F. M. Rabel, J. Chromatogr. Sci., 23:101 (1985).
28. U. A. Th. Brinkman and D. Kamminga, J. Chromatogr. 330: 375 (1985).
29. W. Jost, H. E. Hauck, and W. Fischer, Chromatographia, 21: 375 (1986).
30. T. E. Beesley, J. Chromatogr. Sci., 23:525 (1985).
31. D. E. Jaenchen and H. J. Issaq, J. Liq. Chromatogr., 11: 1941 (1988).
32. D. C. Fenimore and C. J. Meyers, J. Chromatogr., 186:555 (1979).
33. G. Malikin, S. Lam, and A. Karman, Chromatographia, 18:253 (1984).
34. D. E. Janchen, in Proceedings of the Second International Symposium on Instrumental HPTLC (R. E. Kaiser, ed.), Institute for Chromatographia, Bad Durkheim, F. R. G., p. 220, 1982.
35. K. Burger and H. Tengler, in Planar Chromatography, Volume 1 (R. E. Kaiser, ed.), Huethig, Heidelberg, p. 193, 1986.
36. K. Burger, Fresenius Z. Anal. Chem., 318:228 (1984).
37. D. E. Jaenchen, Amer. Labor., March:66 (1988).
38. I. D. Wilson and S. Lewis, J. Chromatogr., 408:445 (1987).
39. P. Buncak, Fresenius' Z. Anal. Chem., 318:289 (1984).
40. K. Kalasz, Chromatographia, 18:628 (1984).
41. Z. Witkiewicz and J. Bladek, J. Chromatogr., 373:111 (1986).
42. E. Tyihak and E. Mincsovics, J. Planar Chromatogr.-Modern TLC, 1:6 (1988).

43. C. F. Poole, J. Planar Chromatogr. - Modern TLC, 2:95 (1989).
44. Sz. Nyiredy, C. A. J. Erdelmeier, and O. Sticher, in Planar Chromatography, Volume 1 (R. E. Kaiser, ed.), Huethig, Heidelberg, p. 119, 1986.
45. E. Tyihak, E. Mincsovics, and H. Kalasz, J. Chromatogr., 174:75 (1979).
46. R. E. Kaiser and R. I. Reider, in Planar Chromatography, Volume 1 (R. E. Kaiser, ed.), Huethig, Heidelberg, p. 165, 1986.
47. R. E. Kaiser, Einführung in die HPPLC, Huethig, Heidelberg, 1987.
48. E. Stahl and J. Muller, Chromatographia, 15:493 (1982).
49. Sz. Nyiredy, C. A. J. Erdelmeier, and O. Sticher, HRC & CC, 8:73 (1985).
50. Sz. Nyiredy, S. Y. Meszaros, K. Dallenbach-Tolke, K. Nyiredy-Mikita, and O. Sticher, J. Planar Chromatogr. - Modern TLC, 1:54, 1988.
51. R. E. Kaiser, HRC & CC, 1:164 (1978).
52. V. Pollock, Adv. Chromatogr., 17:1 (1979).
53. I. M. Bohrer, Topics Current Chem., 126:95 (1984).
54. L. R. Treiber, ed., Quantitative Thin-Layer Chromatography and Its Industrial Applications, Marcel Dekker, Inc., New York, 1987.
55. G. Guiochon and A. Siouffi, J. Chromatogr. Sci., 16:598 (1978).
56. G. Guiochon, G. Korosi, and A. Siouffi, J. Chromatogr. Sci., 18:324 (1980).
57. U. A. Th. Brinkman, Trends Anal. Chem., 5:178 (1986).
58. G. Guiochon, A. Siouffi, H. Engelhardt, and I. Halasz, J. Chromatogr. Sci., 16:152 (1978).
59. G. Guiochon and A. Siouffi, J. Chromatogr. Sci., 16:470 (1978).
60. C. F. Poole, J. Planar Chromatogr. - Modern TLC, 1:373 (1988).
61. C. F. Poole and S. K. Poole, J. Planar Chromatogr. - Modern TLC, 2:166 (1989).
62. G. Guiochon and A. M. Siouffi, J. Chromatogr., 245:1 (1982).
63. H. F. Hauck and W. Jost, J. Chromatogr., 262:113 (1980).
64. E. Mincsovics, E. Tyihak, and H. Kalasz, J. Chromatogr., 191:293 (1980).
65. G. Guiochon, L. A. Beaver, M.-F. Gonnord, A. M. Siouffi, and M. Zakaria, J. Chromatogr., 255:415 (1983).
66. E. Soczewinski, in Planar Chromatography, Volume 1 (R. E. Kaiser, ed.), Huethig, Heidelberg, p. 79, 1986.
67. J. A. Perry, J. Chromatogr., 165:117 (1979).

68. S. Fater and E. Mincsovics, J. Chromatogr., 298:534 (1984).
69. H. Gulyas, G. Kemeny, I. Hollosi, and J. Pucsok, J. Chromatogr., 291:471 (1984).
70. J. Vajda, L. Leisztner, J. Pick, and N. Ahn-Tuan, Chromatographia, 21:152 (1986).
71. D. Nurok, R. E. Tecklenburg, and B. L. Maidak, Anal. Chem., 56:293 (1984).
72. D. Nurok, R. M. Becker, and K. A. Sassic, Anal. Chem., 54:1955 (1982).
73. R. E. Tecklenburg, B. L. Maidak, and D. Nurok, HRC & CC, 6:627 (1983).
74. R. E. Tecklenburg and D. Nurok, Chromatographia, 18:249 (1984).
75. R. E. Tecklenburg, G. H. Fricke, and D. Nurok, J. Chromatogr., 290:75 (1984).
76. S. A. Schuette and C. F. Poole, J. Chromatogr., 239:251 (1982).
77. A. Jeans, C. S. Wise, and R. J. Dimler, Anal. Chem., 23:425 (1951).
78. H. P. Lenk, Fresenius' Z. Anal. Chem., 184:107 (1961).
79. G. Goldstein, Anal. Chem., 42:140 (1970).
80. J. A. Thoma, J. Chromatogr., 12:441 (1963).
81. J. A. Thoma, Anal. Chem., 35:214 (1963).
82. D. Nurok, Anal. Chem., 53:714 (1981).
83. T. H. Jupille and J. A. Perry, J. Chromatogr., 99:231 (1974).
84. J. A. Perry, T. H. Jupille, and L. J. Glunz, Sepn. Purifn. Methods, 4:97 (1975).
85. T. H. Jupille and J. A. Perry, J. Chromatogr. Sci., 13:163 (1975).
86. J. A. Perry, T. H. Jupille, and L. J. Glunz, Anal. Chem. 47:65A (1975).
87. J. A. Perry, J. Chromatogr. 113:267 (1975).
88. T. H. Jupille, J. Assoc. Am. Oil Chem. Soc., 54:179 (1976).
89. R. von Wandruska and F. Gottschalk, Rev. Sci. Instrum., 57:119 (1986).
90. G. Guiochon, M.-F. Gonnord, A. Siouffi, and M. Zakaria, J. Chromatogr., 250:1 (1982).
91. I. D. Wilson, J. Chromatogr., 287:183 (1984).
92. N. Grinberg and S. Weinstein, J. Chromatogr., 303:251 (1984).
93. M. H. Jee and A. S. Ritchie, J. Chromatogr., 299:460 (1984).
94. M. Bakavolia, V. Navaratnam, and N. K. Nair, J. Chromatogr., 299:465 (1984).
95. G. Guiochon, L. A. Beaver, M.-F. Gonnord, A. M. Siouffi, and M. Zakaria, J. Chromatogr., 255:415 (1983).
96. G. Guiochon, M.-F. Gonnord, L. A. Beaver, and A. M. Siouffi, Chromatographia, 17:121.

97. G. Guiochon, A. Krstulovic, and H. Colin, J. Chromatogr.,
     265:159 (1983).
98. D. Nurok, Chem. Revs., in press (1989).
99. D. Nurok, LC/GC Magzn., 6:310 (1988).
100. S. Habibi-Goudarzi, K. J. Ruterbories, J. E. Steinbrunner,
     and D. Nurok, J. Planar Chromatogr. - Modern TLC, 1:161
     (1988).
101. E. K. Johnson and D. Nurok, J. Chromatogr., 302:135 (1984).
102. B. De Spiegeleer, W. Van den Bossche, P. De Moerloose, and
     D. Massart, Chromatographia, 23:407 (1987).
103. M.-F. Gonnord, F. Levi, and G. Guiochon, J. Chromatogr.,
     264:1 (1983).
104. J. E. Steinbrunner, E. K. Johnson, S. Habibi-Goudariz, and
     N. Nurok, in Planar Chromatography, Vol. 1 (R. E. Kaiser,
     ed.), Huethig, Heidelberg, p. 239, 1986.
105. J. E. Steinbrunner, D. J. Malik, and D. Nurok, HRC & CC,
     10:560 (1987).
106. D. Nurok, S. Habibi-Goudarzi, and R. Kleyle, Anal. Chem.,
     59:2424 (1987).
107. D. H. Burns, J. B. Callis, and G. D. Christian, Trends
     Anal. Chem., 5:50 (1986).
108. V. A. Pollak and J. Schulze-Clewing, J. Chromatogr., 437:97
     (1988).
109. V. A. Pollak and J. Schulze-Clewing, J. Liq. Chromatogr.,
     11:1387 (1988).
110. M. L. Gianelli, J. B. Callis, N. H. Andersen, and G. D.
     Christian, Anal. Chem., 53:1357 (1981).
111. M. L. Gianelli, D. H. Burns, J. B. Callis, G. D. Christian,
     and N. H. Andersen, Anal. Chem., 55:1858 (1983).
112. P. Elodi and T. Karasi, J. Liq. Chromatogr., 3:809 (1980).
113. S. Pongor, J. Liq. Chromatogr., 5:1583 (1982).
114. W. E. Neeley, D. Epstein, and A. Zettner, Clin. Chem., 27:
     1665 (1981).
115. T. S. Ford-Holevinski, B. W. Agranoff, and N. S. Radin,
     Anal. Biochem., 132:132 (1983).
116. D. D. Reese, K. E. Fogarty, L. K. Levy, and F. S. Fay,
     Anal. Biochem., 144:461 (1984).
117. D. H. Burns, J. B. Callis, and G. D. Christian, Anal. Chem.,
     58:1415 (1986).
118. R. M. Belchamber, H. Read, and J. D. M. Roberts, in Planar
     Chromatography, Vol. 1 (R. E. Kaiser, ed.), Huethig, Heidel-
     berg, p. 207 (1986).
119. M. Prosek, M. Medja, J. Korsic, M. Pristav, and R. E. Kaiser,
     in Planar Chromatography, Vol. 1 (R. E. Kaiser, ed.),
     Huethig, Heidelberg, p. 221 (1986).
120. R. E. Kaiser, Fresenius' Z. Anal. Chem., 205:284 (1964).

121. H. C. Curtius and M. Muller, J. Chromatogr., 32:222 (1968).
122. J. H. P. Tyman and V. Tychopoulos, J. Planar Chromatogr. -
      Modern TLC, 1:227 (1988).
123. J. H. van Dijk, Fresenius' Z. Anal. Chem., 247:262 (1969).
124. P. R. Boshoff, B. J. Hopkins, and V. Pretorius, J. Chroma-
      togr., 126:35 (1976).
125. A. Karmen, G. Malikin, and S. Lam, J. Chromatogr., 302:31
      (1984).
126. A. Karmen, G. Malikin, L. Freundlich, and S. Lam, J. Chro-
      matogr., 349:2677 (1985).
127. J. W. Hofstraat, M. Engelsma, R. J. van de Neese, C.
      Gooijer, N. H. Velthost, and U. A. Th. Brinkman, Anal.
      Chim. Acta, 186:247 (1986).
128. J. W. Hofstraat, M. Engelsma, R. J. van de Nesse, U. A.
      Th. Brinkman, G. Gooiger, and N. H. Velthorst, Anal. Chim.
      Acta. 193:193 (1987).
129. J. W. Hofstraat, S. Griffioen, R. J. van de Nesse, U. A.
      Th. Brinkman, C. Gooijer, and N. H. Velthorst, J. Planar
      Chromatogr. - Modern TLC, 1:220 (1988).
130. E. Stahl and W. Schilz, Fresenius' Z. Anal. Chem., 280:99
      (1976).
131. E. Stahl, J. Chromatogr., 142:15 (1977).
132. C. F. Poole and S. K. Poole, Anal. Chim. Acta, 216:109
      (1989).

3

Multidimensional Gas Chromatography

WOLFGANG BERTSCH *The University of Alabama, Tuscaloosa, Alabama*

Advances in two-dimensional gas chromatography are highlighted in
this chapter. It is not the intention of the author to provide a historical overview but rather to focus on recent developments in the
field, especially advances relating to capillary columns. To keep
this chapter reasonably short, practical aspects have been emphasized. No attempt has been made to incorporate elements of information theory.

I. INTRODUCTION

A. Chromatography and Information

Within all chromatographic methods, gas chromatography is the most
widely used technique. It is primarily a tool for analysis, in particular for the quantitative determination of individual components
in mixtures. A general feature of analysis is that it is an information gathering process that permits a description of the sampled system. Sometimes, the analytical problem at hand does not require
optimization, or the cost of optimization outweighs the benefits expected. There are, however, many instances where a fair degree
of optimization has to be performed just to get some results at all.
Information theory (1) deals with obtaining information about the
qualitative or quantitative makeup of a sample, based on probability.
Kaiser (2) related the information power of an analytical method to
the signal-to-noise ratio and resolution achieved. Producing information can be considered as an act where the uncertainty in respect
to sample identity, quantity, of composition is reduced (3,4). Two
performance characteristics have been proposed for identification of

a substance by gas chromatography: information content and discriminating power (4,5).   The elimination of uncertainty is the information, not the event itself.   It is expressed in bits.

In gas chromatography, substance identity can be established
on the basis of retention indices or some other measure of retention.
The quality of the information and the degree of separation that can
be produced depends on the discriminating power, e.g., selectivity,
of the stationary phase for the solute or solutes under investigation
and the experimental error.   Huber (5) correlated the information
content of multidimensional gas chromatography and mass spectrometry for drug identification.   As expected, data from polar phases
have a higher information content than from nonpolar ones.   The
correlation of retention data from different phases has the highest
discriminating value if the stationary phases are as dissimilar as
possible.

In gas chromatography, at least three pieces of information can
be obtained for a substance with some degree of certainty, even
for the simplest case: a measure of retention such as retention
index, a description of peak dispersion such as peak width and a
measure of quantity as provided by the detector output signal.   In
simple cases, it is possible to establish the presence or absence of
a substance with a high degree of certainty by carrying out qualitative analysis on the basis of retention data.   In practical situations, ambiguities often arise, forcing the chromatographer to employ complementary methods.   When retention data are produced
with care and interpreted with caution, even difficult analytical
problems may be solved without referring to ancillary methods such
as mass spectrometry.   It is not widely known that retention data
in gas chromatography are reproducible to better than 0.1 index
units if special precautions are taken (6-8).   The inherent precision possible for gas-chromatographic retention data can rival that
of spectroscopic methods.   Unfortunately there are significant practical obstacles to be overcome.

The power of retention data in qualitative analysis (9) has only
recently been rediscovered and given a significant boost.   Progress
has been made in the understanding of the fundamental variables
that influence retention indices in gas chromatography (10,11), including those obtained under conditions of temperature programming
(12,13).   A commercial system has been introduced that permits retention index matching by computer or 3 different phases (14).
Factors affecting the precision of retention indices are beyond the
scope of this chapter.   A comprehensive review is available (15).

B. Hyphenated Techniques

A common approach to increase the confidence level of an aanalysis
is to use redundant systems where the data gathered is based on
complementary techniques. The information may be obtained in
parallel or sequentially. The term "hyphenated techniques" has
been coined by Hirschfeld (16) to describe systems that consist of
directly interfaced instruments, usually under control of a single
computer. The combination of a gas chromatograph to a mass spectrometer is perhaps the most widely used technique where two instruments that require different operational environments have been
brought together to function as a single unit. Giddings (17) discussed fundamental aspects of two-dimensional separations that en-
compass most, if not all, commonly used separation principles. Most
chromatographic methods can be combined directly in some fashion
but only a few direct combinations between chromatographic separations are currently technically feasible.

  Gas chromatographs have been interfaced to other chromatographic equipment, in particular to liquid chromatographs, from the early
days of chromatography. The technique never really caught on in
spite of attractive features. Only recently has LC-GC been rediscovered as a viable extention to gas chromatographic determinations
especially those where a sample cleanup is required. The resur-
gence of LC-GC is a consequence of improved injection techniques
and advances in column technology in gas chromatography. It is
now possible to directly introduce the effluent from a microbore
HPLC column into a capillary column that is installed in a gas chromatograph (18–20). Such analyses are becoming common. An entire session at the 8th Internation Symposium on Capillary Chromatography that took place in Riva del Garda, Italy in 1987 was de-
voted to LC-GC. LC-GC will mature quickly and it is only a matter
of time before instrument manufacturers will introduce dedicated and
integrated LC-GC equipment for routine analysis.

  Chromatographs are increasingly used as inlets for instruments
that measure a variety of physical and chemical properties. The
organizers of the Pittsburgh Conference and Exhibition have organ-
ized annual symposia on "hyphenated techniques" that focus on on-
line interfacing of chromatographic equipment.

C. Approaches to the Analysis of Complex Samples

Retention indexes are very useful in qualitative analysis if they are
measured with care. They can be determined with high precision,
especially when modern data systems are available. To be useful in
practice, another requirement must be met before substances can be
identified or verified solely on the basis of precision indices. It is
necessary that the level of interference from the sample matrix be

kept at a low level. The substance of interest must be isolated
from components eluting in close vicinity. This is not only necessary because of potential overlap with interfering components (21),
but it is well known that substances eluting before the compound
of interest temporarily modify the liquid stationary phase, that is,
by swelling (22). These effects cause uncertainty about peak maxima and retention time shifts. There are many analytical problems
where the substances to be determined cannot be clearly resolved
from interferences. Trace analysis, in particular, is one of the
areas where the analyst usually has to deal with significant background contributions. In this case, it is impossible to establish
chromatographic conditions where the substance or substances of
interest can be clearly and reproducible isolated from the matrix.
This is especially true in cases where the same components are to
be determined in different types of matrixes, that is, the analysis of
dioxins in fly ash, foods, or soil. In practice, the analyst is often
forced to combine the information produced from independent systems, that is, chromatographic retention and selective detection.
The correlation of two or more sets of independent information
sharply increases the probability of correctly determining a particular substance. Two-dimensional chromatography inherently provides
two sets of retention data for one sample and thus is superior over
chromatography which uses only a single channel.

Other levels of information unrelated to chromatographic retention data can also be added. Sometimes the substance or group of
substances may be further characterized by subjecting the sample
to specific chemical reactions such as derivatization, hydrogenation,
adduct formation or ozonization (23,24). Chemical derivatization, in
particular, is an extremely important method to extend the scope of
gas chromatography (25). Derivatization procedures can have several benefits. Proper selection of the derivatizing agent may result
in fairly selective reaction with the substance of interest. In addition, a functional group may be introduced into the molecule which
permits selective detection. By judicious choice of the derivatizing
agent, the retention behavior of the derivatized substance can also
be manipulated to some extent. Independently, each of these methods enhances analysis, but it is the combination of retention data,
chemical selectivity and detector response that has the most significant effect on noise suppression. The analyst not only simplifies
the system by elimination of interferences but also makes it more
specific. In difficult applications, such as biomedical analyses, it
is often necessary to include all of these steps into the overall analytical scheme.

Up to this point, emphasis has been placed on concepts where
additional information is provided by subjecting an analyte to a combination of different chromatographic systems and/or chemical reac-

tions, and/or selective detectors. Most chromatographers, who make
use of multidimensional systems in gas chromatography, think of it
in different terms, namely the physical transfer of partially resolved
sample components from one column to another. This process is
called "heartcutting." In principle, partial transfer of components
from one column to another could be considered as a part of sample
cleanup, with the cleanup system being the first column. Heartcut
operations comprise the majority of all multidimensional systems in
GC practice. The term GC-GC is increasingly gaining acceptance.

Analysis of a sample by simultaneous multicolumn chromatography
is relatively simple and straightforward. Two columns of different
selectivity can be connected via a two-hole ferrule to a single inlet.
If each column is connected to a separate detector, two sets of retention data are obtained simultaneously. For unknown components,
the difficulty often lies in the correlation of the two chromatograms.
If known substances are to be analyzed, it is relatively easy to establish retention indexes and their errors on each column. In order
for substances to be verified, they must appear in a given retention
time window on each column and the response ratio for a given substance must fall within a given range.

D. The Scope of Multidimensional Chromatography

The term "two-dimensional chromatography" has its origin in flat-
bed chromatography where it has been defined as chromatographic
migration in one direction that is followed by a second development
in perpendicular direction (26). The overall separation obtained is
superior if the interactive forces which bring about retention are
different for the two consecutive developments. Herman et al. (27)
demonstrated that overlap of components is far more serious than
commonly thought even with chromatographic systems having high
peak capacity. Davis and Giddings showed that only 37% of all randomly distributed components are fully resolved in a system that has
twice the minimum peak capacity (28). The situation can be drastically improved by expansion into a second dimension. If one considers a two-dimensional system as consisting of boxes that reflect
the peak capacity of the system in each dimension, then the overall peak capacity is roughly the product of the peak capacities of
the individual dimensions (29). Figures 1a and 1b schematically
illustrate the principle for a very simple system. This multiplicative
law for two dimensional peak capacity points out the tremendous increase in resolving power that can be achieved. If one has a two-
dimensional system with a relatively modest peak capacity of 50 in
each axis, a total peak capacity of above 2000 is obtained. An
equivalent one-dimensional separation would require about 10 million
theoretical plates. This is clearly not feasible at this time.

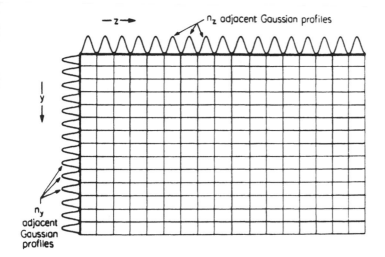

FIGURE 1a   The peak capacity of a two-dimensional system, represented by the number of boxes, is approximately equal to the product of the peak capacities n_z and n_y generated along the two individual axes, as represented by the number of adjacent Gaussian
profiles.   From Ref. 29.

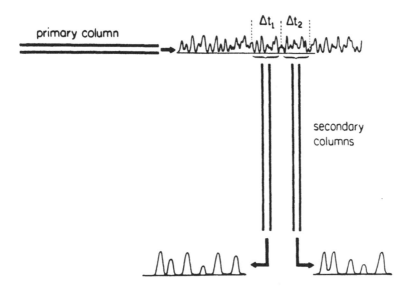

FIGURE 1b   Illustration of a coupled column system consisting of a
primary column and two secondary columns.

The debate on what constitutes multidimensional chromatography has been going on in the chromatographic community for quite some time without arriving at a general concensus (30). Hais (31) has taken the unusual step of throwing into the open the question of how multidimensional chromatography should be defined. In a letter to the editor of a chromatographic journal, he has pointed out that the term multidimensional chromatography is usually chosen to describe sequential combination of chromatographic procedures whereas the same term could also be used to describe "the mathematical treatment of data by applying information theory in a virtual multidimensional space."

In the context of this chapter, the definition of the term two-dimensional chromatography may appear trivial to the casual observer and only the purist may be offended by the misuse of a term. It is, however, important that some agreement is reached as to what techniques constitute two-dimensional chromatography.

The following definition is proposed:

1. A sample is analyzed simultaneously on two columns of different selectivity. Retention data on the two columns are correlated for the same sample components.
2. Two columns of different selectivity are combined such that eluate fractions can be transferred from one column to another.

If one accepts these fairly narrow definitions, some techniques considered by some as two-dimensional no longer fall into this category. A chromatographic system based on the response ratio of a solute in two detectors of different selectivity is unidimensional. An analysis where an eluate fraction that has been trapped at the outlet of one column is reinjected into a second column is carried out in a two-dimensional mode, even if two separate instruments are involved in the off-line mode.

Several review articles and a book chapter dealing with multidimensional gas chromatography have appeared recently. Chizhkov and Yushina (32) have described column switching systems for packed columns, while Deans (33), Miller (34), Schomburg (35,36), Gordon et al. (37), and Mueller (38), reviewed heartcutting techniques. Bertsch (39–41) summarized two-dimensional methods in a three-part review. This chapter will place primary emphases on recent developments.

II. A SHORT SURVEY OF TWO-DIMENSIONAL GAS CHROMATOGRAPHY

Some of the earliest multicolumn systems incorporating complicated valving systems and intermediate traps have been applied in the

petroleum industry for the analysis of crude oil and refinery products (42). In this industry it is important to determine not only
the boiling-point distribution of feedstock and refined fuel, but also
to characterize these materials in terms of alkanes, saturated cyclic
hydrocarbons (naphthenes), aromatic hydrocarbons and olefins.
Some of the most elegant applications of multidimensional GC have
originated from the early work on crude oils (43—45). The gas
chromatographs used for this purpose were bulky and difficult to
operate since multiple cuts had to be timed precisely. Microprocess-
or-controlled instruments incorporating intermediate traps and flow-
switching systems (PONA analyzers) are now routinely used in this
industry.

Process engineers and industrial chemists who are often interested in rapid on-line analysis have routinely used two-dimensional
chromatographic systems for quite some time. Some industrial processes require the analysis of only a single compound or a limited
group of substances. If these are to be determined in a complex
matrix, it is almost imperative to take advantage of some kind of a
prefractionating inlet or precolumn before final chromatography takes
place in the analytical column. This restriction ordinarily applies
even in cases where a highly selective detector is available to record
directly the compound of interest from an interfering matrix. The
reason is as follows: Samples of practical interest often contain
large amounts of high molecular weight materials. Exceedingly long
analysis times are required before a stable baseline is reestablished.
An example of such an analysis is the determination of volatiles in cig-
arette smoke condensate. It is obvious that sample turnaround can
be shortened considerably if only the fraction of the sample that
contains the substances of interest is introduced into the analytical
column and the remainder is backflushed or vented. Another important consideration is the limited lifetime of fragile, low-capacity
capillary columns that are subjected to large sample loads. It is
well known that a deposit of even relatively small amounts of heavy
materials in capillary column inlets adversely affects chromatographic
performance resulting in irreversible absorption of some components.
While such losses caused by the sample matrix can not principally
be solved by a two-dimensional approach alone, a packed precolumn
or a packed precolumn insert can greatly expand the useable lifetime of a capillary column. The distinction between these two arrangements is perhaps rather arbitrary since both fulfill the same
basic function, i.e., to protect the capillary column from unnecessary and unwanted materials (46—48). In any case, it is necessary
to adapt these components to each other. A detailed study on basic
requirements has recently been conducted (49—51). It was concluded that a short column packed with small particle size packing does
not adversely affect the efficiency of a capillary column. Better

reproducibility is claimed for capillary columns that are preceded by packed precolumns (52). Packed inserts that are substituted for injector port liners (46) and solvent bypass inlets (53,54) or programmed temperature vaporizers (55–58) should not be considered components of two-dimensional gas chromatographs, even though they are designed to remove a part of the sample.

Two-dimensional gas chromatography has remained a specialized technique that has not generally spread into the average laboratory. Advances brought by fused silica tubing have been incorporated into some commercial instruments and switching components. Progress has been made toward a better understanding of the principles and operation of cryogenic devices and microtraps that are often used with heartcutting.

There are currently 3 major instrument manufacturers, Siemens, Shimadzu and Dani, that market versatile two-dimensional gas chromatographs. An automated cyclically operated instrument for preparative separations has also been introduced (59). At least two other manufacturers have offered two-dimensional instruments in the past but have withdrawn from the market presumably due to a lack of interest. Add-on accessories and conversion kits to carry out switching operations are marketed bu Valco, SGE, Chrompack, and AC Analytical Controls. Information on installation and applications examples are available from these manufacturers. Technical sessions on multidimensional GC are now routinely included at major meetings such as the Symposia on Advances in Capillary Chromatography, the Pittsburgh Conference and the Eastern Analytical Symposium.

III.   THE PRACTICE OF TWO-DIMENSIONAL GAS CHROMATOGRAPHY

A.   Single column vs. Dual Column Systems

Deans (33) has stated that the primary use of two-dimensional gas chromatography resides in the analysis of complex mixtures when analytical requirements are known. Unquestionably, the majority of all practical applications can be solved with single column systems and in most analytical applications it would be impractical to apply two-dimensional GC. Separations in single column systems employing high resolution capillary columns are primarily based on column efficiency, whereas in two-dimensional systems the source of separation power is the difference in selectivity between two or more stationary phases. In both cases, an increase in peak capacity is sought. The logical first step toward the analysis of unknown samples is the use of short nonpolar columns of limited efficiency (60). The information gained in such a preliminary experiment permits assessment of the scope of the analytical problem. The analyst gets an idea of the boiling point distribution of the sample, its complexity, and whether selective polar phases may be beneficial.

Two-dimensional gas chromatography based on heartcutting techniques invariably uses only a part of the sample. It ultimately gains its power by rejecting the portion of the sample that is not of interest to the analyst. Sometimes only a few substances in a complex sample are targeted. Carrying out only partial analysis may appear to be a very serious limitation. On the other hand, heartcut methods are rarely used when it is necessary to completely analyze every single component in a sample. In principle, it is possible to perform a total analysis by taking consecutive cuts until the enture chromatogram has eventually been "sliced up" and transferred to the second column. The price to be paid is in analysis time. Table 1 illustrates some properties of single column and multicolumn systems.

B. Heartcutting in the On-Line and Off-Line Mode

Two-dimensional gas chromatography, in its simplest version, can be carried out in the off-line mode. The most elementary procedure involves manual collection of effluent from a column, followed by reinjection into another column of a different selectivity (61—64). Details of a particularly simple design that requires only a slight adaptation of standard instrumentation have been described by Sandra et al. (65). Figure 2a presents constructional details, while Figure 2b demonstrates some of the types of analytical problems that can solved with this comparatively simple system. Off-line methods are straightforward and often work quite well but are hardly feasible in routine situations. The collection of effluent fractions from one column for reinjection into a second column is not as trivial as it may appear (64). Many practical problems are encountered, such as the introduction of artifacts, especially if concentration levels are low. The situation may be comparable to the difficulties encountered with GC-MS analysis of complex samples at trace levels, a technique that is particularly difficult in the off-line mode.

Routine analysis requires reproducible conditions. Heartcuts, that is, valve operations, have to be timed precisely. A dedicated two-dimensional GC can carry out heartcutting, foreflushing and backflushing, intermediate trapping, and addition of a standard to the heartcut fraction. All of these basic functions can be time programmed, for example, controlled by the microprocessor of the gas chromatograph. In the following section, properties and performance characteristics of some components that are of critical importance to column switching and heartcutting will be discussed. Some reflections on the future trends in two-dimensional instrumentation will also be presented.

TABLE 1   Single vs. Multicolumn Systems

	Single Capillary Column System	Multicolumn System	
		Heartcut Mode	Parallel Columns
SOURCE OF SEPARATION POWER	primarily column efficiency; stationary phase selectivity to a less extent	difference in stationary phase selectivity; sample capacity of precolumn	difference in stationary phase selectivity, peak correlation by data system.
RESOLUTION POTENTIAL FOR DIFFICULT SAMPLES, I.E., ISOMERS IN TRACE QUANTITY IN A COMPLEX MATRIX.	limited by peak capacity of column	excellent only method that can do this on-line	slightly better than with single column.
DIRECT ANALYSIS OF AQUEOUS SAMPLES	relatively difficult, especially for trace analysis	routine, especially if a porous polymer packing can be used in precolumn	about the same as with single column
QUANTITATION POTENTIAL	excellent if substances of interest can be fully resolved	can be a problem to the need of sample transfer and reinjection; superior over single column if full resolution on single column is not feasible	in most cases slightly better than with single column systems
QUALITATIVE ANALYSIS	of limited scope	two independent retention data sets available; important with identification systems which required relatively pure compounds, i.e., MS, IR	two independent retention data sets available
AVAILABILITY AND COST	relatively inexpensive; widely available	fairly expensive; limited number of commercial sources	same as single column system but relatively sophisticated dual channel data system necessary

TABLE 1 (Continued)

	Single Capillary Column System	Multicolumn System	
		Heartcut Mode	Parallel Columns
OPERATOR EXPERTISE REQUIRED	simple and straightforward	some skill required specific training necessary may be	about same as with single column systems
MAJOR ADVANTAGES	simplicity in operation; low cost; wide availability and choice; easy troubleshooting	two independent retention data sets available; ultimate resolving power; complete separation of closely similar compounds almost always possible regardless of matrix complexity; fast turnaround time; possibility to combine selectivity of gas solid chromatography (packed columns) with capillary column GLC elimination of disturbances caused by major peaks; resolution of minor components coeluting with major components, most suitable for target analysis of trace substances in complex matrixes.	powerful in routine analysis; especially if peaks can be correlated with appropriate user specified soft- ware; can be important where the absence rather than the the pressure of a particular substance has to be demonstrate can compete in certain cases with information provided by selective detectors such MS or IR
MAJOR DISADVANTAGES	limited resolving power; selectivity offered by some gas-solid packing can not be realized	difficult to operate and troubleshoot; setting up operational conditions is time- consuming relatively complicated and expensive few commercial sources	data system with complex samples peak assignment between 2 columns can be ambiguous; not in wide- spread use

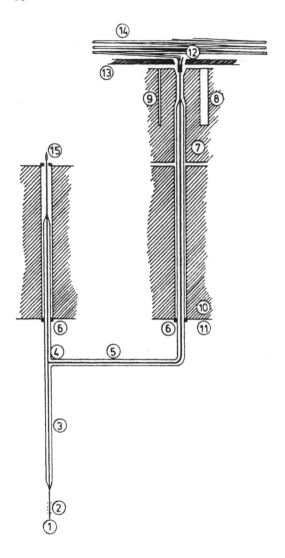

FIGURE 2a Splitter assembly for manual fraction collection. 1.
capillary column; 2. Teflon shrinkable tubing; 3. capillary tube
6 mm OD, 0.3 mm i.d.; 4. capillary tube 6 mm OD, 0.3 mm i.d.;
5. capillary tube 6 mm OD, 0.8 mm i.d.; 6. swagelok union;
7. aluminum block; 8. cartridge heater (100 watt); 9. thermo-
couple; 10. detector base; 11. column oven; 12. conically-shaped
Teflon, shrinkable tubing; 13. asbestos plate; 14. microtrap
(glass capillary); 15. FID. From Ref. 65.

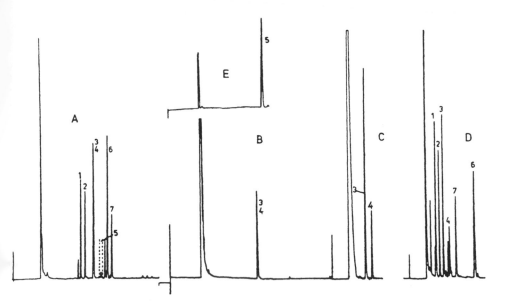

FIGURE 2b   Illustration of the possibilities of microtrap collection with a monoterpene test mixture.   Substances: 1. α-pinene, 2. camphene, 3. β-pinene, 4. myrcene, 5. α-phellandrene, 6. p-cymene, 7. limonene. Conditions: Column I: 80 m × 0.5 mm i.d., coated with SE-30. Analysis at 100°C with a carrier gas (hydrogen) flow rate of 8 ml min^{-1}.   Column II: 25 m × 0.5 mm i.d., coated with Carbowax 20 M. Analysis at 70°C, carrier gas flow rate 4 ml min^{-1}. A. total analysis (column I); B. collection of peak 3.4 (column I); C. collection of peak 3.4 (column II); D. total analysis (column II); E. enrichment of peak 5 (column I).

C. Effluent Switching by the Deans Principle

Basic operations such as backflushing, switching carrier gas between different columns, venting of effluent, etc., can be carried out by either placing one or more mechanical valves in the effluent stream or by manipulating pressures at carrier gas junctions. Both methods have proponents. Pressure switching avoids contact of sample components with the mechanical parts of a valve. Deans (66) introduced the basic principle of pressure switching in 1968 and was awarded a patent on the invention.

The principles of heartcutting, venting and backflushing via Deans switching are demonstrated in a simple two-column system. The basic steps may best be illustrated in a series of figures. The example shown in Figure 3 represents the most general design of a

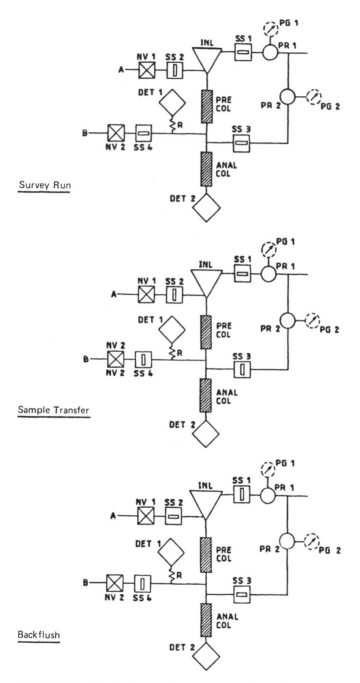

FIGURE 3 Basic layout for a two-dimensional gas chromatograph according to Deans' principle. For details, see Table 2.

Deans type dual capillary arrangement. Table 2 describes the elements shown in Figure 3 and discusses their function. Flow switching is effected by pressure balancing at junctions. The flow of carrier gas through each part of the system is controlled by synchronized opening and closing of off-line solenoid switches. The magnitude of the changes effected by such switches is determined by the inlet and outlet pressures of the interacting components and their flow resistances. Active components which can be adjusted to only a limited extent are the columns (PRE COL and ANAL COL) and the restrictor to detector 1 (DET 1). Needle valves (NV 1 through NV 5) on the other hand act as readily adjustable flow restrictors. The solenoid switches (SS 1 to SS 4, and 3WSS) have two operational modes: (1) to allow passage of carrier gas through a line; or (2) to shut it off. Practical systems are supplemented with slow bypass flow restrictors (fixed or adjustable) and, sometimes, in-line cryogenic traps. The function of the forward bypass is to prevent back diffusion into the tee pieces and keep the solenoid switches free from condensed materials.

Three phases can be distinguished in a heartcutting operation:

1. Initial run on the precolumn (sometimes called prefractionation column or inlet column) to survey the sample and make some basic decisions. The sample is introduced into the precolumn (PRE COL) in the split mode. The positions of the solenoid switches (SS 1 through SS 4) indicate transport of the sample to DET 1. SS 2 stays open for the duration of the sample introduction and may be closed during the run. Carrier flow provided to the lower half of the double tee prevents sample from entering the analytical column (ANAL COL) and provides carrier gas flow at the same time.

2. Sample transfer from the precolumn to the analytical column (heartcut) and venting of substances that are not of interest. Effluent eluting from PRE COL enters ANAL COL. SS 3 and SS 4 are closed during the operation. A small amount of effluent is split to DET 1.

3. Analysis of cut fraction in ANAL COL and backflush of PRE COL. SS 1 is closed, SS 2 and SS 3 are open. Carrier gas through SS 3 provides flow for both ANAL COL and PRE COL.

There have been only a few studies to determine variables such as switching speed and precision that can be obtained for heartcutting in the Deans mode (67–70). It appears that switching speeds in the order of 100 milliseconds and peak area reproducibilities of better than 1% RSD are attainable (69,70). It should be pointed out that effluent switching for the system shown in Figure 3 can also be carried out with in-line valves. There has been considerable debate whether the current generation of microvalves introduce artifacts, i.e., remove components from the effluent stream by absorption or catalysis. The subject will be discussed in the next paragraph.

TABLE 2   Elements and Functions in a Deans Type System

Element	Function	Operation
INL	Sample Inlet	introduction of sample into precolumn
PRE COL	Precolumn	typically a short nonpolar capillary column, i.e. OV 1
ANAL COL	Analytical Column	often longer than the precolumn and usually coated with a polar phase, i.e. CW 20 M
DET 1	Detector 1	sometimes called "monitor", usually FID
DET 2	Detector 2	usually FID, sometimes a selective detector
PR 1/PG 1	Pressure Regulator Pressure Gauge for Precolumn	pressure corresponds to pressure drop across both columns
PR 2/PG 2	Pressure Regulator and Pressure Gauge for Analytical Column	pressure corresponds to pressure at junction
SS 1	Solenoid Switch 1	"off" position means cut-off for precolumn
SS 2	Solenoid Switch 2	"off" position means cut-off for splitter
SS 3	Solenoid Switch 3	"on" position means backflush for precolumn, prevention of solute transfer to analytical column, carrier gas supply for analytical column
SS 4	Solenoid Switch 4	"on" position means venting of effluent from precolumn, provided that SS 3 is also "on"
R	Restrictor	chosen such that a small fraction of precolumn effluent is passed on to detector 1
NV 1	Needle Valve 1	sets split ratio
NV 2	Needle Valve 2	should approximately balance the restriction of the analytical column

D. Mechanical Valves and Recycle Chromatography

As shown in the previous chapter, Deans type switches (66) in
principle avoid contact of the sample with materials other than the
column and its connection tubing. It is thus possible to operate
with "all glass systems." Table 3 compares the relative merits of
Deans type switches and mechanical valves. Some progress has
been made towards suitable low dead-volume valves over the last
few years/ Jenning et al. (71) described a microvalve which incorporates short pieces of fused silica tubing. Another valve has been
introduced by Valco Engineering (72), after years of development.
This valve uses a rotor that is manufactured from a high temperature polymer composite. There are several conflicting reports on
the performance of this valve (73—76). Two aspects have to be
considered in this context: Inertness of surfaces in the valve toward polar substances at trace levels and band broadening effects.
The introduction of the microvalve has spurred another development
that does not in itself constitute two-dimensional chromatography
but was easily incorporated into a two-dimensional system. Recycle
chromatography with capillary columns is an attractive proposition.
A large number of theoretical plates per unit time can be generated
without having to pay the usual price that accompanies very high
efficiency—a large inlet pressure. The underlying principles have
been treated theoretically by Chizhkov et al. (77) and others (78,
79). Practical systems based on packed columns (80—83) are widespread but applications with capillary columns (76,84—86) are sparse.
Recycle chromatography critically depends on the quality of the in-
line valve. Jennings (85,86) was the first to seriously consider this
approach for capillary columns. Figure 4 shows an adaptation of
recycling methodology to multidimensional gas chromatography. The
two-stage recycle unit, as shown in Figure 4, serves as inlet for
the analytical column.

Only moderate progress has been made with capillary column recycle systems and very little has published over the last 8 years.
It appears that a certain degree of peak distortion at each passage
through the valve is unavoidable. Effluent splitting after each valve
passage can minimize this effect but the amount of sample available
after each passage decreases (76). Work is currently in progress
to describe these effects quantitatively (87).

E. Intermediate Trapping vs. Direct Connection in Serially Coupled Columns

The retention behavior of solutes that move through two serially
coupled columns is determined by the relative contribution from both
phases. It would appear that the coupling of 2 columns with dissimilar phases would always ben an advantage, but this is not neces-

TABLE 3   Deans Switch vs. Mechanical Valve

	DEANS	MECHANICAL
Principle	Difference in junction pressures	Positioning of a mechanical component in the carrier gas stream
Ease of Installation and operation	Takes some experience	Simple and straightforward
Memory effects and activity toward sample components	None expected	Minor problems are possible*
Contribution to band broadening	None expected	Insignificant**
Suitability for high temperature operation	Excellent	Good
Cost	More expensive***	Less expensive
Availability	Comparable	
Speed of switching	Comparable	
Potential toward automation	Comparable	

*   inconclusive data.

**   refers to conventional-bore columns; for very small-bore columns, i.e., <100 μm, valves are not likely to be useable.

***   solenoids driven by a system that permits the programming of events are necessary for meaningful operation.

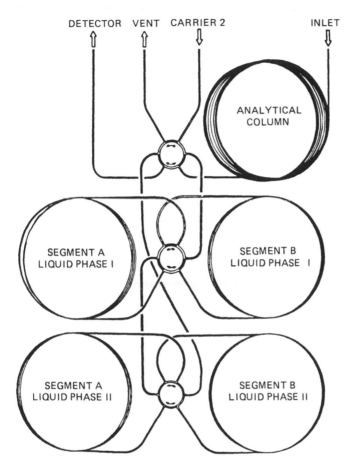

FIGURE 4 Schematic representation of two recycle units (containing different liquid stationary phases) in conjunction with an analytical column. From Ref. 86.

sarily so. A solute pair that may have been resolved in column 1 may merge again in column 2. This effect can be avoided by in- stallation of an intermediate trap. Intermediate trapping offers at- tractive advantages over direct coupling. Table 4 summarizes ad- vantages and disadvantages of the two designs. According to the table, performance characteristics are heavily stacked in favor of instruments equipped with a facility for intermediate trapping. Volkov et al. (88) carried out a theoretical investigation of single- stage preparative chromatography with and without intermediate trapping. They concluded that the use of an intermediate trap

TABLE 4   Intermediate Trap vs. Direct Combination

Advantages, intermediate trap	–	Exact starting point of chromatographic development in second column is known.
	–	Several heartcuts from different section of a precolumn separation can be taken. Chromatographic development in the second column starts at the same time for all fractions collected.
	–	It is feasible to connect a high capacity packed precolumn to a capillary column without split. The gain is in sensitivity for minor components.
	–	Multiple enrichment is possible. Trace components from a particular eluate fraction of the precolumn can be accumulated repetitively in the trap.
	–	Bands broadened by chromatography in the precolumn can be refocussed. The precolumn can be overloaded on purpose to allow enrichment of trace components.
	–	Installation of an auxillary inlet upstream of the trap permits addition of standards to the heartcut fraction. Without trap, an auxillary inlet between the columns is of little use.
	–	With an auxillary inlet, it is possible to operate the analytical column with a carrier gas that is different from the precolumn. A special detector (i.e., negative ion mass spectrometer, photoisomization detector) can be used as detector 2.

TABLE 4   (Continued)

Disadvantages, intermediate trap	-	Relatively complex and expensive.
	-	Proper design is important. Cold spots and inadequate connections must be avoided.
	-	Trapping efficiency not always quantitative, especially for subtances of high volatility.
	-	Heating rate of trap may be inadequate and cause band broadening. Decomposition of thermally labile components is also possible.
	-	Impurities in system, i.e., in the carrier gas, accumulate in cold trap.
Advantages, direct combination	-	Simple.
	-	Inexpensive.
	-	Easily maintained.
Disadvantages, direct combination	-	The major disadvantage is that the sample is subjected to a mixed retention mechanism. Components that have been resolved in the first column may merge in the second column.

brings significant advantages even though a system becomes technically more complex. Intermediate traps seem to provide extraordinary flexibility. In practice, they generate problems that are not
always easy to deal with. Factors that enter into the design of
these devices will be discussed in Section III.H.

F. Serially Coupled Columns Without Intermediate Trap

Instruments that are not equipped with an intermediate trap are less
versatile than instruments that have this device. For certain applications, there are benefits to direct coupling. Direct combinations
are preferable in cases where only one particular problem (preferably, involving a single substance) is to be solved and an instrument
can be dedicated to that special task. The analysis of trace components that elute immediately after a major component is a good example. Figure 5 shows the determination of 10 ppm of ethanol in
methanol (89). Adequate resolution is only possible after the bulk
of the solvent has been removed. It should be noted that the stationary phases of both columns are identical. The instrument is
thus only used to vent a major interfering compound. The degree
of overall band broadening in a directly coupled system is surprisingly small with a properly designed interface (90). Exceedingly
difficult separations such as the analysis of drug metabolites in
human serum have been carried out without the use of an intermediate trap. Figure 6 shows an example in which 2 columns in 2 separate gas chromatographs were linked directly through a heated interface (91). Switching operations were carried by a Deans type
system that was controlled by an external timer. Factors such as
switching speed and extra-column contributions were studied as a
function of several operational parameters. No adverse effects were
found. Figures 6(a) and 6(b) provide instrumental details, while
Figure 6(c) describes the operation of the device. Figure 6(d)
shows two chromatograms, one presenting the direct profile of a derivatized serum extract, and the other showing the results obtained
after a direct heartcut operation. It should be noted that the top
chromatogram in Figure 6(a) was produced from an extract which
had been subjected to a cleanup in the form of acid/base extraction,
whereas the bottom chromatogram in 6(d) (labeled B) was obtained
without prior cleanup.

G. Thermal Focussing in Cryogenic Traps

Advantages and disadvantages of sequentially coupled columns that
are operated with and without intermediate trap have been discussed
in Section III.F. The performances of these devices is of critical
importance and deserves special consideration.

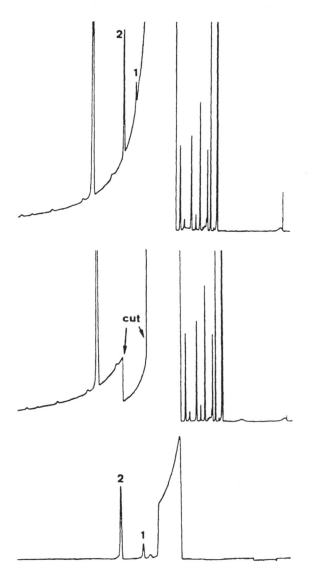

FIGURE 5 Determination of low concentrations of ethanol (10 ppm) in methanol. Sample: methanol (1), internal standard 2-propanol (2). Columns and conditions: Pre-: 56 m polypropyleneglycol, temperature: 37°C, 1.0 bar H_2, u = 28 cm/s. Main: 60 m polypropyleneglycol, temperature: 47°C, 0.5 bar H_2, u = 36 cm/s. Injector temperature 150°C.

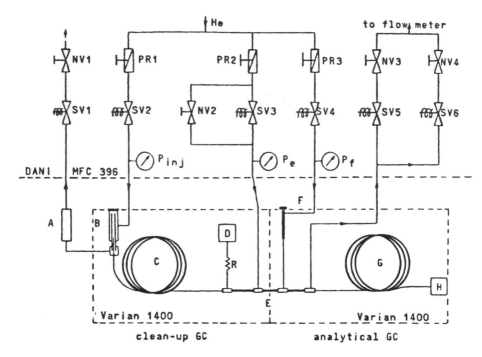

FIGURE 6a Schematic flow diagram of a dual oven dual column gas
chromatograph. A. adsorption filter; B. split-splitless injector
(220°C); C. cleanup capillary column (168°C); D. monitor flame
ionization detector (230°C); E. column connection manifold (172°C);
F. low-volume injector (162°C); G. analytical capillary column
(163°C); H. electron capture detector (270°C), NV1—4: needle
valves; P: pressure gauges; PR1—3: pressure regulators; R:
restriction; SV1—6: solenoid valves. From Ref. 91.

     Cold traps are used in gas chromatography for condensation of
solvent vapors from the carrier gas. It has been shown that cryogenic condensation in capillary tubes is not an easy matter even at
the temperature of liquid nitrogen (67,92—94). This fact is well-
known from preparative-scale gas chromatography. An abundance
of data is available on cryogenic trap designs (96—98), much of it
in conjunction with thermal desorption of adsorbed volatiles from
polymers, a technique widely applied in headspace analysis (99—
101). Trapping efficiencies reported in the literature for comparable conditions vary widely. It has been observed that the presence
of a solvent film in an open tube greatly improves condensation efficiency (64,102). This effect forms the basis of the solvent effect,

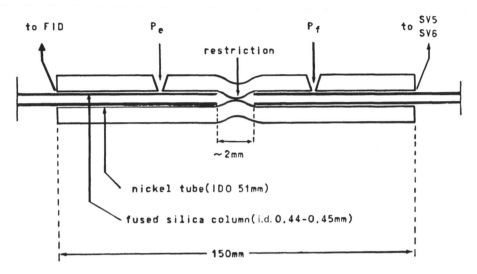

FIGURE 6b   Column connection manifold (T-pieces are not shown).
P_e, P_f, and SV5, SV6 refer to Figure 6a.

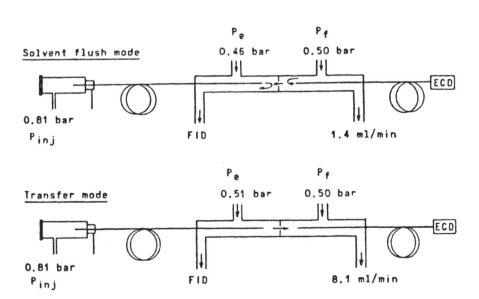

FIGURE 6c   Flow modes used for analysis of KABI 2128. P_e, P_f,
and P_{inj} refer to Figure 6a.

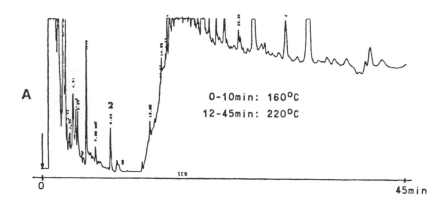

A

0-10min: 160°C

12-45min: 220°C

0   45min

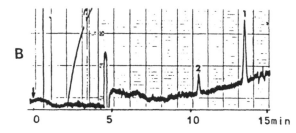

B

0   5   10   15min

FIGURE 6d Typical ECD chromatograms using different workup procedures and different GC instrumentation. (A) Serum level of KABI 2128 was 1.8 ng/ml using an acid/base extraction as a part of the sample preparation. A single 45 m OV-17 glass capillary column was used. (B) Serum level of KABI 2128 was 2.0 ng/ml using a direct derivatization procedure and a two-dimensional GC system (10 m OV 1 + 10 m OV 1701). Peak 1: bis-TFA-KABI 2128. Peak 2: bis-TFA-OAA 647 (internal standard).

a phenomenon that has been under intensive study in conjunction with capillary column inlet systems (103–105).

It has also been noted that solute breakthrough in uncoated cryogenic traps is particularly severe in cases where an abrupt change in temperature occurs, i.e., when a trap is directly dipped into liquid nitrogen (67,97). It has been suggested that open tubular traps should be coated with liquids for efficient trapping. Fortunately, the polysiloxane fluids or gums that are favored by so many chromatographers do not crystallize at subambient temperatures but remain viscous liquids. Paknow (105) found that temperatures of at least −80°C are required for effective trapping of volatile components such as trihalomethanes on thin-film capillaries. Polyethyleneglycol type phases (106) do not produce the same results presumably because of phase solidification.

Trapping efficiencies in packed beds are much higher than in open tubes, even when the packing is uncoated. This is especially true when a temperature gradient is applied along the bed. In this case one does not only achieve more complete retention, but also performs a preliminary separation step (107). Jacobson (108) was able to quantitatively trap volatiles down to ethane with a microtrap, packed with glass beads. Unfortunately, packed-bed geometries are usually not compatible with capillary columns. Rapid heating and/or effluent splitting can overcome some of the problems (70).

To function effectively in two-dimensional chromatography, traps must quantitatively retain solutes and also allow rapid reinjection of the trapped components. Instant heating of the trap content to temperatures typically encountered in injector ports would seem to be a necessary precondition for achievement of a sharp, plug-like profile in the second column. It is important that the condensate in the trap be deposited in the form of a fairly homogeneous film. A number of investigators have examined peak profiles generated by rapid heating techniques (97,109—114). Numerous devices have been described to raise the temperature of traps, usually by electrical resistance heating (67,70,111). Flushing with hot gases (67,115,116) or hot liquids (117,118) has also been done. Electrical resistance heating can be carried out at a very high heating rate. Final temperature can be reached in less than one second, using appropriate electrical circuitry (70,111). It is obvious that traps consisting of glass experience a larger thermal lag than traps made out of metal tubing. Aluminum clad fused silica tubing (119) can, in principle, be heated directly by electrical resistance heating. No information on such methods is available at this time.

H. Serially Coupled Columns With Intermediate Cryogenic Trap

The use of intermediate traps in two-dimensional gas chromatographs has been advocated at a very early stage (43). Incorporation of such devices brings many advantages and, if designed properly, no disadvantages with the exception of the necessary financial investment. The operator of a two-dimensional instrument equipped with intermediate trap has the option of not using the trap. If trapping is not applied, it is only necessary to raise the temperature of the trap, making its function comparable to a transfer line in a GC/MS combination. A modified coupling device based on Deans' principle has been introduced by an instrument manufacturer (38). The system essentially consists of 2 tees that are arranged back to back. Figure 7 (68) is a simplified representation of an instrument using T-junction between the two columns. The lower half of this figure indicates the pressures operative in various parts of the system. It is evident that the direction of flow in the system's components

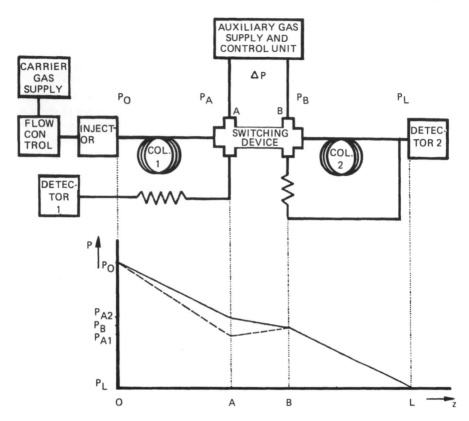

FIGURE 7 Schematic of a gas chromatography for column switching. p = pressure; z = length; p_O = pressure at the inlet of the first column (z = 0); p_A, p_B = pressures at points A and B, respectively, of the switching device; p_L = pressure at the outlet of the second column (z = L). The pressure difference, p, is adjusted with the aid of the auxillary gas supply unit and equals p_{A1} − p_B for single-stage operation and p_{A2} − p_B for two-stage operation. From Ref. 68.

is controlled by the inlet pressure and the pressure on both sides of the switching device. Obviously, the inlet pressure at column 1 corresponds essentially to the sum of the pressure drops of both columns at the given flow rate.

Analysis of aqueous solutions is a particularly important applications area. Direct analysis is usually preferable over methods that require prior solvent extraction. Porous polymers such as Tenax are essentially hydrophobic and elute water before organic materials. A packed porous polymer precolumn is thus a logical choice for di-

rect analysis of aqueous solutions by heartcutting. Unfortunately,
the flows in packed columns and capillary columns are usually incompatible. When directly connected, the overall efficiency is limited
by the packed column, unless a method can be found that overcomes
its limitations. Solutions to the problem are effluent splitting between precolumn and capillary columns and cryogenic focussing.
Figure 8(a) shows a packed/capillary column hybrid system (120).
A thermal conductivity detector is used to monitor the venting of
water. Up to 200 μl of aqueous solution can be injected. The intermediate trap refocusses the trace components of interest. Figure
8(b) shows a typical application. The minimum detectable concentrations are in the order of 100 ppb.

Figure 7 points out a practical limitation of a system where two
columns are connected in series. It is obviously difficult to operate
two columns of high restriction. This problem has been overcome
in an arrangement in which the junction pressure between both columns is held essentially at atmospheric pressure (97). Figure 9(a)
is a schematic diagram of the laboratory assembled system. The
novelty of the design lies with the modification of detector 1 and incorporation of a 3-way solenoid controlled switch. Effluent enters
the transfer manifold line from column 1 (PRECOL). Whether it is
switched toward the detector (DET 1) or the trap depends entirely
on the position of the 3-way solenoid valve 3WSS. The fixed restrictor R is approximately balanced with the adjustable restrictor (needle valve) NV 3, located behind the trap. Conditions are chosen
such that a pressure of around 1—3 psi can build up in the manifold line at a suitable flow rate of the switching gas. Admission of
the gas on either side of 3WSS results in almost instantaneous effluent switching at near atmospheric pressure. The trap accumu-
lates the fractions selected for heartcut. To achieve a suitable pressure buildup in the manifold line to regenerate the trapped components into ANALCOL, DET 1 is mechanically blocked off and the adjustable leak behind the trap is also sealed off by switching solenoid
valve SS 2. One of two methods can be used to prevent effluent
from escaping through detector 1 (DET 1). The detector can either
be internally modified in such a way that it is leak tight and can be
shut off externally or a low dead volume in-line shutoff valve can
be placed in the line leading to the detector. Figure 9(b) shows a
practical application. Three heartcuts were taken, comprising what
appeared to be 2 large individual peaks and 1 partially resolved
doublet. Chromatography in the second dimension on two phases
of different selectivity showed that additional components had coeluted in the three windows.

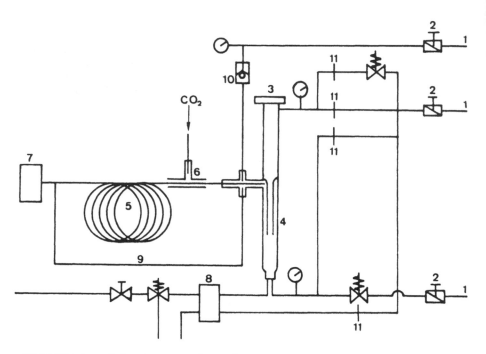

FIGURE 8a Selective sampling of polar trace components (cresols) from aqueous solutions onto capillary columns by using a packed precolumn. 1. carrier gas; 2. pressure regulator; 3. syringe split injector; 4. packed pre-column (Tenax); 5. capillary column; 6. cold trap cooled by liquid CO_2; 7. flame ionization detector (FID); 8. thermal conductivity detector (TCD); 8. direct split connection line to FID; 10. check valve; 11. micro diaphragm. From Ref. 120.

I. Double-Oven Instruments

There are significant advantages to two-dimensional GC in which the columns are located in independently controlled ovens (121). Several instrument manufacturers now offer such instruments. Fenimore et al. (122) was probably the first investigator to use this approach. His application dealt with the determination of low levels of tetrahydrocannabinol and its metabolites in blood serum. The method was based on two-dimensional gas chromatography of suitable derivatives. A packed precolumn and a metal capillary column were operated in separate ovens. A mechanical valve was installed between the columns as a switching device. The overall system efficiency was relatively low, but complete resolution of the target analytes was achieved nevertheless. The double-oven approach provided a stable environ-

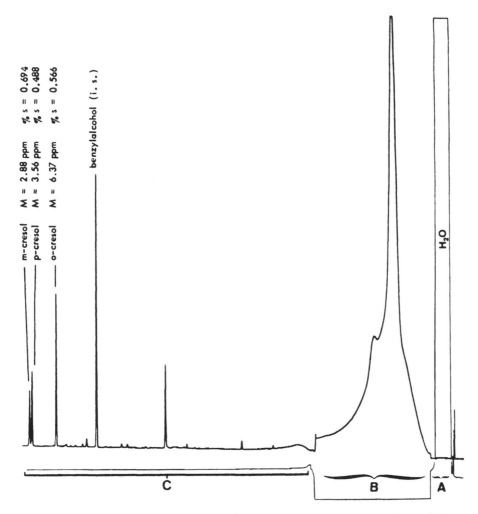

FIGURE 8b Selective sampling of polar trace components (cresols) from aqueous solutions onto capillary columns by using a packed precolumn. Water peak recorded from TCD (A). Cresol peaks obtained by displacement from precolumn and direct transfer to detector without trapping and separation (B). Cresol peaks obtained after trapping and main separation (C). Sample: 10 μl of aqueous solution of cresols, benzylalcohol as internal standard. Columns: A 1 m packed pre-column (Tenax), 0.7 bar, He, 100°C, 2 min iso, 20°C/min to 160°C. B 25 m polyethylene glycol (CW 20 M) on fused silica 80°C, 6 min iso, 8°C/min to 180°C. Detectors: FID and TCD. Instrument: Double oven, Siemens Sichromat 2.

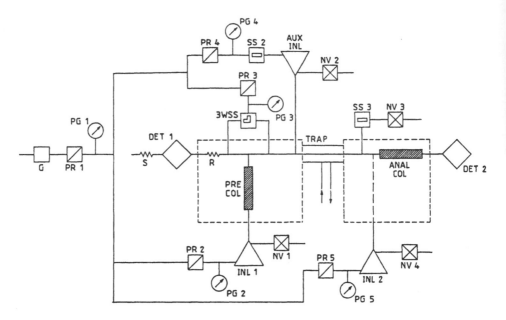

FIGURE 9a Simplified flow diagram of a dual oven system with switching junction at near atmospheric pressure. G = carrier gas supply, INL 1,2 = inlet 1,2 (split/splitless), AUX INL = auxillary inlet, PRE COL = precolumn, ANAL COL = analytical column, DET 1 = detector 1 (monitor), DET 2 = detector 2 (FID), TRAP = cryogenic trap (cooled with liquid, N_2, heated electrically), PR 1—PR 5 = pressure regulators, PG 1—PG 5 = pressure gauges, SS 1—SS 2 = solenoid switches, NV 1—NV 4 = needle valves, R = fixed restrictor, S = external shut-off (provided with small leak). From Ref. 97.

ment for the analytical column and the detector. It was crucial to operate the electron capture detector at the capillary column outlet at the highest sensitivity possible, e.g., under isothermal conditions. Instruments with separate ovens had been used for special applications even earlier, but a truly versatile instrument incorporating two capillary columns, a Deans' type switch, and an intermediate trap was not built until 1976 (67,98).

Multioven systems have been used in the petroleum industry for many years to carry out group type separations of crude oils. A modern example of such an instrument is shown in the applications part of this chapter. Further developments that required gas chromatographs with more than one oven were slow to follow. Experimental work on stationary phase optimization, starting with the introduction of Kaiser's SECAT principle (123), briefly opened up this

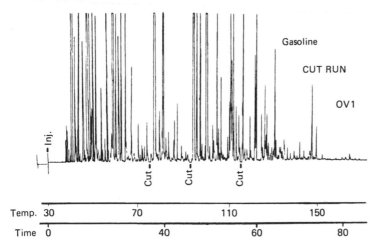

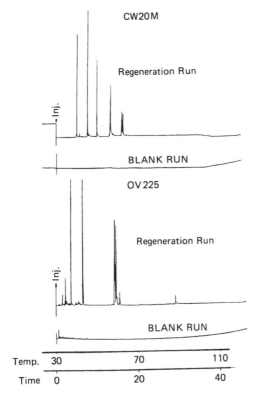

FIGURE 9b Gasoline with 3 heartcuts, ad indicated (top). Development of heartcuts on 2 columns of different selectivity (bottom).

interesting applications area of dual column gas chromatography with double oven instruments. Unfortunately no further developments were reported for many years.

At the present time, much attention is being focused on stationary phase optimization based on two serially coupled columns. The groundwork to stationary phase optimization was laid by Laub and Purnell (124) who showed that selectivity optimization using mixed phases is a relatively straightforward matter. The retention of a substance on a mixed stationary phase is directly proportional to the mole fraction of each phase and the infinite dilution partition coefficients of the substance for each phase. A significant amount of work has been done on mixed phases. The discovery has lead to the well known window diagram technique, a method that is now widely used. The window diagram technique is used to determine the optimal ratio in which two stationary phases should be mixed to achieve optimal selectivity for a given separation problem. Once this ratio has been determined, the analyst may just combine two columns of appropriate length and film thickness to achieve the desired effect. In practice, the method is not quite as straightforward since other factors such as gas compressibility have to be taken into account (125). Other more elegant methods of phase selectivity optimization based on serially coupled columns, are now available (126,127). Deans and Scott pointed out in 1973 (128) that selectivity adjustment can be achieved by manipulating flowrates and/ or temperatures in serially coupled columns. This approach is superior to all other forms of stationary phase adjustment since selectivity optimization can be carried out in a continuous fashion without having to physically manipulate columns. Although phase optimization per se is not a component of multidimensional chromatography, it can be combined with it very easily and effectively. Double oven instruments provide the best of both worlds and allow full flexibility.

Heartcutting and stationary phase optimization can be carried out simultaneously in one system, if necessary. The quantitation of ethyleneglycol in wine at ppm levels is a particularly difficult analysis problem. Mass epectrometry is of limited use in this type of application. A method based on chromatographic isolation of the analyte, rather than its selective detection has been developed. Heartcutting and stationary phase optimization were combined into a unified concept (129). Figure 10(a) represents a diagram of the hardware. A total of four columns are arranged in series. The first two columns, a packed inlet insert and a capillary column, were used to heartcut the sample fraction of interest into the analytical column. The analytical column actually consists of two coupled columns that were optimized in respect to selectivity by adjustment of relative flow rates. Figure 10(b) demonstrates the routine

analysis of ethyleneglycol by direct aqueous injection. Analysis
time was less than 8 minutes.

Advantages of double oven instruments over single-oven instruments can be listed as follows:

o Columns with high temperature phases can be combined with columns of limited thermal stability. The overall temperature limit
  of the system is no longer determined by the phase with the lower
  temperature limit.
o Independent isothermal operation of two or more columns is feasible. Isothermal conditions permit routine determination of high
  precision retention indices.
o Optimization of stationary phase selectivity by temperature adjustment of two sequentially coupled columns, coated with different
  stationary phases, is easily accomplished.
o The precolumn as well as the analytical column can be operated
  independently at different temperatures and/or temperature program rates.
o The oven housing the precolumn can accomodate auxilliary devices
  such as a microhydrogenator or a microozonator.

The superiority of double oven instruments has been demonstrated in a number of recent applications. Some of these will be discussed in the applications section of this chapter.

J. Optimization of Sample Quantity

One of the most important applications of heartcutting is the isolation of pure substances for structure elucidation. The most suitable
and widely used ancillary devices in gas chromatography are infrared spectrometers and mass spectrometers, although other instruments are occasionally used as well. Spectrometers usually require
pure substances. Deconvolution of mixed spectra by mathematical
methods is possible but difficult when spectral features have a high
degree of similarity. Structural isomers fall into this category.
Consequently, isolation of pure substances is always desirable.
This is not an easy task when trace components elute in the vicinity
of major peaks.

To introduce into the system a sufficient amount of the trace
materials of interest, it sometimes becomes necessary to purposely
overload a column. Overloading deteriorates the resolution of minor
substances in a sample. Ligon and May have introduced an effective method of automatically attenuating major components in a sample that are not of interest. Figure 11(a) shows a schematic diagram of a two-dimensional gas chromatograph designed for quantity
optimization. A flame ionization detector situated at the end of the

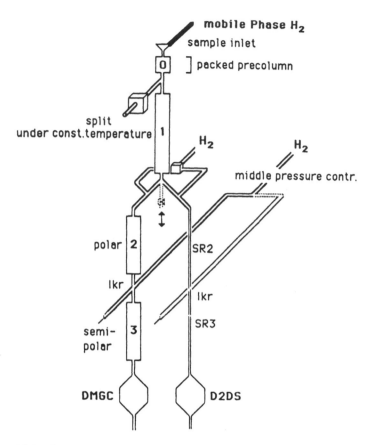

FIGURE 10a Principle of Deans heartcut conditions on-line MGC.
0,1,2,3 are the separation columns or capillaries. "Middle pressure
contr." is an electronically controlled very precise pressure regulator. It controls the pressure in the middle between the multi-GC-
column tandem. It does not mean we are working at a middle high
pressure level. ikr is the so called "IfC-cross," a completely dead
volume free capillary connecting device. "SR2, SR3" are flow restrictors. "DMGC, D2Ds" are the two detectors connected at the
multi-gas-chromatographic part or at the two-dimensional Deans-
switching part of the system. From Ref. 129.

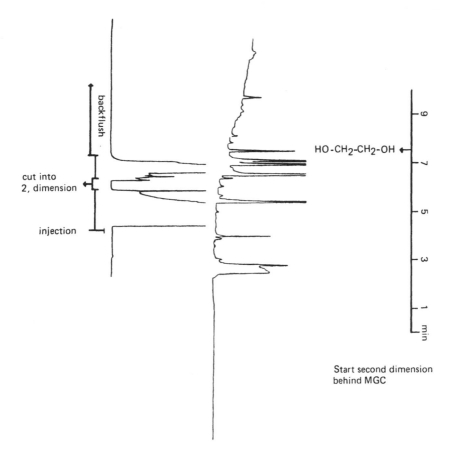

FIGURE 10b The chromatogram CUT and the second dimension analysis, on-line MGC, behind three capillaries and a precolumn. Sample as example: Champagne Deutz, Maison Deutz & Geldermann, Demi-Sec Champagne 12% VOL, Produce of France. Result: 14 mg/L MEG (HO—CH$_2$—CH$_2$—OH).

packed precolumn records the signal produced by the effluent. If the detector response exceeds a preset threshold level, a valve leading to a vacuum system opens and removes up to 98% of the effluent. Major components are largely eliminated from the system, whereas minor substances that do not exceed the threshold limit directly pass into the cryogenic trap. Figure 11(b) demonstrates the performance of the device using a hydrocarbon sample. The top chromatogram shows the major sample components, labeled as A, B, D, G, H and F at high attenuation. The column is clearly

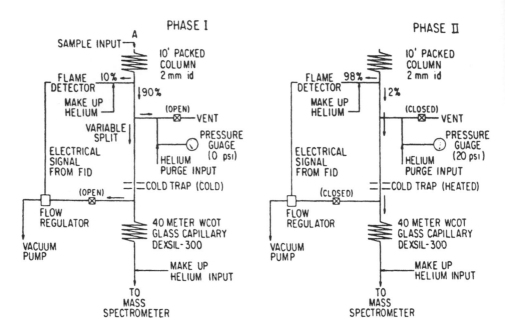

FIGURE 11a Schematic diagram of a two-dimensional gas chromatograph for quantity optimization in the primary and secondary phases of operation. From Ref. 130.

overloaded, resulting in relatively poor resolution for the minor substances marked C, E, F, I and J, as shown in the middle chromatogram. The bottom chromatogram demonstrates enhanced resolution for the minor compounds after quantity optimization, for example, removal of the major peaks. The system seems to work well and several applications were presented by the author in difficult areas of analysis (131—133). It is surprising that this elegant concept has not found more widespread use in the chromatographic analysis of trace components. The technical approach is relatively complex and is not easily implemented. The method appears to be ideally suited for trace analysis in fields like biochemical analysis or in the flavor industry where it is often necessary to isolate and quantify trace substances in the presence of large interferences.

K. Simultaneous Multicolumn Chromatography

The preceding chapters focused exclusively on two-dimensional methods of the heartcutting type. As discussed in Sections I.B and I.C, simultaneous multicolumn chromatography must also be considered a version of two-dimensional chromatography. Heartcutting, a method

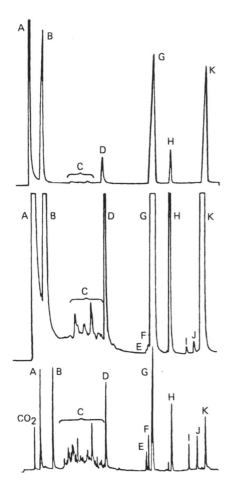

FIGURE 11b Quantity optimization. Hydrocarbon sample (X 1) (top); hydrocarbon sample (X 10) (middle); quantity optimized total ion current chromatogram (bottom). (A) n-C_6H_{14}, (B) n-C_8H_{18}, (C) n-$C_{12}H_{26}$, (G) n-$C_{16}H_{34}$, (H) n-$C_{18}H_{38}$, (K) n-$C_{22}H_{46}$.

that is based on sequential rather than parallel chromatographic approach, retention indexes of concurrently produced data are correlated. Many examples of gas chromatographic separations that use two or more columns in parallel have been described in the literature. There are very few examples where the data obtained from such multicolumn chromatography has been subjected to postanalysis treatment with the aim of correlation. The importance of such methods in chromatography has been recognized early (134) and attempts

have been made to express the information content of multidimensional chromatography, using principles of information theory. Sevcik
(135) provided a detailed discussion and showed that the information
from two inefficient packed columns of different polarity can be superior to that obtained from a very efficient capillary column.

The introduction of small powerful computers, even with low-cost
instruments, enables the chromatographer to carry out postrun calculations in a routine fashion. In 1977 a main frame computer was
used to correlate data obtained from three parallel capillary columns
(136). Modern personal computers or laboratory data systems have
recently been used to carry out such postrun calculations (137–141).
Instruments for simultaneous multicolumn chromatography are readily
available from a variety of manufacturers. The required components
are two or more capillary columns, connected to the same inlet but
to separate detectors, and an appropriate data system that can keep
track of the individual channels. Gas chromatographs programmable
in BASIC can be set up relatively easily for this type of analysis.
Software for specific applications is sometimes available from manufacturers. Figure 12 shows an early application of this type of anal-
sis for the analysis of priority pollutants (142). Retention times
for the target components are established on two columns of different selectivity and retention times windows are set that are sufficiently narrow to exclude most interferences. Component identity is established when two conditions are met independently: a) A signal
must appear in the defined retention time window and b) the response ratio for the target compound must also fall within a defined
limit. In the case where two detectors of the same type are being
used, the response ratio reflects the split ratio between the two columns, that is, for a 1:1 split, the signals should be of about equal
intensity. Response ratio for detectors of different selectivity, i.e.,
between a nitrogen/phosphorus detector and a flame ionization detector can also be used as a criterion. An application based on
this approach is presented in Section IV.B.

It is rather surprising that multidimensional chromatography
based on concurrent development and data linkage is not a common
technique in routine analysis. It appears that additional information
on a sample can be gained at very little additional cost to the analyst. It can provide a critical margin of confidence in difficult applications. Neu et al. (143) recently discussed a comprehensive system that has been in use in a large industrial laboratory for several
years. A laboratory data system was used to control each stage of
the analysis, from sample management to the final analysis report.
Correlation was conducted in four stages that were related to the
probability for correct identification. A short description of the
search routine may be helpful.

In stage 1, a search is made in the reference file of a previously specifiec chromatogram for a retention time that includes that of

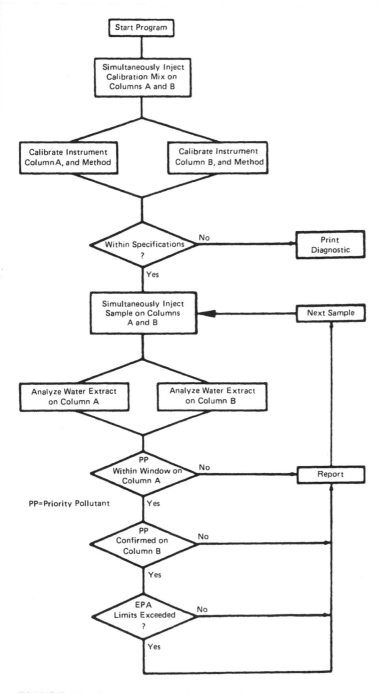

FIGURE 12   Flow diagram for simultaneous dual column chromatography.   The sequence is under BASIC program control.   From Ref. 142.

the first peak.   If a matching retention time is found in the refer-ence file, a search for the same compound is made in the appropri-ate retention time window of the second column.   If the expected signals on both channels fall within preset specifications, the sub-stance is considered to be present.   The system treats all data points in this manner and then moves on to the second, third and fourth stage.   With each successive stage, requirements are relaxed, i.e., the signal ratio from both elements may depart further and further.   These provisions were built in to deal with unresolved peaks and other disturbances.   Table 5 shows an example of such

TABLE 5   Analysis Report for a Wastewater Sample Analyzed by Simultaneous Multicolumn GC

Part of a dual column analysis report of a waste water sample analyzed by headspace gas chromatography. The total run time is 40 min with 117 respectively 116 peaks in both chromatograms exceeding a value of 1000 area counts. Unknown components of A/D 104 are printed in sequence of their retention times bottom right at the end of the report.

Sample Number: 07232
A/D-Channel: 103/104
Reference files: MTAB3/MTAB4

Journal Number: 3168
Files: 69/70

Date: 7. Apr. 87
Last update: 5. Apr. 87

Columns: No. 123   DB 1, 30 m, i.d. 0.25 mm, film 1.0 μm
No. 98   DB 1701, 30 m, i.d. 0.25 mm, film 1.0 μm
Temp. program: 40 °C – 6 min. – 5°/min → 220 °C – 15 min.

Compound name	Result mg/l	*	Rt	A/D 103, col. 123 dRt	area	conc.	A/D 104, col. 98 Rt	dRt	area	conc.
Acetone	9.130		2.13	.00	271254	8.771	2.91	.00	295416	9.448
unknown			2.42		10137					
tert-Butylamine	.549	$	2.54	.06	56575	0.726	2.61	.04	40046	0.549
unknown			2.84		1421					
Meth.-tert-Butyl ether	.286	$	3.38	–.01	22446	0.286	3.22	–.02	33818	0.381
1,1-Dichloroethane	.028	#	3.38	.03	22446		4.15	.04	917	0.028
Methyl ethyl ketone	.360	$	3.64	–.01	21639	0.360	5.73	–.03	117737	1.811
Chloroform	1.238		4.29	–.03	11813	1.185	6.20	–.02	14754	1.292
iso-Butanol	11.987		4.76	.02	793165	12.448	8.08	.05	768751	11.526
n-Butanol	2.110		6.11	–.02	84480	2.191	9.75	–.02	79631	2.029
Benzene	.182	#	6.11	.04	84480		6.92	–.04	26633	0.182
Cyclohexane	.257		6.45	–.03	12998	0.259	5.32	–.03	14585	0.254
1,2-Dichloropropane	.031	$	7.35	.03	2085	0.047	9.50	–.02	1573	0.031
Triethylamine	.562		7.52	–.02	54249	0.583	6.57	–.02	49960	0.541
1,4-Dioxane	.385	$	7.70	–.02	7244	0.385	10.12	.02	28107	1.266
unknown			9.92		900					
Toluene	.453		10.86	–.03	54196	0.451	11.76	–.02	61963	0.455
Cyclopentanone	.109	$	11.07	–.00	10738	0.300	15.39	–.00	4541	0.109
unknown			11.83		1153					
unknown			12.54		1236					
Tetrachloroethene	.633	$	12.94	–.05	9521	0.886	12.80	.02	7824	0.633
4-Methylpiperidine	.064	$	13.32	–.01	18166	0.236	14.67	–.03	899	0.064
Chlorobenzene	.240	$	14.28	–.03	18166	0.237	15.80	–.02	21276	0.243
unknown			14.79		1730					
Ethylbenzene	.108	$	15.13	–.02	14730	0.146	15.89	–.02	12226	0.108
Cyclohexanone	.392		15.70	–.02	32863	0.387	19.77	–.00	37956	0.397
Cyclohexanol	1.222		15.81	–.03	61924	1.266	19.17	–.01	64405	1.179
Styrene	.093	$	16.27	–.03	9579	0.093	17.72	–.00	16227	0.138
:	:		:	:	:	:	:	:	:	:
:	:		:	:	:	:	:	:	:	:

* Classification stages: $ = results differing by more than 10 %,
# = One peak of this pair mainly due to other identified component.

an analysis report. Tens of thousands of samples have been analyzed in this fashion (143,144).

One of the disadvantages of systems that rely on comparison to
stored data is the frequent need of recalibration. Slight shifts in
retention time during an analysis cannot be compensated for and retention time windows must be adjusted periodically to keep up with
these variations. External calibration methods are inherently undesirable. It would be advantageous to use an internal series of standards that are concurrently chromatographed with the sample. System variations would be greatly reduced with such internal calibration methods. On the other hand, the standard has to be added to
each sample, making an already complex sample even more complex.
An elegant method has been devised that largely overcomes this potential disadvantage (145). Figure 13 presents a system where only
trace amounts of the internal calibration standard need to be added.
A wide range internal calibration standard is chosen that is only detected by the selective calibration detector such as an ECD but not
by the detector that registers the effluent from the analytical column.
It is clear that slight variations in chromatographic conditions are
automatically compensated for since all determinations are based on
the use of retention indexes. The system does not need to rely on
stored calibration data but uses calibration data from standards that
were obtained concurrently with the sample. An analogy may be
drawn to high resolution mass spectrometry where accurate mass determinations are made by comparison of the signals from the sample
to a standard that is continuously being bled into the ion source.

IV. SELECTED APPLICATIONS

Selected applications of two-dimensional gas chromatography have
been presented in the preceding text to illustrate principles. This
last section is a summary of some recent examples which may have
some significance because of wide interest and/or a unique approach.

A. Petroleum Industry and Geochemistry

Many two-dimensional instruments dedicated to routine analysis are
currently in use in the petroleum industry. Crude oil, geochemical
materials and coal-derived substances are some of the most complex
samples which have to be dealt with on a large scale. Mass spectrometry, in particular high resolution mass spectrometry, has been
extensively applied in this field but methods based on chromatography are more commonly used.

It is important to determine the composition of petroleum products to provide the petrochemical engineer with the data required

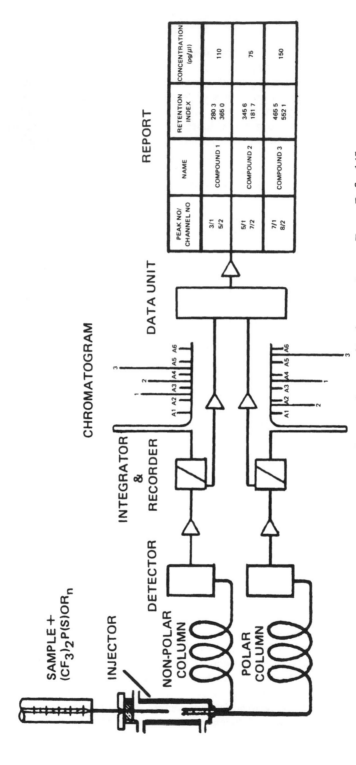

FIGURE 13   Scheme of the basic two-channel retention monitoring system.   From Ref. 145.

118

for process control.   The oil exploration chemist is in need of predicting geochemical parameters relating to origin and maturity of
crude oils, kerogens and similar materials.   The analytical methods
applied very much depend on the nature of the substances and their
matrix.   Both liquid and gas chromatography are being used but gas
chromatography is clearly more important.   Crude oils and source
rocks require different procedures even though some of the same
materials are analyzed.   The analysis of petroleum products has
spurred some of the most comprehensive research efforts on any single subject.   Research Project No. 6, initiated by the American Petroleum Institute, is a good example.

Sample characterization is carried out with two objectives: boiling point distribution and classification according to structural groups.
The former is done by distillation (146) or simulated distillation, a
gas chromatographic procedure (147,148).   Structural information in
terms of paraffins, olefins, naphthenes and aromatics (PONA analysis) may be obtained by one of two methods:

1.   Complete chromatographic resolution and characterization of in-
     dividual components on a single high-performance capillary col-
     umn.   The term "high performance" is related not only to high
     efficiency but also to high reproducibility, i.e., column stability.
2.   Group-type separations on several columns of different selectivity
     using intermediate trapping and heartcutting techniques.

The relative merits of these two approaches have recently been
discussed.   Johansen et al. (149) have provided a comprehensive
review on the subject.   He demonstrated the power of high resolution GC in conjunction with modern data systems in the analysis of
such complex chromatograms.   Disadvantages of the single-column
method are the need of identification of essentially every peak in
the chromatogram and the relatively long time necessary for analysis
and data presentation.   The major disadvantage of group type separations that rely on multidimensional chromatography is primarily
the complexity of the necessary instrumentation.   Both methods have
proponents.   It appears that the direct approach is favored for the
analysis of light end petroleum products where sample complexity is
manageable.   With increasing carbon number, the number of possible isomers to be resolved increases sharply.   Considering only
alkanes, naphthenes and aromatics, more than 10,000 structures are
possible for a carbon number up to only C_{12}, making the direct approach less attractive.   Group type separations are clearly preferable for such samples.   Examples for the analysis of natural gas and
of crude oil by group type separations, i.e., multidimensional gas
chromatography, are presented below.

In natural gas, inorganic gases such as O_2, N_2, CO, and CO_2
need to be determined in addition to hydrocarbons, ranging from

approximately C_1 to C_{10}. The analysis of fixed gases requires gas-solic chromatography on packed columns with TCD detection whereas capillary column GC with FID is favored for the hydrocarbons. Figure 14a shows a valve based system and Figure 14b illustrates the analysis sequence (150). The fixed gases are resolved by packed column GC on Porapak and molecular sieve adsorbents. The hydrocarbons are introduced into the capillary column, after backflushing of the packed columns. The system is under control of the GC microprocessor, for example, valves are actuated by a proper sequencing program. Typical relative standard deviations are in the order of 1%. This general approach to gas analysis seems to be widely accepted. Several systems have recently been described, with relatively minor variations. Both valve type (150–152) and Deans type switches (153) have been used.

Crude oil and its products are commonly subjected to PONA analysis. Figure 15a presents a state of the art PONA analyzer (154). The system shown relies on packed columns. Advantage is taken of stationary phase selectivities that can distinguish between alkanes, naphthenes, olefins, and aromatics. The sample is sliced up into fractions and each fraction is then sequentially subjected to the three separations systems. Figure 15b presents the sequence in which fractions are switched between the 3 columns and Figure 15c shows the final result. It is obvious that the process is quite complex. It appears that this type of instrumentation is now used worldwide but a study group of committee D-2 of ASTM is currently also

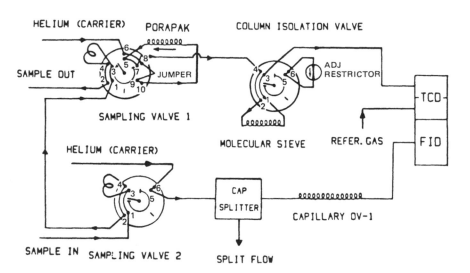

FIGURE 14a Schematic diagram of a gas chromatographic system for analysis of natural gas. From Ref. 150.

STEP   COLUMNS   DETECTORS   PEAKS

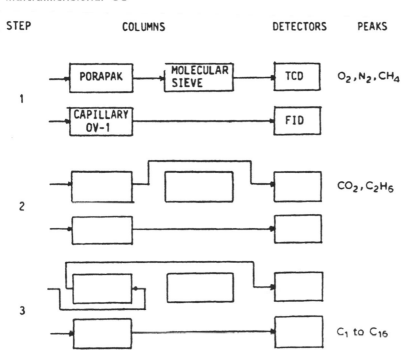

FIGURE 14b   Analysis sequence.

investigating the direct approach.   PONA analysis on instrumentation
less complex than that shown in Figure 15a has also been presented
(155).   One instrument uses a support coated capillary column prepared with molecular sieve 13X.   Figure 16 shows a chromatogram
obtained from that system.   The result is comparable to Figure 15c
but no data are provided on quantitative aspects.

Organolead compounds in gasoline are gradually being phased
out in the U.S.   Oxygenates such as methyl-t-butylether (MTBE)
and diisopropylether (DIPE) are added to gasoline instead.   The
analysis of oxygenates represents a difficult task, due to the extremely complex nature of gasoline.   Multidimensional GC is ideally
suited for this type of analysis since polar phases provide excellent
selectivity for oxygenates (138,156−158).   A system has been proposed to isolate these substances and, at the same time, provide optimal conditions for general analysis.   Gasoline contains a large number of highly volatile substances in the C_4 to C_7 range, that require
thick film columns and/or low temperatures.   Unfortunately, thick
film columns are unsuitable for the high boiling range fraction of
gasoline, for example, substituted naphthalenes.   Both requirements
can be met by switching the high volatility part of the sample to a

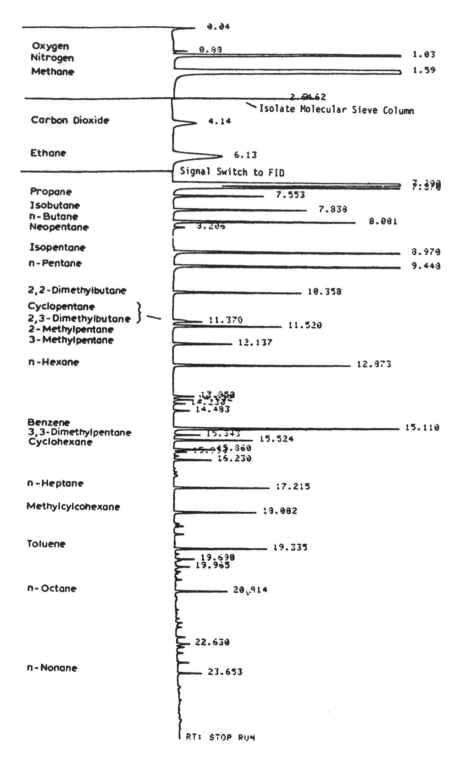

FIGURE 14c Chromatogram of a natural gas sample.

thick film column and carrying out the remainder of the analysis on
a column of standard film thickness (159).

Analysis of sedimentary organic matter provides important information for the assessment of maturity and other parameters that are
of interest in oil exploration. The composition of methylphenanthrene
isomers is an important indicator to determine the thermal history of
a crude oil. Analysis of the isomers has been carried out by two-
dimensional GC on cyanopropyl- and methylpolysiloxane phases (160).
High temperature liquid crystal phases that provide excellent selectivity for structural isomers have recently been introduced (161).
They should be superior to cyanopropyl type phases such as SILAR
10C. Kerogen, the precursor of crude oil, is of very high molecular weight and thus inaccessible by gas chromatography. Pyrolysis
GC is often used to study thermally labile products derived from
kerogen. Two-dimensional systems have been introduced to deal
with the abundance of products in the kerogen pyrolysate (162,163).
The thermal desorbate from small rock samples has been studied
with a sophisticated two-dimensional instrument that included an on-
line desorption unit (164). The results help the petroleum chemist
and geologist to gain insight into the critical steps in the petroleum
generation process. Biomarkers such as the pristane/phytane and
methylphenanthrene isomers are of special importance to the geochem-
ist. Rapid quantitation of these substances is important. The chromatographic resolution of the individual isomers in pristane and phy-
tane has been a real challenge to some chromatographers (165).
The separation is very difficult since it has to be based on efficiency. Selective phases are unavailable for these closely related isomers. Columns generating in excess of 750,000 theoretical plates
are required for baseline separation (166).

The analysis of geochemical polymers of crude oil and petroleum
is likely to remain a challenge for the application of multidimensional
GC.

B. Environmental Analysis

Many of the substances of interest in environmental analysis, in particular, air (167,168) and water pollutants (169,170) are related to
petroleum products or materials derived from combustion processes.
Polynuclear aromatic hydrocarbons and their derivatives have been
under special scrutiny (171–172). Elaborate two-dimensional systems have been described that process collected samples on-line
(173–176). Another important class of chemicals of environmental
interest are organohalogen compounds, especially pesticides, PCBs
and dioxins. GC-MS is often a preferred method in pesticide analysis since mass fragments are usually very characteristic. Analysis
is thus based primarily on detector selectivity and less on chromato-

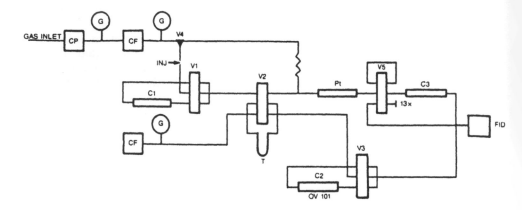

FIGURE 15a PNA analyzer flow scheme: G = gauge; CP = constant pressure controller; CF = constant flow controller; V1,2,3,5 = six-port sliding valve; V4 = threeport solenoid valfe; INJ = injection port; FID = flame ionization detector; T = trap, 40 cm, 1/8 in. o.d., 2.5 mm i.d. filled with Tenax; Pt = hydrogenator, 5% Pt on $BaSO_4$ mixed with Chromosorb 750; Cl = polar column, 3 m ss 1/8 in o.d., 2 mm i.d., 20% OV-275 on Chromosorb P-AM-DMCS, 60/80 mesh; C2 = nonpolar column, 4 m ss 1/8 in. o.d., 2 mm i.d., 5% OV-101 on Chromosorb W-AW-DMCS; C3 = column for paraffin/naphthene separation, 1.7 m ss 2 mm o.d., 1.6 mm i.d., molecular sieve 13x. Temperatures: column oven 120°C; injector 150°C, detector 150°C; trap, absorption was ambient, desorption was 300°C; hydrogenator 180°C; mol. sieve column 100°C, raised 6°C/min to 400°C. Flowrates: for hydrogen carrier gas in the polar column 15 mL/min., in the nonpolar column 15 mL/min., for nitrogen make-up gas in the FID 30 mL/min.; air in the FID 400 mL/min. From Ref. 154.

graphic resolution. Although a very large number of pesticides may potentially be present in a sample, only a very limited number is of interest in practical analysis. It rarely happens that more than a few pesticides are applied simultaneously in an agricultural setting. Electron capture detectors are still in widespread use in pesticide analysis.

Stan (177–180) and others in 1983 reported on pesticide residue analysis in food matrices using two-dimensional gas chromatography. The method is based on an internal calibration technique that includes peak recognition with name annotation. Simultaneous analysis on two fused silica capillary columns of high efficiency, coated with stationary phases of different selectivity provided unequivocal identification for some 50 possible pesticides. The retention behavior on different phases and the response ratio on two selective detec-

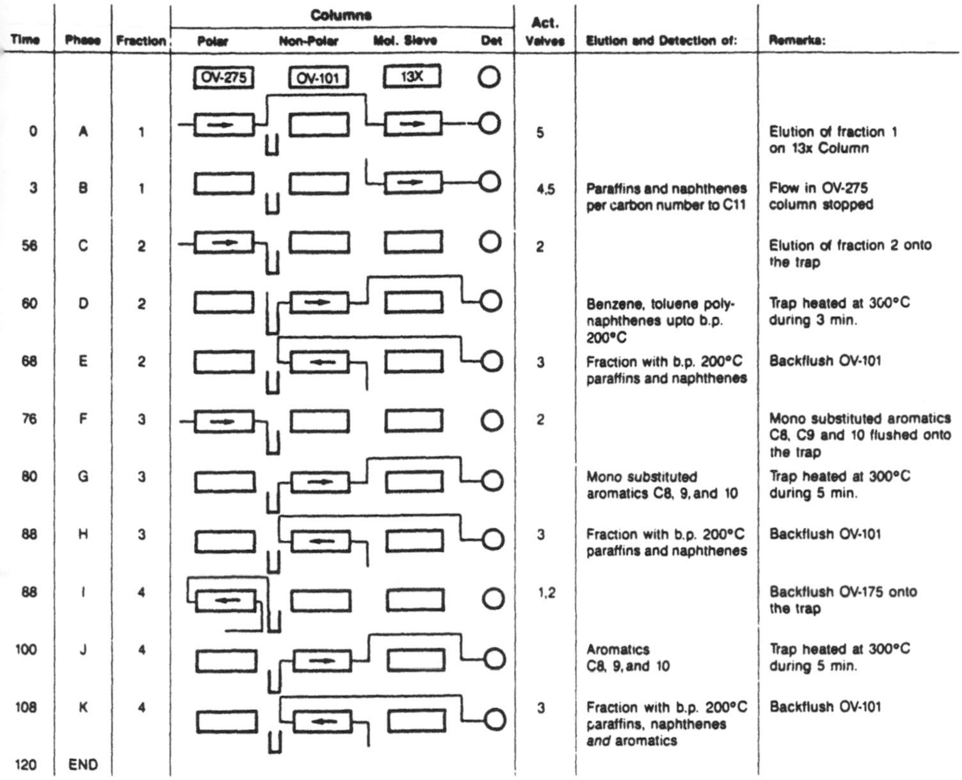

Time	Phase	Fraction	Columns: Polar (OV-275)	Non-Polar (OV-101)	Mol. Sieve (13X)	Det	Act. Valves	Elution and Detection of:	Remarks:
0	A	1				○	5		Elution of fraction 1 on 13x Column
3	B	1				○	4.5	Paraffins and naphthenes per carbon number to C11	Flow in OV-275 column stopped
56	C	2				○	2		Elution of fraction 2 onto the trap
60	D	2				○		Benzene, toluene poly-naphthenes upto b.p. 200°C	Trap heated at 300°C during 3 min.
68	E	2				○	3	Fraction with b.p. 200°C paraffins and naphthenes	Backflush OV-101
76	F	3				○	2		Mono substituted aromatics C8, C9 and 10 flushed onto the trap
80	G	3				○		Mono substituted aromatics C8, 9, and 10	Trap heated at 300°C during 5 min.
88	H	3				○	3	Fraction with b.p. 200°C paraffins and naphthenes	Backflush OV-101
88	I	4				○	1,2		Backflush OV-175 onto the trap
100	J	4				○		Aromatics C8, 9, and 10	Trap heated at 300°C during 5 min.
108	K	4				○	3	Fraction with b.p. 200°C paraffins, naphthenes and aromatics	Backflush OV-101
120	END								

FIGURE 15b   PNA analysis sequence.

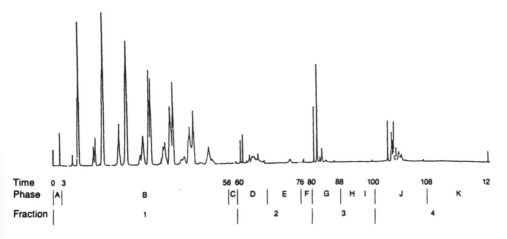

Time	0 3		56 60		76 80	88	100	108	12
Phase	A	B	C D	E	F G	H I	J	K	
Fraction		1		2		3		4	

FIGURE 15c   Operational phases (A–K) and hydrocarbon fractions (1–4) obtained in PNA analysis.

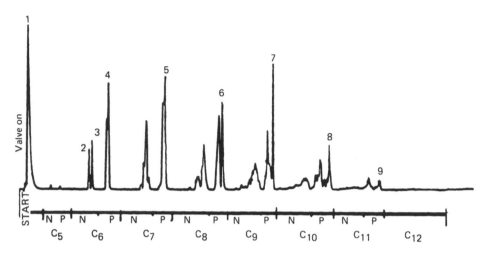

FIGURE 16 Chromatogram of naphtha analysis. Precolumn temperature: 430°C; 13 × MS PLOT. column temperature: 160°C–450°C,
20°C/min. Carrier gas: Hydrogen (A) 4.5 mL/min., (B) 4.5 mL/
min. 1. Total aromatics, 2. cyclohexane, 3. methylcyclopentane,
4. n-hexane, 5. n-heptane, 6. n-octane, 7. n-nonane, 8. n-
decane, 9. n-undecane. Samples size 0.16 μl. Split ratio 20:1.
From Ref. 155.

on two selective detectors, e.g., an FPD and an ECD were determined simultaneously for each substance. The data were linked,
i.e., retention times and/or detector signals had to fall within specified limits. Table 6 presents an example of a final report (178).
In practical samples that produce a significant level of interferences,
even highly efficient capillary columns may not always provide the
resolution necessary at such trace levels. Heartcutting or extensive
precolumn cleanup has to be performed in such cases (180).

The analysis of polychlorinated biphenyls (PCBs) and of polychlorinated dibenzodioxins/dibenzofurans (PCDDs and PCDFs) is a
particularly challenging problem in environmental analysis. PCDDs
and PCDFs contain isomers that are some of the most toxic materials
known to man. It is therefore necessary to detect these substances
at very low levels and in many different many matrixes. In environmental samples, dioxins are often overshadowed by PCBs that interfere with their analysis. Detection by mass spectrometry is commonly used to circumvent some of these interferences at the required
concentration levels (181,182). This approach is very difficult when
PCBs are present in large excess. A method based on parallel FID-
ECD detection and heartcutting has recently been proposed (183).
Figure 17 shows a chromatogram in which 1,2,3,4 TCCD has been

TABLE 6 Relative Retention Times of Organophosphorus Pesticides in Two-Dimensional Gas Chromatography[a]

No.	Pesticide	Test (ng/l)	First column RRT	First column Response	First column NAME	Second column RRT	Second column Response	Second column NAME
1	trichlorphon	5		5		0.108	0.82	TRICFON
2	dimefox	2	0.216	3.28	DMEFOX	1.057	4.75	DMEFOX
3	dichlorvos	2	0.371	2.29	DCVOS	0.212	2.38	DCVOS
4	mevinphos	2	0.497	2.70	MEVINF	0.294	1.79	MEVINF
			0.497	2.70	MEVINF	0.307	0.53	MEVINF
5	acephate	3	0.556	1.67	ACEPHAT		n.n.	
6	demephion	5	0.567	0.82	DEMEFIO	0.389	0.73	DEMEFIO
X	.....							
	ISTD 1	5	0.518	0.56	ISTD.1	0.323	0.67	ISTD.1
	ISTD 2	5	1	1	ISTD.2	1	1	ISTD.2

[a]Compounds are listed according to their retention times on the first column (screening run). Column 1: fused silica, SD 2100. Column 2: soda-lime glass, OV 225. IST.1 and IST.2 refer to internal standards. Partial table from Ref. 178.

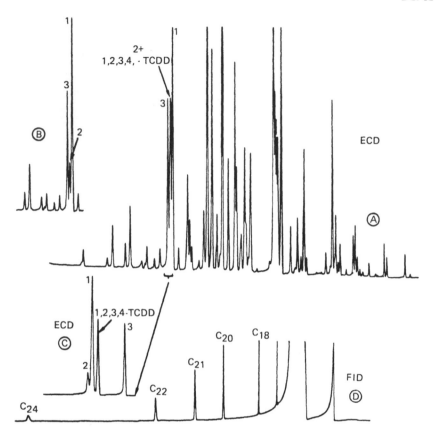

FIGURE 17 MDGC separation of chlorinated aromatic hydrocarbons using detection by paralleled ECD and FID. Identification of 1,2,3,4-TCDD by Kovats index. Samples: (A) 0.4 µl Clophen A 50, solution spiked with 1,2,3,4-TCDD; (B) Clophen A 50; (C) cut of (A); (D) 0.4 µl n-alkane solution. Columns: (A), (B) 24 m SILAR 7CP on soft glass, 0.24 mm i.d. (C), (C) 25 m methyl polysiloxane OV-1 on FS, 0.32 mm i.d. Temperatures: (A), (B) 50—130°C ball., 130—260°C, 4°/min; (C), (D) 45 min. Detectors: ECD for chromatograms (A), (B), (C), FID for chromatogram (D). From Ref. 183.

added to a PCB standard. It is apparent that heartcutting is necessary for adequate resolution of this isomer. Confidence in analysis results can further be substantiated by carrying out precise retention index measurements. Retention indexes can routinely be determined to better than 1 index unit, especially under isothermal conditions (90). A double oven instrument that permits temperature programming of the precolumn but allows isothermal chromatography in the analytical column is a clear asset. Retention index vs. temperature data provide still an additional measure of verification. Table 7 presents some retention data at different temperatures.

Environmental analysts often use liquid-solid chromatography as a cleanup step to remove major interferences before gas chromatographic analysis. Two-dimensional GC can eliminate this extra step and thus provides a more direct approach. The chemist must weigh carefully the potential risks involved in adding additional steps to the analysis and balance these considerations against the additional expense incurred by two-dimensional GC.

C. Biochemical and Clinical Applications

Ervard's extensive dissertation (184) on the application of two-dimensional gas chromatography to biochemiccal problems was the first serious attempt to introduce this technique into the clinical laboratory. One of the few areas where two-dimensionas gas chromatography has gained some significance in clinical chemistry is in the area of drug analysis. This type of analysis is of interest to the forensic chemist as well. Drug screen methods have been described that are based on simultaneous multicolumn chromatography (139, 140). Since the presence or absence of drugs usually has far reaching implications, most analysts feel compelled to rely on additional information such as provided by spectrometry. It can be expected that capillary columns with selective detectors, especially mass spectrometry, will be favored for the foreseeable future over two-dimensional GC.

TABLE 7   Retention Indices of the Most Toxic
2,3,7,8-Substituted PCDD- and PCDF Isomers

	OV-1 220°C	OV-1 240°C	OV-1 280°C	OV-1 300°C
2,3,7,8-tetra-CDF		2309		
2,3,7,8-tetra-CDD	2307	2336		
1,2,3,7,8-penta-CDD		2510	2574	
1,2,3,4,7,8-hexa-CDD			2744	
1,2,3,6,7,8-hexa-CDD			2749	
1,2,3,7,8,9-hexa-CDD			2764	2800
1,2,3,4,6,7,8-hepta-CDD			2930	2971
1,2,3,4,6,7,8,9-octa-CDD				3150

Most substances of biological interest contain chiral atoms. It would appear that verification of enantiometic purity would be of major concern in the manufacture of drugs. Stationary phases based on polysiloxanes have been synthesized that provide rapid analysis of optically active materials without a need for diasteroisomer formation (185). A large body of information has been accumulated in that area but only few practical applications have been presented that require the use of multidimensional GC (117,186).

Two-dimensional gas chromatography with capillary columns has been applied to the analysis of steroids in body fluids (187). A heartcutting system in conjunction with a radioactive monitor has been suggested for the analysis of carbon-14 labeled materials (188, 189), but this approach should be viewed as an exception rather than the rule. Clinical chemists still prefer methods based on liquid chromatography. It is doubtful that two-dimensional gas chromatography will find widespread use in routine clinical laboratories.

D. Food and Flavor Industry

Analysis of natural products presents formidable problems to the chromatographer. Frequently, samples are extremely complex and trace components are often overlapped by major components. Consequently, direct identification of minor substances by mass spectrometry or infrared spectrometry is a difficult task. In many cases, both of these complementary techniques must be applied. Sample complexity of natural products requires the use of highly efficient columns, for example, small bore columns. This is in conflict with another requirement. To obtain enough material for a useable spectrum, the maximum permissable amount of sample must be introduced into a column. This is one of the reasons that widebore capillary columns are still popular with many flavor chemists, even though they produce only moderate efficiencies.

Since trace constituents are often the most important components in flavors, complete physical resolution of individual compounds is necessary when organoleptic properties are to be correlated to chromatographic profiles (190,191). Two-dimensional gas chromatography using off-line methods has also been applied in the flavor industry. The results which can be obtained with seemingly simple systems are impressive. The magnitude of sample complexity in natural products may best be illustrated by an example involving organoleptic evaluation of the active components in a common spice. Figure 18 shows a chromatographic profile of the volatiles of black pepper (65). The shaded area refers to the properties of the chromatogram which impart the typical pungent odor of pepper. It is quite clear that the shaded area consists of many unresolved substances. Subsequent re-injection of the isolated fraction into a column of different

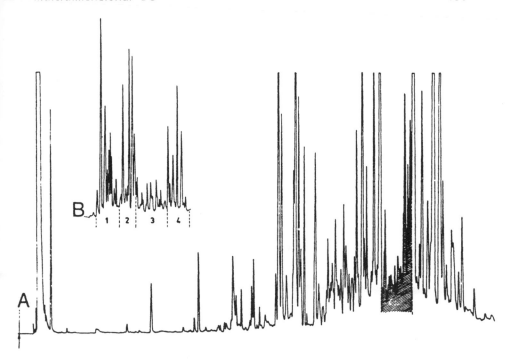

FIGURE 18 Analysis of an organoleptic interesting part of the oxy-gen fraction of pepper essential oil, obtained by preparative GC. (A) Column 80 m × 0.5 mm W.W.C.O.T., P.M.P.E.   Temperature programmed 70° to 220° at 2°/min.   Carrier gas flow, hydrogen 8 ml/min.   (B) Collection of fraction (shaded part in Figure A).   Several collections washed out with dichloromethane, concentrated and injected onto an FFAP column: 40 m × 0.5 mm.   Temperature programmed 70° to 180° at 2°/min.   Carrier gas flow, hydrogen 4 ml/min.   From Ref. 65.

selectivity (insert) showed that components in fractions 1 and 3 imparted the typical flavor.

Correlation of organoleptic properties with chromatographic data is a particularly difficult endeavor.   The human nose responds very selectively to chemical stimuli, covering a concentration range of at least 8 orders of magnitude.   Heartcutting is often the only alternative.   The analysis of volatiles in wine and alcoholic beverages is not as simple as it may appear to the casual observer (192).   The major problems with wine volatiles is their very low concentration, the delicacy of some individual substances and the fact that one has to deal with an aqueous matrix containing salts and other nonvolatile materials as well.   Figure 19 shows chromatograms of wine volatiles

obtained under different conditions (193). Figure 19a represents
an efficient capillary column whereas Figure 19b was obtained on a
packed column coated with a polar stationary phase. The three
chromatograms shown in Figure 19c represent the cuts taken from
the polar column. It is clear that some overlap does occur between
the selected fractions. The major advantage of this particular system is the rejection of water and other undesirable components from
the sample.

The last example presented in this book chapter deals with a
particular difficult problem in flavor research, that of tobacco smoke
analysis. In 1975, Grob and Grob published a chromatogram of the
so called semivolatile fraction that contained about 1,000 reproducible
signals (194). Tobacco smoke analysis can be described as an analyst's nightmare. The range of substances stretches from gases to
solids; a concentration range of many orders of magnitude needs to
be covered; a large variety of different functional groups are encountered, that is, strongly acidic or basic substances may be pres-

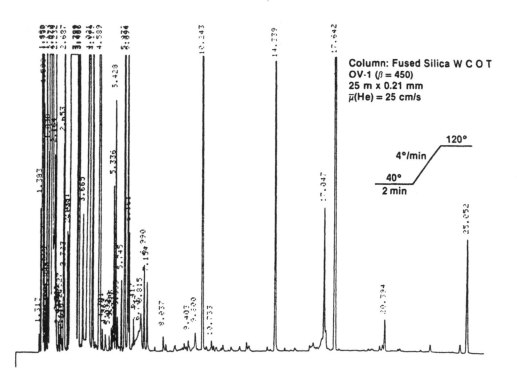

FIGURE 19a   Capillary column analysis of wine volatiles.   From Ref.
193.

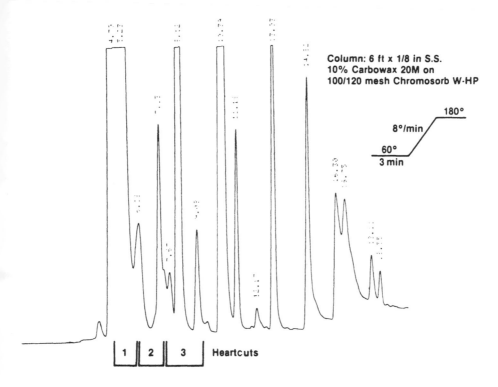

Column: 6 ft x 1/8 in S.S.
10% Carbowax 20M on
100/120 mesh Chromosorb W-HP

180°
8°/min
60°
3 min

1 | 2 | 3 | Heartcuts

FIGURE 19b   Packed column separation of wine volatiles.

sent at the same time; many substances are reactive and/or thermally labile.

In spite of these obstacles, hundreds of components have been
identified over the years. Figure 20a, top, shows a part of a chromatogram for a tobacco smoke sample (190).   The bottom chromatogram is an expansion of a section of the above chromatogram.   All
peaks are severely distorted and can be assumed to represent mixtures.   Not surprisingly, mass spectra taken at points to E proved
inconclusive.   The situation improved only after the fraction corresponding to the scanned region in the bottom chromatogram was
heartcut to a different column.   The resulting chromatogram is
shown in Figure 20b.   It consists of at least 9 components that
were well resolved and thus amenable to mass spectral elucidation.
This example points out the major disadvantage of heartcutting.
Only a small fraction, about 1% of the chromatogram shown on Figure 20a has been analyzed in this heartcut operation.   On the other
hand it is also clear that almost all separations amenable to gas chromatography can be solved in practice, regardless of sample complex-

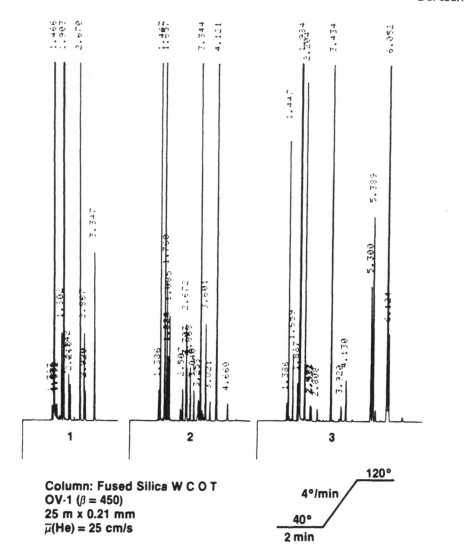

Column: Fused Silica W C O T
OV-1 (β = 450)
25 m x 0.21 mm
$\bar{\mu}$(He) = 25 cm/s

120°
4°/min
40°
2 min

FIGURE 19c   Capillary column analysis of fractions cut from packed
column, Figure 19b.

ity. If the resolution at hand is insufficient, it can be improved in
two ways: the width of the heartcut fraction can be narrowed or
another stationary phase may be selected. If necessary, a heartcut
of a heartcut can be taken. The only requirement is that the stationary phase shows some selectivity for a critical pair of substances.
     Analysis of flavors is probably one of the most fruitful and pleas-
ant areas for the application of multidimensional GC. The fact that
only a few applications of multidimensional GC are reported for applications from the food industry may reflect the preoccupation of
many flavor chemists to solve problems by profiling techniques, including pattern recognition. This does not diminish the significance
of multidimensional gas chromatography as the ultimate analytical
tool for applications where highest resolving power is required.

V. OUTLOOK

Two-dimensional gas chromatography will certainly retain a special
niche in chromatography, at least within the near future. This
means that it probably will not experience strong growth as it translates into wider use. Unquestionably, specialized laboratories will
continue to refine and improve the technique, simply because it is
the only realistic approach that is presently available toward the
solution of extremely difficult separations problems.
     In a certain sense, competition to two-dimensional GC arises
from the successful interfacing of very sensitive and selective detectors to capillary columns. A system that can simultaneously provide infrared spectra, mass spectra, and precise retention data, all
on the fly, considerably eases chromatographic requirements. Complete chromatographic resolution is often no longer required. Integrated systems based on capillary columns in conjunction with selective detectors and postrun data processing will continue to effectively compete with two-dimensional GC. With the introduction of low
cost mass spectrometers as dedicated detectors for gas chromatography many analytical problems that previously required a high degree
of chromatographic resolution can now be tackled in routine analysis. Other, very powerful detectors such as tandem mass spectrometry further expand the horizon. A GC-GC-MS-MS system has recently been described (195). It is expected that further advances
in detector technology and data processing will enable analysts that
have only a moderate level of training to routinely carry out analyses that were difficult to accomplish only a few years ago. The
trend to turnkey systems will continue. Two-dimensional high resolution gas chromatography is not a technique of the past, however.
There are many avenues that have barely been touched. Systems
that are capable of simultaneous heartcutting and stationary phase

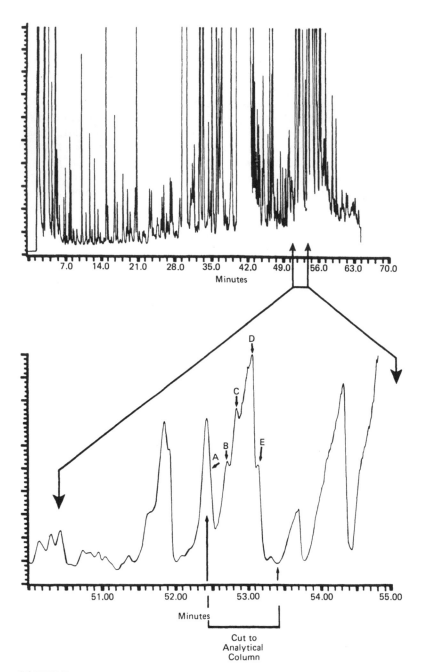

FIGURE 20a Chromatogram generated by tobacco smoke sample. A normal chromatogram generated on the DB-Wax precolumn (top). An expanded view of the section indicated on the top chromatogram (bottom). The section indicated on the bottom chromatogram was transferred to the analytical column. The areas indicated (A—E) were scanned by the mass spectrometer. From Ref. 190.

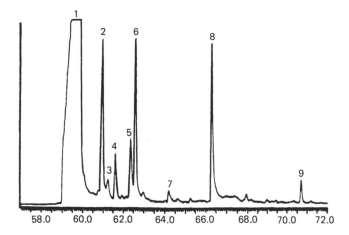

FIGURE 20b Analytical column chromatogram generated from the heartcult (indicated on Figure 20a) from a tobacco smoke sample. Peak numbers indicate ares where mass spectral data were acquired.

optimization have already been described (77,196). If one adds selective detectors such as IR/MS or MS/MS to multidimensional GC and correlates the wealth of data from these many resources, new frontiers will be opened up.

ACKNOWLEDGMENT

I would like to thank D. Roberts for proofreading and S. Shirley for typing the manuscript, under trying conditions.

REFERENCES

1. E. C. Shannon and W. Weaver, The Mathematical Theory of Communication, University of Illinois Press, Urbana, IL, 1949.
2. H. Kaiser, Anal. Chem., 42:24A (1970).
3. J. F. K. Huber and H. C. Smit, Z. Anal. Chem., 245:84 (1969).
4. A. C. Moffat, K. W. Smalldon, and C. Brown, J. Chromatogr., 90:1 (1974).
5. J. F. K. Huber, E. Kenndler, and G. Reich, J. Chromatogr., 172:15 (1979).
6. J. A. Rijks and C. A. Cramers, Chromatographia, 7:99 (1974).

7. J. A. Rijks, Dissertation, Technical University Eindhoven, NL, 1973.
8. M. Goedert and G. Guiochon, Anal. Chem., 42:962 (1970).
9. E. Kovats, Helv. Chim. Acta., 41:1915 (1958).
10. C. A. Cramers, J. A. Rijks, V. Pacakova, and I. de Andrade, J. Chromatogr., 51:13 (1970).
11. J. A. Rijks, and C. A. Cramers, Chromatographia, 7:99 (1974).
12. J. Curvers, J. Rijks, C. A. Cramers, K. Knauss, and P. Larson, HRC & CC, 8:607 (1985).
13. J. Curvers, J. Rijks, C. A. Cramers, K. Knauss, and P. Larson, HRC & CC, 8:611 (1985).
14. The Sadtler Standard Gas Chromatographic Retention Index Library. Sadtler Research Laboratories, Philadelphia.
15. J. F. K. Huber, and R. G. Gerritse, J. Chromatogr., 80:25 (1973).
16. T. Hirschfeld, Anal. Chem., 52:297A (1980).
17. J. C. Giddings, Anal. Chem., 56:1238A (1984).
18. T. V. Raglione, N. Sagliano, Jr., T. R. Floyd, and R. A. Hartwick, LC-GC, 4:328 (1986).
19. H. J. Cortes, C. D. Pfeiffer, and B. E. Richter, HRC & CC, 8:469 (1985).
20. B. Pacciarelli, E. Mueller, R. Schneider, K. Grob, W. Steiner, and D. Froehlich, HRC & CC, 1:135 (1988).
21. H. T. Neu, and R. Zinburg, HRC & CC, 2:395 (1979).
22. K. Grob, Jr., On-column Injection in Capillary Gas Chromatography, Heuthig, 1987, Chapter 3.
23. T. Toth, J. Chromatogr., 279:157 (1983).
24. G. Schomburg, E. Huebinger, H. Husmann, and F. Weeke, Chromatographia, 16:228 (1982).
25. C. F. Poole and S. A. Schuette, Contemporary Practice of Chromatography, Elsevier, 1984, p. 485.
26. R. Consden, A. H. Gordon, and A. J. P. Martin, Biochem. J., 38:244 (1944).
27. D. P. Herman, M. F. Gonnord, and G. Guiochon, Anal. Chem., 56:995 (1984).
28. J. M. Davis, and J. C. Giddings, Anal. Chem., 55:418 (1983).
29. J. C. Giddings, HRC & CC, 10:319 (1987).
30. W. V. Ligon and R. T. May, J. Chromatogr., 294:77 (1984). Footnote by the editor.
31. I. M. Hais, J. Chromatogr., 187:466 (1980).
32. Y. P. Chizhkov and G. A. Yushina, J. Anal. Chem. USSR, 31:1, Chapter 10 (1976).
33. D. R. Deans, J. Chromatogr., 203:19 (1981).
34. R. Miller, in High Resolution Gas Chromatography, 2nd ed. (R. R. Freeman, ed.), Hewlett-Packard, Avondale, Pennsylvania, 1981, Chapter 4.

35. G. Schomburg, LC-GC, 5:303 (1987).
36. G. Schomburg, in Sample Introduction in Capillary Gas Chromatography (P. Sandra, ed.), Huethig, 1985, p. 235.
37. B. M. Gordon, C. E. Rix, and M. F. Borgerding, J. Chrom. Sci., 23 (1985).
38. F. Mueller, Am. Lab., 10:15 (1983).
39. W. Bertsch, HRC & CC, 1:85 (1978).
40. W. Bertsch, HRC & CC, 1:187 (1978).
41. W. Bertsch, HRC & CC, 1:289 (1978).
42. J. V. Brunnok and L. A. Luke, Anal. Chem., 40:215B (1968).
43. H. Boer, in Gas Chromatography (S. G. Perry, ed.), Applied Science Publishers, Barking, 1983, p. 109.
44. II. Boer and P. Van Arkel, Chromatographia, 4:300 (1971).
45. N. G. McTaggart and L. A. Luke, Erdoel Kohle, 24:586 (1982).
46. R. E. Kaiser, HRC & CC, 2:95 (1979).
47. A. S. Christophersen and K. E. Rasmussen, J. Chromatogr., 243:57 (1982).
48. G. Schomburg, HRC & CC, 2:461 (1979).
49. Jianying Zhang, Bingchen Lin, and Peichang Lu, Chromatographia, 23:487 (1987).
50. Feng Ye, Bingchen Lin, and Peichang Lu, Chromatographia, 23:492 (1987).
51. Daixin Tong, Fangbao Xu, and Peichang Lu, Chromatographia, 23:499 (1987).
52. V. G. Berezkin, T. P. Popova, A. A. Korolev, V. E. Shriyaeva, L. S. Vasin, and V. N. Lipavskii, HRC & CC, 11:41 (1988).
53: P. R. McCullough and W. A. Aue, J. Chromatogr., 82:269 (1982).
54. C. R. Warner, M. C. Johnson, D. G. Prue, and B. T. Kho, J. Chromatogr., 82:263 (1963).
55. G. Schomburg, H. Husmann, F. Schulz, G. Teller, and M. Bender, in Proceedings 5th International Symposium on Capillary Chromatography, Riva del Garda, 1983 (J. Rijks, ed.), Elsevier, Amsterdam, 1983, p. 452.
56. W. Vogt, K. Jacob, A. B. Ohnesorge, and H. W. Obwexer, J. Chromatogr., 235:225 (1981).
57. F. Poy, Chromatographia, 16:345 (1982).
58. F. Poy and L. Cobelli, J. Chromatogr., 279:689 (1983).
59. Gerstel, Inc., Muelheim/Ruhr, FRG. Brochure.
60. W. G. Jennings, Gas Chromatography with Glass Capillary Columns, 2nd ed., Academic Press, New York, 1980, Chapter 4.
61. W. Bertsch, F. Hsu, and A. Zlatkis, Anal. Chem., 48:928 (1976).
62. D. A. Leathard and B. C. Shurlock, Identification Techniques in Gas Chromatography, Wiley-Interscience, New York, 1970, p. 7.

63. A. Zlatkis, J. W. Anderson, and G. Holzer, J. Chromatogr.,
     142:127 (1977).
64. I. Klimes, W. Stuenzi, and D. Lamparsky, J. Chromatogr.,
     136:23 (1977).
65. P. Sandra, T. Saeed, G. Redant, M. Godefroot, M. Verstappe,
     and M. Verzele, HRC & CC, 3:107 (1980).
66. R. R. Deans, Chromatographia, 1:18 (1968).
67. E. A. Anderson, Multidimensional High Resolution Gas Liquid
     Chromatographia in Trace Organic Analysis: Heartcutting
     with Intermediate Trapping. Dissertation, University of Ala-
     bama, 1978.
68. J. F. K. Huber, E. Kenndler, W. Nyiri, and M. Oreans, J.
     Chromatogr., 247:211 (1982).
69. G. L. Johnson and A. Tipler, Proceedings of the 8th Inter-
     national Symposium on Capillary Chromatography (P. Sandra,
     ed.), Huethig, 1987.
70. A. Hagman, and S. Jacobson, HRC & CC, 8:332 (1985).
71. W. G. Jennings, J. A. Settlage, D. F. Ingraham, N. Wohlers,
     and R. J. Miller, in Proceedings of the 4th International Sym-
     posium on Capillary Chromatography. Hindelang, 1981 (R. E.
     Kaiser, ed.), Huethig, 1981, p. 563.
72. Valco Engineering, Houston, TX. Product information.
73. I. C. M. Wessels, and R. P. M. Dooper, in Proceedings of
     the 5th International Symposium on Capillary Chromatography,
     Riva del Garda, Italy, 1983 (J. Rijks, ed.), Elsevier, 1983,
     p. 452.
74. W. Jennings, J. Chromatogr. Sci., 22:129 (1984).
75. S. T. Adams, HRC & CC, 11:85 (1988).
76. R. J. Miller, HRC & CC, 10:457 (1987).
77. V. P. Chizhkov, G. A. Yushina, L. A. Sinitzina, and B. A.
     Rudenko, J. Chromatogr., 120:35 (1976).
78. M. D. Zabokvitsky, V. P. Chizhkov, and B. A. Rudenko,
     HRC & CC, 6:170 (1983).
79. M. D. Zabokvitsky, V. P. Chizhkov, and B. A. Rudenko,
     HRC & CC, 6:460 (1983).
80. R. E. Pauls, A. T. Shepard, J. E. Phelps, J. E. Davis, and
     L. B. Rogers, Sep. Sci., 12:289 (1977).
81. A. M. Reid, J. Chromatogr. Sci., 14:203 (1976).
82. J. W. Root, E. K. C. Lee, and F. S. Rowland, Science, 143:
     676 (1967).
83. R. Yoder, and R. Sacks, J. Chromatogr. Sci., 25:21 (1987).
84. J. Sevcik, in Proceedings 5th International Symposium on Cap-
     illary Chromatography, Riva del Garda, Italy, 1983 (J. Rijks,
     ed.), Elsevier, 1983, p. 771.
85. W. Jennings, J. A. Settlage, and R. J. Miller, HRC & CC, 2:
     441 (1979).

86. W. G. Jennings, J. A. Settlage, R. J. Miller, and W. G.
    Raabe, J. Chromatogr., 186:189 (1979).
87. D. Roberts. The University of Alabama. Dissertation
    (1989).
88. S. A. Volkov, V. I. Rezuikov, V. Y. Zenvenskii, and K. I.
    Sadokinskii, Chromatographia, 15:765 (1982).
89. G. Schomburg, F. Weeke, F. Mueller, and M. Oreans, Chro-
    matographia, 16:87 (1983).
90. J. A. Rijks, J. H. M. Van Den Berg, and J. P. Piependaal,
    H. Chromatogr., 91:603 (1974).
91. H. Broetell, G. Rietz, S. Sandquist, M. Berg, and H. Ehrs-
    son, HRC & CC, 5:596 (1983).
92. J. W. Graydon and K. Grob, J. Chromatogr., 253:265 (1983).
93. D. R. Rushneck, J. Gas Chromatogr., 1:318 (1965).
94. D. E. Willis and R. M. Engelbrecht, J. Gas Chromatogr., 3:
    536 (1967).
95. S. Adam, HRC & CC, 5:36 (1983).
96. J. F. Pankow, HRC & CC, 6:292 (1983).
97. E. L. Anderson, M. M. Thomason, H. T. Mayfield, and W.
    Bertsch, HRC & CC, 2:235 (1979).
98. H. U. Buser, R. Soder, and H. M. Widmer, HRC & CC, 5:
    159 (1982).
99. N. Schmidbauer, and M. Oehme, HRC & CC, 10:398 (1987).
100. B. Versino, H. Knoeppel, M. DeGroot, A. Peil, J. Poelmann,
    H. Schauenburg, H. Vissers, and F. Geiss, J. Chromatogr.,
    122:373 (1976).
101. K. H. Bergert, V. Betz, and D. Pruggmayer, Chromatographia,
    7:115 (1974).
102. K. Grob and K. Grob, Jr., HRC & CC, 1:57 (1978).
103. W. E. Harris, J. Chromatogr. Sci., 11:184 (1973).
104. V. Pretorius, C. S. G. Phillips, and W. Bertsch, HRC & CC,
    6:232 (1983).
105. J. Pankow, HRC & CC, 6:292 (1983).
106. M. Young, J. Chromatogr., 214:197 (1981).
107. R. E. Kaiser, in Proceedings 8th International Symposium on
    Advances in Chromatography (A. Zlatkis, ed.), University
    of Houston, 1973, p. 215.
108. S. Jacobson and S. Berg, HRC & CC, 5:236 (1982).
109. G. Holzer, private communication.
110. M. M. Thomason. Dissertation, The University of Alabama
    (1985).
111. B. Hopkins and V. Pretorius, J. Chromatogr., 158:465 (1978).
112. H. R. Buser, HRC & C, 5:154 (1982).
113. J. Sevcik, HRC & CC, 2:436 (1979).
114. J. A. Settlage and W. G. Jennings, HRC & CC, 4:146 (1980).
115. G. Schomburg, H. Husmann, and F. Weeke, J. Chromatogr.,
    112:205 (1975).

116. G. Schomburg, H. Husmann, and F. Weeke, J. Chromatogr.,
      99:63 (1974).
117. D. C. Fenimore, R. R. Freeman, and P. R. Loy, Anal. Chem.,
      45:2331 (1973).
118. W. G. Jennings, Comparison of Fused Silica and Other Glass
      Columns in Gas Chromatography, Huethig, 1981.
119. S. R. Lipsky, M. L. Duffy, HRC & CC, 9:376 (1986).
120. G. Schomburg, E. Bastian, H. Behlau, H. Husmann, F. Weeke,
      M. Oreans, and F. Mueller, HRC & CC, 7:4 (1984).
121. W. Bertsch, E. Anderson, and G. Holzer, Chromatographia,
      10:449 (1977).
122. D. C. Fenimore, R. R. Freeman, and P. R. Loy, Anal. Chem.,
      45:2331 (1973).
123. R. E. Kaiser and R. I. Rieder, HRC & CC, 2:416 (1979).
124. R. J. Laub and J. H. Purnell, J. Chromatogr., 112:71 (1975).
125. J. H. Purnell and P. S. Williams, HRC & CC, 6:507 (1983).
126. R. E. Kaiser, R. I. Rieder, Lin Leming, L. Blomberg, and
      P. Kusz, HRC & CC, 8:580 (1985).
127. T. S. Buys, and T. W. Smuts, HRC & CC, 4:102 (1981).
128. D. R. Deans, and I. Scott, in Proc. 8th Intern. Symp. on
      Advances in Chromatography (A. Zlatkis, ed.), 1973, p. 77.
129. R. E. Kaiser and R. I. Rieder, HRC & CC, 10:240 (1987).
130. W. V. Ligon, Jr. and R. J. May, Anal. Chem., 52:901 (1980).
131. W. V. Ligon and R. J. May, J. Chromatogr., 294:77 (1984).
132. W. V. Ligon and R. J. May, Anal. Chem., 58:558 (1986).
133. W. V. Ligon and R. J. May, J. Chromatogr. Sci., 24:2 (1986).
134. J. F. K. Huber and H. C. Smit, Z. Anal. Chem., 245:84
      (1969).
135. J. Sevcik, J. Chromatogr., 186:129 (1979).
136. K. Kugler, W. Halang, R. Schlenkermann, H. Webel, and R.
      Langlais, Chromatographia, 10:438 (1977).
137. J. M. Levy, J. Chromatogr. Sci., 22:149 (1984).
138. J. M. Levy and J. A. Yancey, HRC & CC, 9:383 (1986).
139. G. W. Hime and L. R. Bednarczyk, J. Anal. Tosicol., 6:247
      (1982).
140. S. Alm, S. Jonson, H. Karlson, and E. G. Sundholm, J.
      Chromatogr., 254:179 (1983).
141. Z. Penton, Varian Instrument, Palo Alto, Applications Note
      No. 56, August 1981.
142. R. R. Freeman, T. A. Rooney, R. M. Przybylski, and L. H.
      Altmayer, Technical Paper No. 83, Hewlett-Packard, Avondale,
      Pennsylvania.
143. H. T. Neu, M. Larsen, H. Panzel, and W. Merz, HRC & CC,
      11:131 (1988).
144. W. Merz, H. J. Neu, and H. Panzel, Vom Wasser, 59:103
      (1982).

145. J. Enqvist, P. Sunila, and U. M. Lakkisto, J. Chromatogr., 279:667 (1983).
146. D. 86-78, Distillation of Petroleum Products: 1981 Book of ASTM Standards, Part 23, American Society for Testing and Materials, Philadelphia, PA, 1981, pp. 8—26.
147. D 2887-73 (1978), Boiling Range Distribution of Petroleum Fractions by Gas Chromatography: 1983 Book of ASTM Standards, Part 24, American Society for Testing and Materials, Philadelphia, PA, 1981, pp. 799—807.
148. A. Rastogi, HRC & CC, 10:479 (1987).
149. N. G. Johansen, L. S. Ettre, and R. L. Miller, J. Chromatogr., 256:393 (1983).
150. L. Huber and H. Obbens, J. Chromatogr., 279:167 (1983).
151. P. Coleman and L. S. Ettre, HRC & CC, 8:112 (1985).
152. E. H. Osjord and D. M. Sorenssen, J. Chromatogr., 279:219 (1983).
153. H. Tani and M. Furuno, HRC & CC, 9:712 (1985).
154. P. van Arkel, J. Beens, H. Spaans, D. Grutterink, and R. Verbeek, J. Chromatogr. Sci., 25:141 (1987).
155. Z. Naizhong and L. E. Green, HRC & CC, 9:400 (1986).
156. S. M. Sonchik and T. Q. Walker, Am. Lab., 2:58 (1985).
157. S. W. S. McCreadie, D. F. K. Swan, G. M. Ogle, and R. Pintus, Proceedings of the 6th International Symposium on Capillary Chromatography (R. Sandra and W. Bertsch, eds.), Huethig, 1985, p. 456.
158. N. J. Johansen, HRC & CC, 8:487 (1984).
159. W. Jennings, J. Chromatogr. Sci., 22:129 (1984).
160. G. Schomburg, F. Weeke, and R. G. Schaefer, HRC & CC, 8:388 (1985).
161. K. E. Markides, H. C. Chang, C. M. Schregenberger, B. J. Tarbet, J. S. Bradshaw, and M. L. Lee, HRC & CC, 8:516 (1985).
162. H. Dembicki, Jr., B. Horsfield, and T. T. Y. Ho., Bull. Am. Assoc. Pet. Geol., 87:1094 (1983).
163. E. L. Colling, B. H. Burda, and D. A. Kelley, J. Chromatogr. Sci., 24:7 (1986).
164. R. G. Schaefer and H. Pooch, Chromatographia, 16:257 (1982).
165. C. P. M. Schutjes, Dissertation, Eindhoven/The Netherlands, 1983.
166. M. Proot, F. David, P. Sandra, and M. Verele, HRC & CC, 8:426 (1985).
167. J. E. Bunch and E. D. Pellizari, J. Chromatogr., 186:811 (1979).
168. W. Bertsch, in Application of Glass Capillary Gas Chromatography (W. J. Jennings, ed.), Marcel Dekker, Inc., 1981.
169. D. C. K. Lin, in Application of Glass Capillary Gas Chromatography (W. J. Jennings, ed.), Marcel Dekker, Inc., 1981.

171. M. L. Lee, M. Novotny, and K. O. Bartle, Analytical Chemistry of Polycyclic Aromatic Compounds. Academic Press, New York (1981).
172. Handbook of Polycyclic Aromatic Hydrocarbons, A. Bjorseth, editor, Marcel Dekker, Inc., New York, 1983.
173. B. J. Tyson and G. C. Carle, Anal. Chem., 46:611 (1974).
174. A. Jonsson and S. Berg, J. Chromatogr., 279:307 (1983).
175. J. Berg and A. Johsson, HRC & CC, 6:687 (1984).
176. R. A. Lunsford and Y. T. Gagnon, HRC & CC, 10:102 (1987).
177. O. Mrowetz and H. J. Stan, in Proc. 5th Intern. Symp. on Capillary Chromatography, Riva del Garda, 1983 (J. Rijks, ed.), Elsevier (1983), p. 203.
178. H. J. Stan and D. Mrowetz, HRC & CC, 6:255 (1983).
179. H. J. Stan and D. Mrowetz, in Proc. 5th Intern. Symp. on Capillary Chromatography, Riva del Garda, 1983 (J. Rijks, ed.), Elsevier, 1983, p. 189.
180. H. J. Stan and D. Mrowetz, J. Chromatogr., 279:173 (1983).
181. D. Schmid and M. D. Mueller, HRC & CC, 10:548 (1987).
182. M. Swerev and K. Ballschmiter, HRC & CC, 10:544 (1987).
183. G. Schomburg, H. Husmann, and E. Huebinger, HRC & CC, 8:395 (1985).
184. E. Evard, Two-Dimensional Gas Chromatography, Dissertation, Universite Catolique de Louvrain, 1975.
185. H. Frank, J. G. Nicholson, and E. Bayer, J. Chromatogr. Sci., 15:174 (1977).
186. W. Chinghai, H. Frank, W. Guanghua, Z. Liangmo, E. Bayer, and Lu Peichang, J. Chromatogr., 262:352 (1983).
187. E. C. Horning, C. D. Pfaffenberger, and A. C. Moffat, in Prod. 7th Intern. Symp. on Advances in Chromatography (A. Zlatkis, ed.), 1971, p. 119.
188. K. Herkner, and W. Swoboda, in Proc. 4th Intern. Symp. on Capillary Chromatography, Hindelang 1981 (R. E. Kaiser, ed.), Huethig, 1981, p. 429.
189. K. Herkner, Chromatographia, 16:39 (1982).
190. D. W. Wright, K. O. Mahler, and L. B. Ballard, J. Chromatog. Sci., 24:60 (1986).
191. P. A. Rodriguez and C. L. Eddy, J. Chromatogr. Sci., 24:18 (1986).
192. R. H. M. van Ingen, L. M. Nijssen, D. van der Berg, and H. Maarse, HRC & CC, 10:151 (1987).
193. R. J. Phillips, K. A. Knauss, and R. R. Freeman, HRC & CC, 5:546 (1982).
194. K. Grob and G. Grob, Chromatographia, 8:423 (1975).
195. S. G. Claude, R. Tabacchi, in Proceedings of the 8th Int. Symposium on Capillary Chromatography (P. Sandra, ed.), Huethig, 1987, p. 564.
196. P. Sandra, F. David, M. Proot, G. Diricks, M. Verstappe, and M. Verzele, HRC & CC, 9:782 (1985).

4

Selectivity Tuning in Capillary Gas Chromatography

PAT SANDRA* and FRANK DAVID *Research Institute for Chromatography, Wevelgem, Belgium*

I. INTRODUCTION

Thirty years after M. Golay presented his fundamental work (1–3), capillary gas chromatography (CGC) has become one of the most powerful analytical techniques. Characterized by excellent resolving power, high sensitivity and short analysis times, CGC offers the analyst a tool, applicable to a large variety of separation problems.

Although the possibilities of capillary gas chromatography were soon realized, the technique was restricted to the laboratories of some specialized groups. Only recently, a breakthrough to routine analyses was made by the introduction of flexible fused-silica columns (4). Developments such as the static coating technique, high temperature silylation, the introduction of new gum phases and their immobilization, on-column injection, multidimensional CGC, CGC-MS, CGC-FTIR and many others, have catalyzed this breakthrough and have resulted in a widespread use of CGC in all fields of chemical analysis.

As the application of CGC becomes more and more important in routine analysis, capillary gas chromatographic researchers are confronted with an increasing number of new and challenging problems. Many of these problems are related to the high complexity of the mixtures to be analyzed. In such cases, unequivocal identification and accurate quantification of single compounds is only possible if sufficient resolution is obtained. Resolution optimization is therefore the heart of chromatographic development.

The most important factor influencing resolution is the selectivity of the chromatographic system. Selectivity for a given sample is in

---

*Current affiliation: University of Gent, Gent, Belgium

first place determined by the choice of the stationary phase. In packed column GC this principle is well-known and the most suitable stationary phase is carefully selected. For this reason more than 300 stationary phases are available.

On the other hand, due to the much higher plate numbers of capillary columns, many separation problems encountered with packed column GC, can easily be solved on a capillary column coated with an apolar dimethylsilicone stationary phase.

In recent years, however, it was realized that selectivity is also very important in capillary gas chromatography. The recent developments in stationary phase synthesis and the re-investigation of selectivity tuning clearly illustrate the growing interest in selectivity optimization in capillary gas chromatography.

In this chapter the importance of selectivity to the overall resolution in CGC will be demonstrated. Five stationary phases are selected as basic phases. Methods are then described to tune the selectivity between the basic phases and thus to adapt the chromatographic system to a given problem. The different methods are compared and practical examples are given.

We have to emphasize that selectivity tuning is not the panacea of capillary GC. For complex, unknown mixtures, multidimensional chromatography is much more powerful. Selectivity tuning is however, also applicable in combination with multidimensional capillary gas chromatography.

II.   THEORETICAL CONSIDERATIONS ON THE RESOLUTION

The resolution for two compounds can be expressed as

$$R = \frac{\sqrt{n}}{4} \left(\frac{\alpha - 1}{\alpha} \right) \left(\frac{k_2}{k_2 + 1} \right)$$

where n equals the plate number of the column, which is dependent on the length of the column L and on the height equivalent to a theoretical plate h (HETP)

$$n = \frac{L}{h}$$

α equals the selectivity factor, which is given by the relative retention of the two compounds

$$\alpha = \frac{t'_{R_2}}{t'_{R_1}} = \frac{k_2}{k_1}$$

and k equals the capacity factor, which is dependent on the distribution coefficient K and on the phase ratio β (β = volume of mobile phase/volume of stationary phase)

$$k = \frac{K}{\beta}$$

The influence of the different parameters (plate number, selectivity and capacity factor) on the resolution is illustrated in the following example. Let us consider the separation of two compounds (k_1 = 4.76; k_2 = 5.00; α = 1.05) on a 10 m × 0.50 mm i.d. column with n = 20,000. Under these circumstances, the resolution of the two compounds is 1.40. This value is the basis of our comparison. By varying one of the three factors and keeping the other two constant, the resolution will change according to the curves in Figure 1. The following conclusions can be made:

1. The selectivity has the most important influence on the resolution. Increasing α from 1.05 to 1.10 results in an increase of R with a factor 2. A small decrease in α (α = 1.03) results in a drastic decrease of R (R = 0.86).

2. The capacity factor has a limited influence on R. Only for compounds with k<3, the increase of the capacity factor (for example by applying thicker film columns or working at lower temperatures) will result in a considerable increase of R.

3. The influence of the plate number is intermediate to the influences of the other factors. To obtain an increase of the resolution with a factor 2, the plate number must be 4 times larger—square root relationship.

High resolution capillary gas chromatography implies the optimization of the plate number, the selectivity and the capacity ratio. For many years, the resolving power of a capillary column has been related nearly exclusively to the plate number of the column, while adaption of the selectivity and of the capacity factor to the specific analytical need were hardly considered. If the separation of certain compounds of interest was insufficient, one often resorted to longer columns to obtain higher plate numbers. This results however in a proportional increase of the analysis time and a decreased sensitivity, while the resolution will hardly be affected.

Another possibility to increase the plate number of the column, is the reduction of the HETP value. The height equivalent to a theoretical plate is a complex function of the carrier gas velocity,

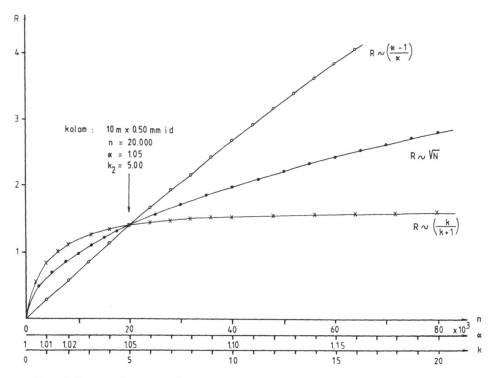

FIGURE 1   Influence of n, α and k on the resolution.

the column inner diameter, the film thickness, the diffusion coefficient
in mobile and stationary phase and the capacity factors (2,3).   In
practice, the column inner diameter is the most important parameter
influencing the efficiency.   By reducing the column inner diameter,
the minimum HETP-value is reduced proportionally.   In this way,
long narrow bore columns (i.d. < 100 μm) can offer extraordinary
plate numbers (over 1,000,000).   However, in comparison to a separation on a 25 m × 0.25 mm i.d. (standard) column, the influence
of the higher plate number on the resolution is small (5).   Moreover,
the applicability of long narrow bore columns is limited by the very
high inlet pressure required and by the increased analysis time.
Short narrow bore capillary columns, on the other hand, are important
for high-speed CGC, since these columns allow the reduction of the
analysis time without simultaneously impairing resolution (6).
    The optimization of the resolution through the capacity factor
has recently be outlined (7–12).   Compounds showing a low capacity
factor (k<2) require very high plate numbers to be separated.   For
the resolution of two compounds with k_1 = 0.49 and k_2 = 0.50 (α =

1.02), for example, 375,000 plates are required. By increasing the capacity factors by a factor of 5, the required plate number is reduced to 58,000. The most important parameter controlling the capacity factor is the stationary phase film thickness. In this respect, the introduction of thick film columns has extended the applicability of CGC to the analysis of very volatile compounds.

The most spectacular progress in high resolution CGC, however, can be expected by the optimization of the selectivity of the chromatographic system. The selectivity is a very complex parameter. The relative retention of the compounds is determined by all interactions between the stationary phase and the solutes. These interactions depend on the chemical structure of the stationary phase and of the solutes and on working temperature. The choice of the stationary phase is therefore the starting point in selectivity optimization.

III. STATIONARY PHASE SELECTION IN CGC

In packed column gas chromatography, a large variety of stationary phases has been used. Most of these stationary phases, however, cannot be coated efficiently on smooth fused-silica or glass capillary surfaces. The selection of the stationary phase for high resolution CGC should therefore primarily be based on other criteria than selectivity. The criteria to which CGC phases should correspond, can be summarized as follows:

1. The stationary phase must allow efficient coating of the smooth capillary surface, without applying roughening techniques.
2. The stationary phase must be of high purity to ensure complete inertness of the column and to prevent decomposition of the film, due to the catalytic activity of residual impurities.
3. A broad useful temperature range of the stationary phase is required and the film must remain stable at elevated temperatures.
4. Preferably, the stationary phase must allow immobilization without the need of the incorporation of groups facilitating immobilization.

In this respect, apolar dimethyl silicones are superior to any other stationary phase. The gum character of these silicones is responsible for a homogeneous covering of the capillary surface, resulting in highly efficient columns. Moreover, the viscosity of the phase is hardly influenced by temperature, guaranteeing high thermostability of the film.

Among the more polar phases, only high molecular weight polyethylene glycols could be used with relative success. The other stationary phase used in packed column GC, are liquids and cannot form stable homogeneous film on smooth surfaces. The great need

for a much wider choice of gum phases was therefore emphasized
(13).   Some low polar gum phases became commercially available
(PS-286, OV-1701, OV-225 vinyl) and different groups started to
synthesize polar silicone gum phases, covering a wide range of polarity and selectivity.

This however introduced another problem.   These stationary
phases all have a slightly different composition, resulting in different
chromatographic behavior.   This leads to nonuniform results in CGC
and interlaboratory exchange of retention data becomes very difficult.
For routine analysis it is therefore necessary to restrict the variety
of applied stationary phases to a few well-defined CGC-compatible
phases.

An important contribution in this direction was made by Stark
et al. (14).   Based on a study on the selectivity variance of packed
column GC phases, six high molecular weight cross-linkable gums,
covering a wide selectivity range were proposed as optimized phases
for CGC.

This philosophy was followed in our laboratory.   Figure 2 summarizes stationary phase selection and selectivity tuning.   In accordance with Stark et al. (14) and based on our own experience in
stationary phase technology, the basic phases for CGC are methyl
silicone, phenylmethyl silicone, trifluoropropyl silicone, cyanopropyl
silcone and HMW polyethylene glycol.

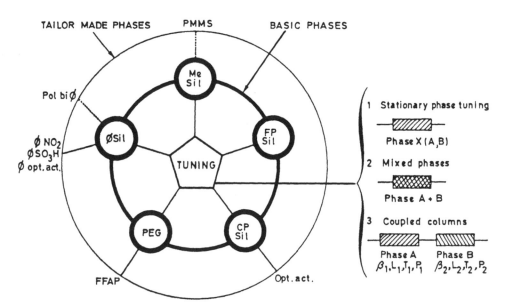

FIGURE 2   Stationary phase selection in CGC.

The apolar dimethyl silicones (possibly containing small percentages of phenyl- or vinyl-groups: OV-1, PS-255, SE-52, SE-54,...) can be coated on capillary columns with a film thickness varying from 0.05 to 10 μm. The efficiency, inertness and thermal stability of methyl silicone coated columns are excellent and immobilization can easily be performed without influencing, in a negative way, the chromatographic performance. Columns coated with apolar dimethyl gum phases are considered as the workhorse in CGC.

Trifluoropropyl silicone phases have a unique selectivity towards electron pairs on carbonyl- and nitro-groups. A phase containing 50% trifluoropropyl-groups (PS-286) can be coated efficiently and immobilized in capillary columns. A higher degree of trifluoropropyl-substitution leads to inferior chromatographic characteristics and therefore a 50% trifluoropropyl stationary phase is considered as a basic stationary phase.

Phenylmethyl silicones are interesting phases due to their selectivity and high thermal stability. The commercial silicones with high (> 25%) phenyl substitution however are liquids and cannot be used successfully. In the last years, much progress has been made in the development of CGC compatible phenylmethyl silicones (15–19). Highly viscous phenylmethyl silicones can be prepared by the hydrolysis of silane mixtures and by further polymerization of the low viscous prepolymers. The OH-terminated silicones are immobilized by heat curing (19). The phenyl-group can be incorporated from 1–100% in the silicone backbone. The phase with 50% phenyl and 50% methyl possesses, as far as efficiency, selectivity and useful temperature range are concerned, the highest chromatographic value. This stationary phase was applied successfully in the analysis of PAH (20), dibenzothiophenes (21), chlorinated phenols (20), steroids (22) and triglycerides (23).

Cyanopropyl silicones possess a permanent dipole and are highly selective towards double bonds. For example, in the analysis of fatty acid methyl esters, dioxins, vitamin E isomers, some PAH isomers, cyanopropyl silicones yield the highest resolution. Various silicones, containing from 25%–100% cyanopropyl, are commercially available. They are characterized by a low viscosity. Capillary columns can be prepared with these phases, but as the viscosity decreases with increasing temperature, droplet formation occurs above 100°C when coated on a smooth deactivated surface. Different groups therefore synthesized cyanopropyl silicones with gum properties (18,24–31). In this way, efficient and thermostable columns were prepared. A high degree of immobilization, however, could only be obtained by incorporation of vinyl-, tolyl- or ditolyl-groups in the silicone matrix. These structural modifications have a considerable effect on the selectivity and the polarity of the columns (31). In this case, a choice has to be made between maximal selectivity

and an immobilized phase. For selectivity optimization a 90% cyanopropyl phase is chosen as basic phase.

Among the nonsilicone phases, high molecular weight polyethylene
glycols are by far the best known. They are characterized by a
unique selectivity and polarity and are therefore also selected as
basic phases for CGC. We have tried several procedures for im-
mobilizing polyethylene glycol films (32,33). Immobilization degrees
of 100% are attainable but we have found all the procedures to give
an increased activity when compared to pure polyethylene glycol
coatings. The specific analytical need is crucial in deciding whether
an immobilized or nonimmobilized film has to be applied. For example,
the analysis of aqueous solutions requires an immobilized film. For
the analysis of essential oils and perfumes a normal PEG column is
preferred.

The five stationary phases cover a wide selectivity range. For
some special separation problems however, the selectivity variance
of the basic phases is insufficient. Therefore, some tailor-made
phases, such as liquid crystal phases or optical phases cannot be
excluded and these phases complete our set of CGC phases.

For many analytical problems, a selectivity intermediate to the
selectivities of the five basic phases can offer a better separation
in a shorter time. This is illustrated with the following example.
On stationary phase X, compounds A and B can be separated from C,
but they cannot be separated from each other. On another stationary
phase Y, A and B are separated, but one of them is now co-eluting
with C. By using a chromatographic system with a selectivity between the selectivities of X and Y, an optimum can be found to
separate the three compounds.

Intermediate selectivities and polarities can be obtained by selectivity tuning. Selectivity tuning means that the selectivity is adapted
to the analytical need by creating a selectivity between two (or more)
extreme (or considered as being extreme) selectivities (34). In
principle, this can be performed in three different ways: 1) by synthesizing a tuned phase with predetermined amounts of monomers
containing the required functional groups; 2) by mixing, in different
ratios, two or three phases in one column and 3) by coupling two
or more columns of extreme selectivities, variables being the β-values
(r,df), the column length, the column temperatures and/or the
pressures.

IV. METHODS OF SELECTIVITY TUNING

A. Synthesis of a Specific Stationary Phase

The present knowledge of silicon chemistry allows us to synthesize
a gum phase with a predetermined polarity and selectivity. For

example, all polarities and selectivities between dimethyl silicone and biscyanopropyl silicone can be obtained by the copolymerisation of dimethyl silane and biscyanopropyl silane. This approach is well-known in packed column GC, as is illustrated by the SILAR series (5CP, 7CP, 8CP, 9CP, 10C) and the methylphenyl series (OV-7, OV-11, OV-17, OV-22, OV-25).

An example of stationary phase tuning in CGC is the development of a substitute for the Ucon and Pluronic stationary phases. The polarity of these phases, and more especially of Ucon LB-550-X, are greatly appreciated in essential oil analysis. However, roughening techniques are required to coat these liquid polypropylene-ethylene glycol phases on capillary walls. Moreover, these industrial polymers suffer from a lack of batch-to-batch reproducibility and the maximal allowable operating temperature is limited to 220°C.

To approach the selectivity and polarity of the Ucons and Pluronics, a methylsilicone-ethylene glycol copolymer was synthesized (35). The stationary phase (RSL-310) can be immobilized and possesses good characteristics (Figure 3). The elution sequence of the compounds of the Grob polarity mixture indicates that the polarity and selectivity of RSL-310 is very similar to that of Ucon LB-550-X and of the Pluronics 61, 101, 121 (36). The thermal stability of RSL-310 reaches 260°C.

For daily practice, however, stationary phase tuning has only a limited value because each specific phase has to be synthesized. This is not practical for the chromatographer in the field. Moreover, the number of applied stationary phases is again increased.

B. Mixed Phase Columns

Selectivity tuning by mixing stationary phases has been investigated from the beginning of the sixties. Maier and Karpathy (37) and Hildebrand and Reilley (38) were the first to investigate mixed phases in packed column GC and to compare this with coupled columns of different length, each coated with pure stationary phase. The retention behavior on the mixed polarities was predicted from the retention on the pure stationary phases and the relative amounts of the phases in the mixutre. From their experiments they concluded that polarity tuning can be realized by:

1. packing a column with solid support coated with a mixture of two phases
2. packing a column with a mixture of two differently coated solid supports
3. serial coupling of predetermined lengths of columns, each packed with support coated with a pure stationary phase

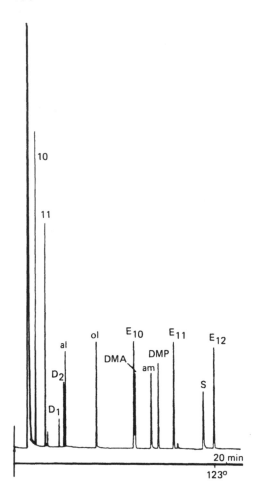

FIGURE 3 Analysis of the Grob polarity test mixture on a 25 m ×
0.32 mm i.d. FSOT column coated with an immobilized methyl silicone -
ethylene glycol copolymer (RSL-310). Temperature program: 65°C
to 140°C at 3°C/min. Compounds: 10: n-decane; 11: n-undecane;
D1: 2,3-butane diol (rac); D2: 2,3-butane diol (meso) al: n-nonanal;
ol: 1-octanol; DMA: 2,6-dimethyl aniline; DMP: 2,6-dimethyl phenol;
am: dicyclohexyl amine; S: 2-ethylhexanoic acid; E10: decanoic acid
methyl ester; E11: undecanoic acid methyl ester; E12: dodecanoic
acid methyl ester.

The technique of mixed phases was further investigated by Laub and Purnell (39–40), who introduced window diagrams to calculate the optimum mixing ratio for a certain problem.

Stationary phase tuning by mixing stationary phases has been claimed to be restricted because certain stationary phases are immiscible (41,42) or a nonideal solution behavior is claimed (41). Until now we have not observed difficulties in mixing different gum silicones or in mixing methyl silicones with high molecular weight ethylene glycols.

Coating capillary columns with a mixture of two stationary phases was described by Sandra and Van Roelenbosch (43). OV-1 and Superox 20 M were mixed in different amounts, approaching in this way the polarity of the Pluronics and the Ucons.

By mixing silicone stationary phases it is also possible to approach or even substitute a commercial available silicon stationary phase with an intermediate polarity and selectivity. However, to obtain the same selectivity of a certain stationary phase, it is not sufficient to have the same amount of functionality in the mixture. The same selectivity can only be reached if the same monomer units are used. This is illustrated with OV-1701, a copolymer consisting of 86.4% dimethyl siloxane units and 13.6% cyanopropylphenyl units. OV-1, a home-made 50% phenylmethyl silicone and a homemade 90% biscyanopropyl silicone were mixed to give 85.3% methyl substitution, 6.7% phenyl substitution, 6.7% cyanopropyl substitution and 1.3% vinyl substitution. The analysis of the Grob polarity test mixture shows that the ternary mixture gives very efficient and inert columns (Figure 4A). The temperature stability of the column is 270°C, corresponding to the thermal stability of the least stable polymer, in this case the biscyanopropyl silicone.

The polarity and selectivity of the mixed phase column can be compared with a OV-1701 column (Figure 4B). Due to the vinyl content some polarity shifts were expected, but not as drastically as illustrated in Figure 4. For example, octanol elutes before undecane on the mixed phase but elutes just before dodecane on OV-1701. The overal polarity thus is lower.

The same decrease in polarity was observed in substituting OV-225, which consists of 50% methylphenyl siloxane units and 50% methylcyanopropyl siloxane units. The mixed phase, consisting of 50% methyl-, 24.3% cyanopropyl-, 24.3% phenyl- and 1.4% vinyl-substitution was considerably less polar than OV-225. This decrease in polarity can only be explained by the difference between a mono- and a di-substituted silicon atom.

The investigation of mixed phases is also a clear illustration of the complexity of the selectivity in CGC and the sensitivity of solute—stationary phase interactions. In this respect, it is also clear that polyethylene glycol stationary phases cannot be substituted

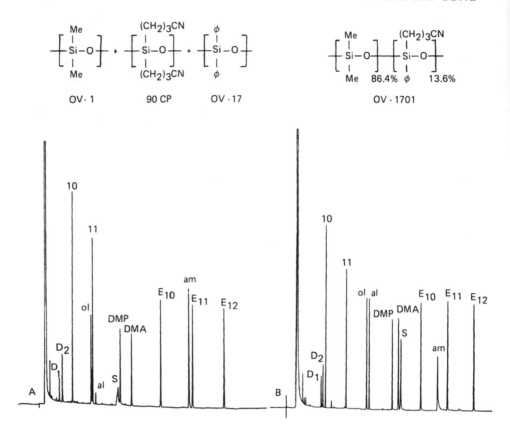

FIGURE 4 Analysis of the Grob polarity test mixture. (A) 25 m ×
0.32 mm i.d. FSOT column coated with a mixed phase consisting of
OV-1, methylphenyl silicone and biscyanopropyl silicone. (B) 20 m
× 0.32 mm i.d. FSOT column coated with OV-1701. Temperature
program: 60°C to 150°C at 3°C/min. Compounds: see Figure 3.

by a silicone stationary phase or a mixed phase (34). Based on
their theoretical study, Stark et al. (14) suggested that a silicone
containing 40–45% cyanopropyl groups and 20–25% phenyl groups
should have a selectivity similar to polyethylene glycol phases. This
was checked by comparing the selectivity of pure HMW-PEG with
the selectivity of different phenylmethyl-biscyanopropyl combinations. Figure 5 shows the diagrams of the relative retention (E10 =
1.00) of the compounds of a polarity test mixture on pure HMW-
PEG (Superox 20M), on a mixed phase, consisting of 45% cyanopropyl,
25% phenyl and 30% methyl substitution and on serial coupled columns,
consisting of different lengths of columns coated with pure phenylmethyl silicone and pure 90% biscyanopropyl silicone. The polarities
of the mixed phase and of the coupled columns approach the polarity
of polyethylene glycol, but the selectivities are different. The combination best correlating with Superox 20M is the coupled column
15 m 90 CP / 2.5 m OV-17 (Figure 6). However, it is clear that
important differences are noted. For example, it is impossible to
elute the octanol peak between the two diol isomers as on Superox
20M. The elution sequence of 2,6 dimethyl aniline (DMA) and 2,6

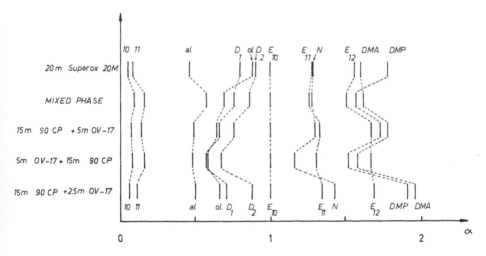

FIGURE 5   Relative retention of the compounds of a polarity test
mixture (E10 = 1.00) on phenylmethyl (OV-17)- biscyanopropyl
(90 CP) combinations in comparison with Superox 20M. Columns:
(A) 20 m × 0.25 mm i.d. Superox 20M. (B) 20 m × 0.25 mm i.d.
mixed phase (45% cyanopropyl: 25% phenyl: 30% methyl). (C) directly
coupled columns coated with OV-17 and 90 CP. (length fractions:
as indicated). Column temperature: 60°C to 150°C at 3°C/min.
Compounds: polarity test mixture similar to Grob test mixture
(Figure 3); N: naphthalene.

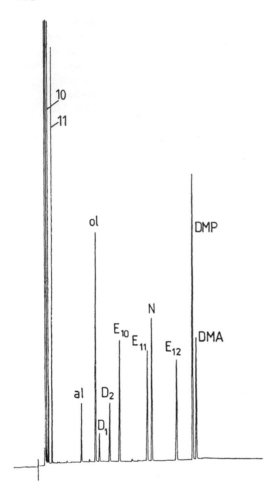

FIGURE 6   Analysis of polarity test mixture on a 15 m × 0.25 mm
i.d. OV-17 + 2.5 m × 0.25 mm i.d. 90 CP serial coupled column
combination.   Conditions: see Figure 5.

dimethyl phenol (DMP) is also an excellent marker for the selectivity. The elution of DMA before DMP is very characteristic for polyethylene glycol phases and cannot be obtained by selectivity tuning. In conclusion, by combining silicone phases, the polarity of polyethylene glycol phases can be approached but the selectivity will never be the same.

C. Serial Coupled Columns With Variation of the Column Characteristics (L,r,df)

In this case, two or more columns are coupled directly and installed in one GC oven. The temperature is the same for each column and only the inlet pressure of the first column can be controlled. The inlet pressure of the following columns is equal to the outlet pressure of the preceding column.

In the direct coupling of capillary columns coated with pure stationary phases, the relative amount of stationary phases determines the retention of the solutes. The relative amount of stationary phases can be adjusted by using different lengths of columns or by combining columns with different β -values, variables being the film thickness and the column radius. For practical reasons, only the variation of the column lengths have been used.

In 1981, Purnell et al. (44) described the coupling of two capillary columns and selectivity tuning by varying the lengths of the columns. The method was applied by several groups (41,42,45-47). The mathematical methods developed for mixed phases in packed column GC were used to predict the column lengths necessary to achieve a desired separation in CGC. These methods did not take gas compressibility into account and empirical corrections were required (41,42,47). In recent contributions the theory of coupled columns was further developed and corrected for gas compressibility effects (48-53).

D. Serial Coupled Columns With Independent Control of Column Temperatures and Flowrates

The techniques of mixed phases and direct serial coupling of columns have serious limitations. The preparation of a column with a mixture of stationary phases or cutting expensive capillary columns into pieces surely are not practical ways to adjust the selectivity for a specific application. Other methods have therefore been developed, allowing continuous adjustment of the selectivity.

In 1973, Deans and Scott (54) demonstrated that the separation characteristics of a multicolumn system can be adjusted by changing the ratio of the gas velocities in the serially coupled columns. Kaiser et al. (55-56) and Toth et al. (57) noticed dramatic selectivity changes in a capillary tandem when the carrier gas flows through each column were changed.

As an alternative, similar effects can be noticed if both columns are operated at different temperatures. Pretorius et al. (58) and Kaider and Rieder (59) described the technique for packed and micropacked columns. The SECAT-mode for capillary columns was introduced by Kaiser and Rieder (60). The technique was successfully applied in the analysis of pesticides (61–62).

Theoretical studies were made to predict the retention behavior in serial coupled columns (63–68). An overview of the most important relationships, controlling the retention in serial coupled columns was given by Hinshaw and Ettre (69–70).

In contrast to the first three methods, the serial coupled columns with independent control of the column temperatures and/or the carrier gas flows allows a fast and reproducible selectivity tuning and the same setup can be applied to various problems. An ideal setup for selectivity tuning is shown in Figure 7.

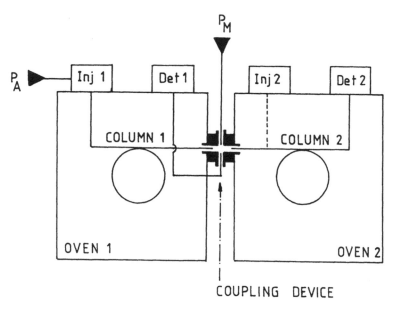

FIGURE 7 Diagram of setup for selectivity tuning using serial coupled capillary columns with independent control of column temperatures and flowrates. P_A = inlet pressure on first column. P_M = inlet pressure on second column = outlet pressure of first column. This system allows monitoring of the effluent from the first column (FID 1) and can contain a second injection port.

V. EXAMPLES OF SELECTIVITY TUNING IN CGC

A. Selectivity Tuning Between Methyl Silicone and High Molecular Weight Polyethylene Glycol

Selectivity tuning between methyl silicone (OV-1) and HMW-PEG (Superox 20M) was applied to approach the selectivity of the Ucon and the Pluronic stationary phases (43). Excellent results were obtained by coating capillary columns with a mixture of OV-1 and Superox 20M (Figure 8). The gum character of OV-1/Superox 20M mixtures is responsible for the high efficiency (UTE > 95%). Moreover, the columns have a high inertness and a temperature stability up to 250°C. Recently, we were able to immobilize the mixed stationary phase film without influencing the characteristics (71).

The polarity and selectivity of the mixed phase columns can be characterized based on the relative retention of the compounds of the Grob test mixture (36). In Figure 9 the retention of the compounds, relative to methyldecanoate (E10), is given as a function of the mixed phase composition. The temperature was programmed from 65° to 150°C at 3°C/min. As the amount of Superox 20M increases, the polar compounds (ol, DMP, DMA) shift to a higher relative retention, while the relative retention of the hydrocarbons (10,11) decreases. It is also clear that a low percentage of Superox 20M is sufficient to reverse the elution sequence of 2,6-dimethyl phenol (DMP) and 2,6-dimethyl aniline (DMA).

The comparison of the selectivities and the elution sequences with the data given by Grob (36), indicates that the binary mixture 50% OV-1/50% Superox 20M corresponds to Ucon 50 HB-5100 and to Pluronic L-64, and that 25% OV-1/75% Superox 20M corresponds to Pluronic F-68. However, this does not imply that the selectivity of the mixed phases toward individual compounds will be exactly the same as on the corresponding Ucon or Pluronic phase. In general, the mixed phases are less polar than the Ucons and Pluronics. For example, on the 50% OV-1/50% Superox 20M phases nonanal elutes between undecane and dodecane, while on Ucon 50 HB-5100 nonanal elutes between tridecane and tetradecane.

OV-1/Superox 20M mixed phases have been successfully applied in essential oil analysis (72). This is illustrated in the analysis of hop oil, spiked with some aroma compounds, on three different mixed phases (Figure 10). The retention times of the terpenes myrcene (peak 1), caryophyllene (peak 5) and humulene (peak 7) are only slightly affected by the Superox 20M content. The polar solutes cis-3-hexenol (peak 2), linalool (peak 3), and carvone (peak 6), on the other hand, shift drastically as a function of the percentage of Superox 20M. The polar interactions are furthermore evident by the increase of the retention times of farnesol (peak 9) and benzyl benzoate (peak 10).

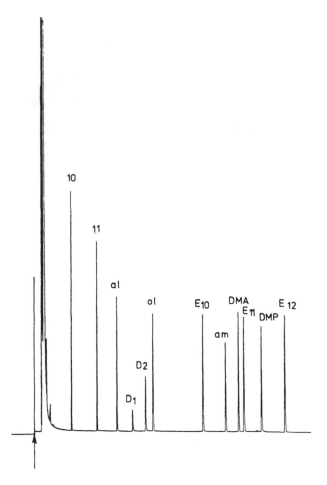

FIGURE 8 Analysis of the Grob polarity test mixture on a 18 m ×
0.32 mm i.d. FSOT column coated with a 50% OV-1: 50% Superox 20M
mixed phase.   Temperature program: 65°C to 150°C at 3°C/min.
Compounds: see Figure 3.

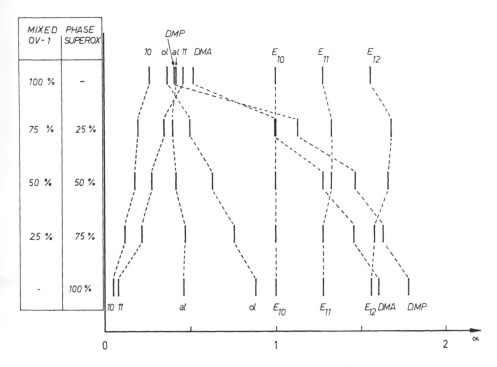

FIGURE 9   Relative retention of the compounds of a polarity test
mixture (E10 = 1.00) in function of the composition of the mixed
phase.   Compounds: see Figure 3.

For practical work, a 50% OV-1/50% Superox 20M mixed phase
offers the most interesting selectivity since monoterpenes, monoter-
penoids, sesquiterpenes and sesquiterpenoids show a minimal overlap.   In the chromatogram of a complex perfume oil, these four groups
can easily be distinguished (Figure 11).

In conclusion, by using the binary mixtures OV-1/HMW-PEG,
the selectivity and the polarity of the Ucons and the Pluronics can
be approached.   Because of the different chemical structure, the
mixed phases cannot exactly duplicate the corresponding Ucons and
Pluronics.   For practical work, however, the real selectivity of the
phase is not always predominant and due to a much higher efficiency
than the polypropylene-ethylene glycol phases, these mixed phase
columns can be of great help in practice.

Another way to obtain an intermediate selectivity between a
methyl silicone and a HMW-polyethylene glycol, is the direct coupling
of two columns, each coated with the pure stationary phase.   Two
20 m × 0.32 mm id columns coated with respectively 0.25 μm OV-1

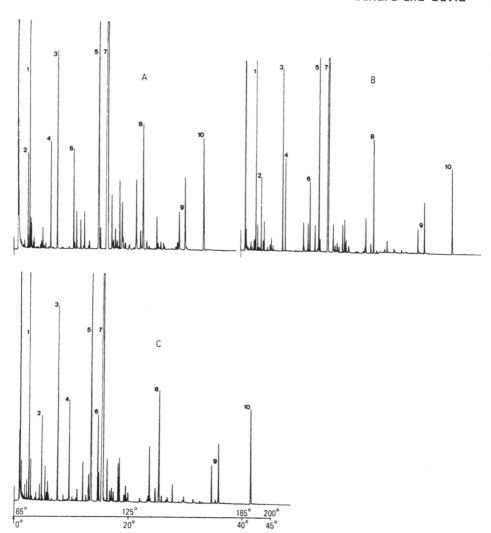

FIGURE 10 Analysis of hop oil spiked with cis-3-hexenol (peak 2), ethyl octanoate (peak 3), linalool (peak 4), carvone (peak 6), farnesol (peak 9) and benzyl benzoate (peak 10). (A) 13 m × 0.22 mm FSOT column coated with a 75% OV-1: 25% Superox 20M mixed phase. (B) 18 m × 0.22 mm FSOT column coated with a 50% OV-1: 50% Superox 20M mixed phase. (C) 14 m × 0.22 mm FSOT column coated with a 25% OV-1: 75% Superox 20M mixed phase. Compounds: 1: myrcene; 5: caryophyllene; 7: humulene; 8: humuladienone. Temperature: 65°C to 200°C at 3°C/min. Flow adjusted so that the retention time of ethyl octanoate remains constant.

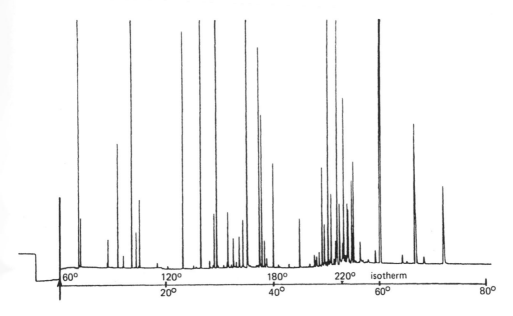

FIGURE 11   Analysis of a perfume oil.   Column: 25 m × 0.25 mm i.d.
FSOT column coated with a 50% OV-1: 50% Superox 20M mixed phase.
Temperature: 50°C to 220°C at 3°C/min.

and 0.25 μm Superox 20M were coupled in a polyimide seal (73).   A
1/32 in. Valco coupling can also be used for this purpose.   The
capillary tandem was evaluated in both directions and the temperature
was programmed from 60°C to 150°C at 3°C/min.   The analysis of a
polarity test mixture demonstrates the high efficiency of the capillary
tandem (Figure 12).   From the elution sequence of the compounds, it
is clear that the selectivity of the tandem is dependent on the column
sequence.   On the OV-1/Superox 20M combination, N elutes before
E10 and DMA elutes before E11.   If the Superox 20M column is
mounted in front, these elution sequences are reversed.   In Figure
13 the relative retention (E10 = 1.00) of the test compounds on the
capillary tandem, either with the OV-1 or with the Superox 20M
first, is compared to the retention on the columns as such.   The
highest polarity for the capillary tandem is observed with the Superox
20M column first.   The dependence of the overall selectivity on the
overall selectivity on the column sequence is due to the effect of gas
compressibility in a capillary column.   Along the column, a carrier
gas velocity gradient is created and the residence time in the first
half of the column (in this case the first column of the tandem) is
longer than in the second half.   Therefore, the polarity and the
selectivity of the first column will have a more important contribution

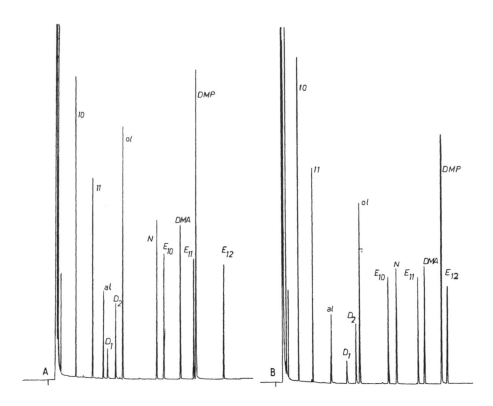

FIGURE 12   Analysis of a polarity test mixture.   (A) 20 m × 0.25 mm i.d. OV-1 (0.2 µm) FSOT column + 20 m × 0.25 mm i.d. Superox 20M (0.2 µm) FSOT column (directly coupled).   (B) The same capillary tandem installed with the Superox 20M column first.   Temperature: 60°C to 150°C at 3°C/min.   Compounds: see Figure 3, N: naphthalene.

to the overall selectivity.   The OV-1/Superox 20M tandem is thus less polar than the Superox 20M/OV-1 tandem.

The overall polarity and selectivity of a 50% OV-1/50% Superox 20M mixed phase column is intermediate to the selectivities of the capillary tandem, evaluated in both directions.   This is demonstrated by the elution of DMP between E11 and E12.   On the mixed phase column, DMP elutes halfway both esters (Figures 8 and 9).   On the OV-1/Superox 20M combination, DMP elutes just after E11 and on the Superox 20M/OV-1 combination DMP elutes just before E12 (Figures 12 and 13).

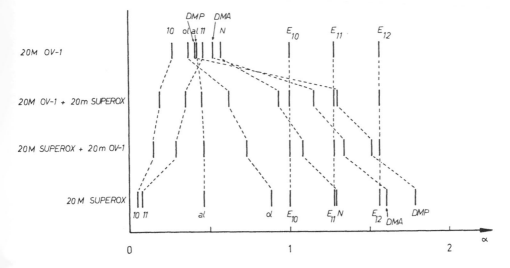

FIGURE 13 Relative retention of the test compounds (E10 = 1.00) on OV-1, Superox 20M and on the OV-1/Superox 20M tandem, operated in both directions. Compounds and temperature program: see Figure 12.

Selectivity tuning by direct coupling of capillary columns has a limited practical value. Because of the gas compressibility effects, the respective column lengths needed to obtain a target selectivity, is more difficult to calculate than the mixing ratio for mixed phase columns. Both mixed phase columns and direct coupled columns also suffer from a lack of flexibility. Once the selectivity of the chromatographic system has been optimized for a given application, it is not possible to change it in a fast and practical way for another application.

The continuous tuning of the selectivity is possible if the carrier gas flow and/or the temperature of each serial coupled column can be selected independently. For the experimental setup, we used a double-oven Siemens Sichromat 2 gas chromatograph. Two capillary columns were coupled via the live-switching device (74). In this way, the pressure on both columns and the column temperatures can be changed.

A 20 m × 0.32 mm i.d. OV-1 column was installed in the first oven and a 20 m × 0.32 mm i.d. Superox 20M column was installed in the second oven. In a first series of experiments, the carrier gas flows through both columns were kept constant and nearly equal (0.4 bar on each column). By changing the temperatures in both ovens, the overall selectivity of the system can be tuned (Figure 14). With

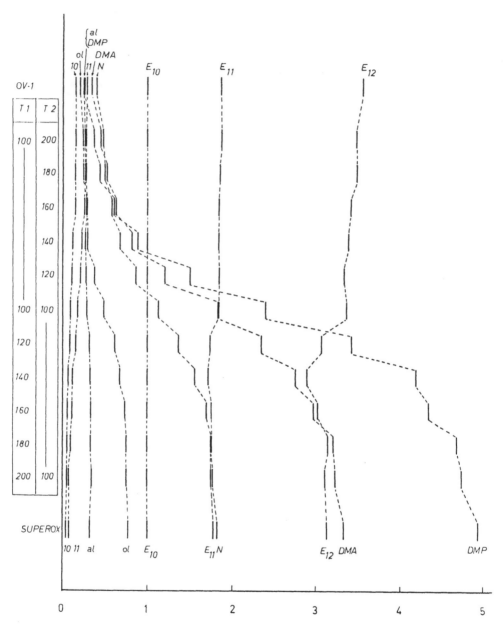

FIGURE 14 Relative retention of the test compounds in function of
the individual column temperatures of an OV-1/Superox 20M combination. Instrument: Siemens Sichromat 2. Column A: 20 m × 0.25 mm
i.d. FSOT coated with 0.2 μm OV-1. Column B: 20 m × 0.25 mm i.d.
FSOT coated with 0.2 μm Superox 20M. Pressure: P_A = 0.80 bar
(hydrogen) P_M(coupling piece) = 0.38 bar. Column temperatures:
isothermal as indicated.

the first column at 100°C and the second column at 200°C, the tandem behaves in an apolar manner. However, the selectivity factors are not the same as for pure OV-1. DMA, having a very high retention on Superox 20M, elutes after undecane. The polarity of Superox 20M is thus still felt at 200°C. Both temperature and pressure adjustment are necessary to obtain the same selectivity as on OV-1.

By heating up the first column and/or cooling down the second column, the polarity and the selectivity shift to the Superox 20M side. At 200°C for the first column, the selectivity of pure Superox 20M is approached. Similar shifts can be noticed when temperature programming is applied in one or in both ovens (Figure 15).

Another possibility to obtain intermediate polarities and selectivities between a methyl silicone and a HMW-polyethylene glycol is the variance of the flow rates. Both columns were kept at 100°C. The inlet pressure and the middle pressure were changed, according to the pressures mentioned in Figure 16. Here, we have to notice that the netto-pressure on the first column equals $P_A - P_M$, because the outlet pressure of the first column is not atmospheric pressure but the inlet pressure of the second column.

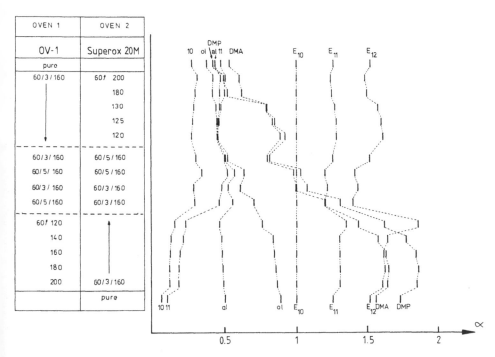

FIGURE 15 Relative retention of the test compounds in function of the column pressures of an OV-1/Superox 20M combination.

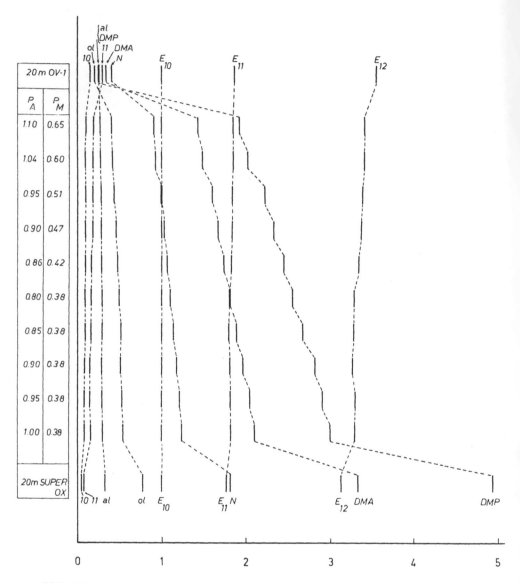

FIGURE 16 Relative retention of the test compounds in function of the column pressures of an OV-1/Superox 20M combination. Columns: see Figure 14. Column temperature: $T_1 = T_2 = 100°C$. Pressure: as indicated. P_A = inlet pressure first column, P_M = inlet pressure second column = outlet pressure first column.

In general, the same shifts can be observed as with temperature variation. The lower the pressure in the second column and the higher the pressure in the first column, the more polar the tandem acts. If the flowrate of each column can be adjusted independently, gas compressibility effects are avoided. This was illustrated by reversing the column sequence. With the Superox 20M column installed in the first oven and the OV-1 column in the second oven, the same elution sequence is observed as with the OV-1 column first. This observation is not only valid for isothermal operation. If temperature programming is applied, the selectivity is still independent of the column sequence. Figure 17 shows the analysis of the polarity test mixture with both setups. It is clear that the same polarity and selectivity is obtained. This selectivity is equal to a 50/50 mixed phase (Figure 8) and is intermediate to the direct coupled column evaluated in both directions (Figure 12). For this comparison, we take the elution of DMP and DMA relative to the esters as a marker.

Further evidence for the fact that the selectivity is independent on the column sequence when separate pressure regulation is used, is given in Table 1, listing the elution temperatures for both setups. For these experiments, the effluent from the first column is split and one part is directed to a FID, while the other part is directed to the second column. In this way it is possible to measure the elution temperatures for the first column and for the total system simultaneously. The total elution temperatures are equal for both setups. For example, DMA elutes at 133°C. If the OV-1 column is installed first, the elution temperature on this column is 84°C and the compound stays in the second column up to 133°C. If the Superox 20M column is installed first, DMA elutes at 127°C from this column, but the compound is hardly retained on the apolar column and the total elution temperature is again 133°C. This behavior is

TABLE 1 Elution temperatures on a "live"-coupled OV-1/Superox 20M combination

Combination	10	11	al	ol	DMP	DMA	N	E_{10}	E_{11}	E_{12}
Column 1: 20 m OV-1										
Column 2: 20 m Superox 20M										
Elution temp. OV-1	73	81	80	77	79	84	86	106	118	131
Elution temp. Superox 20M (Total elution temp.)	77	87	95	106	140	133	122	121	135	148
Column 1: 20 m Superox 20M										
Column 2: 20 m OV-1										
Elution temp. Superox 20M	64	65	81	99	134	127	113	101	113	125
Elution temp. OV-1 (Total elution temp.)	78	87	94	106	140	133	122	121	134	147

Analysis: 60°C to 150°C at 3°/min.

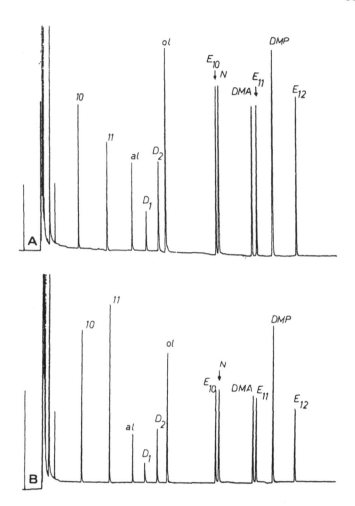

FIGURE 17   Comparison of the elution profiles on the "live" - coupled
OV-1/Superox 20M combination.   (A) OV-1 column in first oven.
(B) Superox 20M column in first oven.   Columns: see Figure 14.
Temperature: both ovens simultaneously programmed 60°C to 150°C
at 3°C/min.   Pressure: P_A = 0.44 bar; P_M = 0.22 bar.

remarkable since, depending on the column sequence, the compounds
are chromatographed at different temperatures. This experiment
suggests that the interactions between a methyl silicone or a polyethylene glycol and the solutes have similar temperature dependence.
For other stationary phases, such as phenylmethyl silicones or cyanopropyl silicones, this is no longer valid since the selectivity of
these phases towards unsaturated compounds increases with increased
temperature (31,34).

   In general, the combination of methyl silicone and HMW-polyethylene glycol offers excellent possibilities for selectivity tuning.
A theoretical and experimental study of this column combination was
also presented by Hinshaw and Ettre (69,70). Mathematical methods
are described to predict the retention behavior on a capillary tandem
and to achieve the best possible resolution of multicomponent mixtures.

B.   Selectivity Tuning Between Methyl Silicone and Biscyanopropyl Silicone

   Methyl silicones and biscyanopropyl silicones are the two extremes
on the polarity scale of the silicone phases (14). In our experiments,
we have combined OV-1 with a homemade cyanopropyl silicone, containing 90% cyanopropyl groups (90 CP). This gumlike stationary
phase shows excellent chromatographic characteristics (32). By
selectivity tuning between these stationary phases a large selectivity
and polarity range is covered. The same methods, described for
the combination of OV-1 and Superox 20M, can hereby be applied.

   The coating of capillary columns with mixed phases can easily
be performed and both phases seem to be perfectly miscible. This
is illustrated by the analysis of a test mixture on a 20 m × 0.25 mm
i.d. column coated with a 50% OV-1/50% 90 CP mixed phase. From
Figure 18, the high efficiency and good inertness of the column is
evident. The mixed phase column can be used up to 250°C, without
any polarity modification.

   Selectivity tuning was also performed by the direct coupling of
different lengths of columns coated with respectively OV-1 and 90
CP. Each combination was evaluated in both directions. The relative
retention (E10 = 1.00) of the compounds of the test mixture on the
different combinations in a temperature programmed run are presented in Figure 19 and compared to the 50/50 mixed phase. The
hydrocarbons are shifted to lower retention and the polar compounds
to higher retention as the length fraction of the polar column increases. The most pronounced shifts are observed for naphthalene,
DMA, and DMP. This is due to the strong formation of π-complexes.
Gas compressibility effects are also noticed here. The chromatograms
on the 10 m OV-1/10 m 90 CP combination are shown in Figure 20.
By comparison with Figure 18, it can be seen that the mixed phase
column takes an intermediate position between the runs on the tandem.

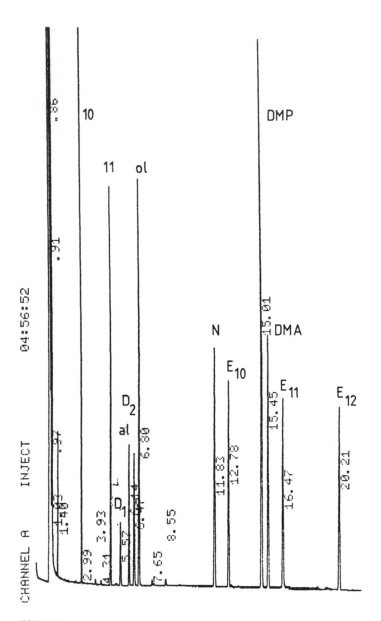

FIGURE 18 Analysis of polarity test mixture on a 20 m × 0.25 mm
i.d. FSOT column coated with 0.2 μm 50% OV-1: 50% 90 CP mixed
phase. Column temperature: 60°C to 150°C at 3°C/min.

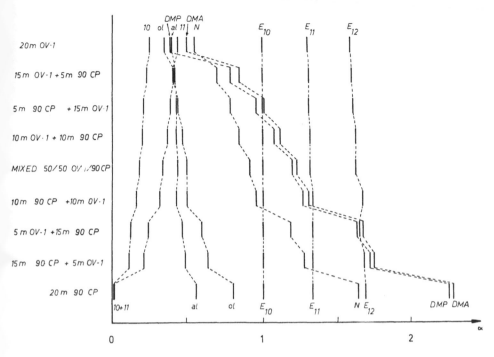

FIGURE 19 Relative retention of the test compounds in function of the length of the two capillary columns of a capillary tandem. Columns: 0.25 mm i.d. FSOT OV-1 (0.2 μm) + 0.25 mm i.d. FSOT 90 CP (0.2 μm). Directly coupled and evaluated in both directions. (Individual column lengths as indicated) Column temperature: 60°C to 150°C at 3°C/min.

The polar compounds (N, DMP, DMA) shift to higher retentions if the polar column is installed in front. The relative elution of the diol isomers (D1, D2) also is a good indication for the overall polarity and confirms the intermediate polarity of the mixed phase column.

 Gas compressibility effects are more pronounced if longer columns are coupled, because the pressure drop over the tandem is higher. To illustrate this, two 25 m × 0.25 mm id columns were coupled and operated in both directions. From Figure 21 we can deduce that the elution sequence again is strongly dependent on the column sequence. When the OV-1 column is installed first, octanol elutes before nonanal and DMP/DMA elute before methyl decanoate. Reversing the tandem results in a reversed elution order of the alcohol and the aldehyde, while DMP/DMA now elute between E10 and E11. The differences are much larger than the differences observed with the 10 m OV-1/ 10 m 90 CP combination (Figure 20). After these analyses, half of

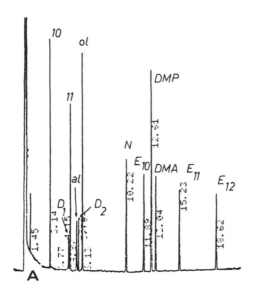

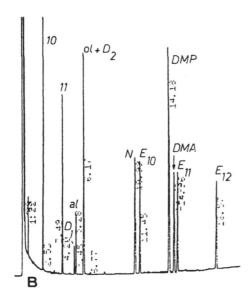

FIGURE 20 Analysis of polarity test mixture on (A) 10 m × 0.25 mm i.d. FSOT OV-1 + 10 m × 0.25 mm i.d. FSOT 90 CP (directly coupled). (B) same capillary tandem operated in reversed direction (90 CP column first). Column temperature: 60°C to 150°C at 3°C/min.

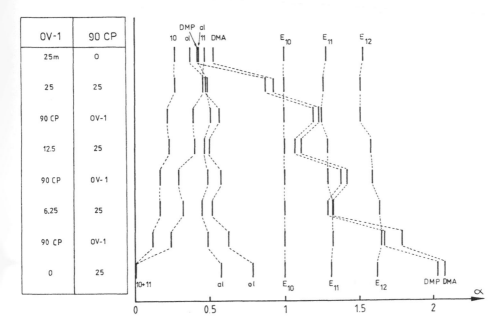

FIGURE 21 Relative retention of the test compounds on directly coupled OV-1/90 CP combinations. Columns: 0.25 mm i.d. FSOT OV-1 (0.2 μm) + 0.25 mm i.d. FSOT 90 CP (0.2 μm). Directly coupled (individual column lengths as indicated). Each capillary tandem operated in both directions. Column temperature: 60°C to 150°C at 3°C/min.

the OV-1 column was removed. The results are remarkable. The 12.5 m OV-1/25 m 90 CP combination with the OV-1 column first, seems to be less polar than the 25 m 90 CP/25 m OV-1 combination with the 90 CP column first. The compounds are shifted to a position between the two analyses on the 25 m combinations. These effects were confirmed by again removing half of the OV-1 column. This experiment emphasizes the importance of pressure drop and compressibility effects in direct coupled systems, whereby the carrier gas flow through each column cannot be adjusted independently. If no gas compressibility effects are considered, a 66% 90 CP combination is found to be less polar than a 50% 90 CP combination. From this point of view, it is easy to understand why trial and error corrections had to be used to reach a desired phase ratio.

   In the same way as described for the OV-1/Superox 20M combination, all intermediate polarities and selectivities between methyl silicone and biscyanopropyl silicone can of course also be obtained by operating both columns at different temperatures or by applying

different carrier gas flowrates in each column. An example is given
to illustrate this. A 10 m × 0.32 mm i.d. OV-1 column and a 25 m ×
0.32 mm i.d. 90 CP column are combined in the Siemens Sichromat 2
gas chromatograph and by using different temperatures or different
temperature programs, the test compounds can be shifted between
the two extreme selectivities (Figure 22).

The combination of OV-1 and 90 CP can be applied to the analysis
of fatty acid methyl esters. On OV-1, FAMEs are separated according to their vapor pressure. On cyanopropyl stationary phases,
separation is achieved according to the number of unsaturation. A
complete separation of methylstearate (C 18:0), methyloleate (C 18:1),
methyllinoleate (C 18:2) and methyllinolenate (C 18:3) is obtained
on a 90 CP column (31), but methylarachidate (C 20:0) elutes between C 18:2 and C 18:3. If a separation according to unsaturation
together with a carbon number separation is desired, selectivity
tuning can provide the answer. Hereby the elution order must be:
C 18:0, C 18:1, C 18:2, C 18:3, C 20:0. This sequence can be obtained by synthesizing a stationary phase with a lower cyanopropyl
content (31). Another possibility is the operation of two serial
coupled columns at different temperatures. A FAME test mixture
was analyzed on the 10 m OV-1/25 m 90 CP combination installed in
the Siemens instrument. The relative retentions (C 18:0 = 1.00)
of the compounds are presented in Figure 23. The best resolution
is achieved with the 90 CP column at 180°C and the OV-1 at 240°C.
If the OV-1 column is at 200°C, C 18:0 coelutes with C 18:1, while
at 280°C, C 18:3 coelutes with C 20:0.

VI. SELECTIVITY TUNING IN MULTIDIMENSIONAL CGC

For highly complex natural samples, the separation power of a single
capillary gas chromatographic system is insufficient to achieve complete resolution for the compounds of interest. Even by optimization of the selectivity, important compounds will still coelute, since
the better separation of one pair of compounds is likely to be counteracted by the overlapping of another pair of compounds present in
the sample. If the number of compounds to be separated is larger
than the peak capacity of the optimized chromatographic system,
only the combination of more than one system can provide a solution.
In multidimensional CGC, a group of compounds, not separated on
a first column, is transferred (heartcut) to a second column, where
complete resolution is achieved.

When using multidimensional CGC, selectivity tuning can esily
be applied, adding an extra dimension to the system. This is illustrated by some examples of essential oil analysis (75). For these
experiments, a Siemens Sichromat 2 gas chromatograph was used
and a 20 m × 0.32 mm i.d. OV-1 column and a 20 m × 0.32 mm i.d.

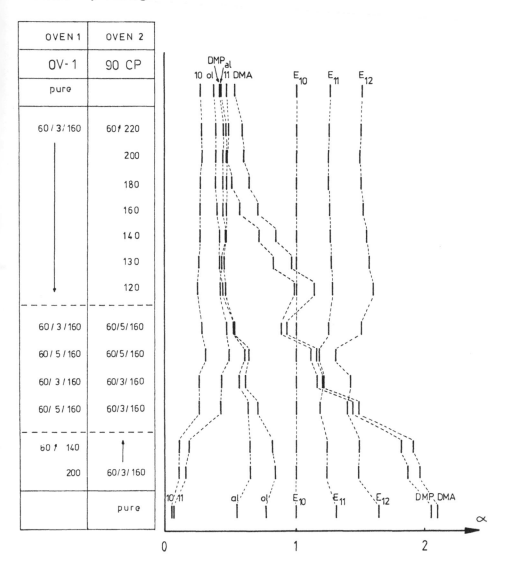

FIGURE 22   Relative retention of the test compounds in function of the individual column temperatures.   Instrument: Siemens Sichromat 2.   Column A: 10 m × 0.32 mm i.d. OV-1 (0.24 μm df).   Column B: 25 m × 0.32 mm i.d. 90 CP (0.5 μm dr).   Pressure: P_A = 1.0 bar (hydrogen); P_M = 0.8 bar.   Column temperatures: as indicated.

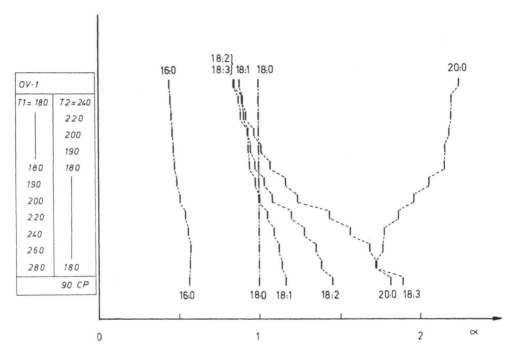

FIGURE 23 Relative retention of fatty acid methyl ester (FAME) test mixture in function of the individual column temperatures of a OV-1/ 90 CP combination. Instrument: Siemens Sichromat 2. Column A: 10 m × 0.25 mm i.d. OV-1 (0.2 μm df). Column B: 10 m × 0.25 mm i.d. 90 CP (0.2 μm df). Pressure: P_A = 0.27 bar; P_M = 0.13 bar. Column temperatures: as indicated.

Superox 20M column were applied. By operating the live-switching system, fast and reproducible heartcuts can be executed (74,75).

Two operation modes can be distinguished. If all transferred compounds are trapped before further analysis on the second column (intermediate trapping), the selectivities of both columns are "decoupled." The elution pattern of the transferred compounds, observed after analysis on the second column, will then be independent on the selectivity of the first column and will be exactly the same as if the pure sample compounds were analyzed on the second column only. Of course, interference with other compounds is now avoided.

Alternatively, when no intermediate trapping is applied, the elution pattern of the transferred compounds, observed after analysis on the second column, will be the same as the elution pattern for an analysis on a mixed phase column or a coupled column system with an intermediate selectivity between the selectivities of both columns.

The elution pattern is determined by the residence times in each column.

Both principles are illustrated in Figure 24. A terpene standard mixture was analyzed on a OV-1 column (Figure 24A), a Superox 20M column (Figure 24E) and OV-1/Superox 20M combination (= 50/50 mixed phase) (Figure 24B). This last analysis was performed by directing the total sample through both columns in the Siemens setup. It is clear that co-elution occurs in each case.

By heartcutting fractions X and Y from the OV-1 separation and analyzing these fractions on the Superox 20M column, complete resolution is obtained. If no intermediate trapping is applied, the total retention times of the individual compounds in fraction X (Figure 24C) and fraction Y (Figure 24D) is exactly the same as their retention times on the OV-1/Superox 20M combination (Figure 24B). If intermediate trapping is applied, the total retention times of the compounds in fraction X (Figure 24F) and fraction Y (Figure 24G) correspond to the analysis of the complete sample on the Superox 20M column alone (Figure 24E).

The practical value of selectivity tuning in MD-CGC is illustrated in the analysis of peppermint oil. On a OV-1 column the separation of the menthone-menthol fraction is incomplete (Figure 25A). This fraction was heartcut (Figure 25B) and further analyzed on a Superox 20M column. With intermediate trapping, compound 3 elutes very close to menthone (peak 1) (Figure 25C). A better separation is obtained without intermediate trapping (Figure 25D).

Another example is the analysis of the essential oil from the magnolia plant. The chromatogram obtained on an apolar column shows a very complex sesquiterpene region (Figure 26). Fraction A was heartcut and, without intermediate trapping, further analyzed on a Superox 20M column. One of the transferred peaks is now separated into two compounds (Figure 26C - peaks 2 and 3). By applying another temperature program for the polar column, the elution sequence is altered (Figure 26D). Because no intermediate trapping was applied, the elution pattern is determined by the selectivity of each column and by the residence time of the compounds in each column. In the second analysis, the temperature of the second column was higher and its contribution to the overall selectivity was smaller. In this way, selectivity tuning can lead to an optimal separation. The relative shifts of the peaks by changing the overall selectivity also gives information about the identity of the compounds. As the retention of peak 2 decreases with decreasing polarity of the system, the compound must be polar.

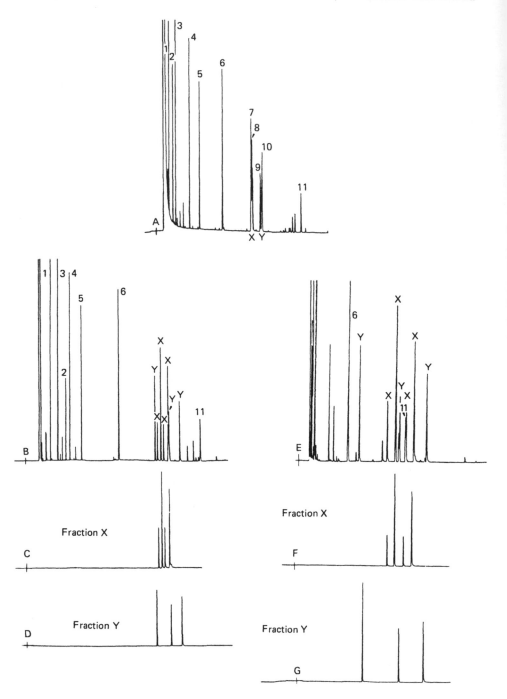

FIGURE 24

VII. CONCLUSION

Since selectivity is the most important factor influencing the resolution in gas chromatography, selectivity optimization is a main road to high resolution CGC. By selectivity tuning, the selectivity of the chromatographic system is adapted to the specific analytical need. Different techniques can be used. For simple routine applications mixed phase columns can offer excellent results. If a continuous adaption of the selectivity is needed, serial coupled columns with independent temperature and pressure control offers the highest versatility and flexibility.

Selectivity tuning also adds an extra dimension to multidimensional CGC. If no intermediate trapping is applied, the polarity and selectivity of both columns are coupled and the selectivity can be tuned to an optimum.

To fully exploit selectivity tuning, further research is needed to understand the solute-stationary phase interactions better and especially their temperature dependence. Artificial intelligence can then be developed to predict the polarity and selectivity; and selectivity tuning can become an important tool in chromatographic analysis.

FIGURE 24 Analysis of terpene standard mixture. Instrument: Siemens Sichromat 2. Column A: 20 m × 0.32 mm i.d. OV-1 (0.24 μm df). Column B: 20 m × 0.32 mm i.d. Superox 20M (0.24 μm df). Column temperatures: 60°C; 2 min isothermal; then to 180°C at 3°C/min. Compounds: 1. 2-heptanone; 2. hexen-3-ol; 3. hexan-3-ol; 4. β-pinene; 5. limonene; 6. camphor; 7. nerol; 8. neral; 9. L-carvone; 10. geraniol; 11. geranyl acetate. A. Analysis of total mixture on OV-1 column; B. Analysis of total mixture on OV-1/ Superox 20M combination (selectivity = 50:50 mixed phase); C. Multidimensional analysis of fraction X without intermediate trapping; D. Multidimensional analysis of fraction Y without intermediate trapping; E. Analysis of total mixture on Superox 20M column; F. Multidimensional analysis of fraction X with intermediate trapping; G. Multidimensional analysis of fraction Y with intermediate trapping.

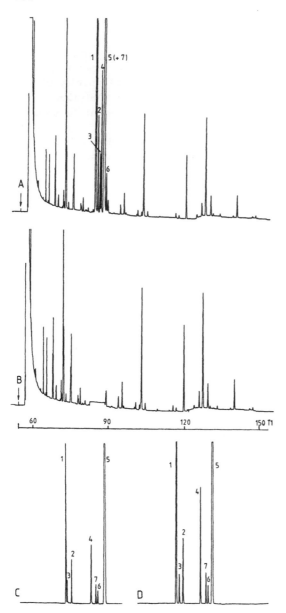

FIGURE 25 Multidimensional analysis of peppermint oil. Chromatographic conditions: see Figure 24. (A) Precolumn separation on
OV-1. (B) Heartcut of menthone-menthol fraction. (C) Analysis of
transferred fraction on Superox 20M with intermediate trapping.
(D) Analysis of transferred fraction on Superox 20M without intermediate trapping.

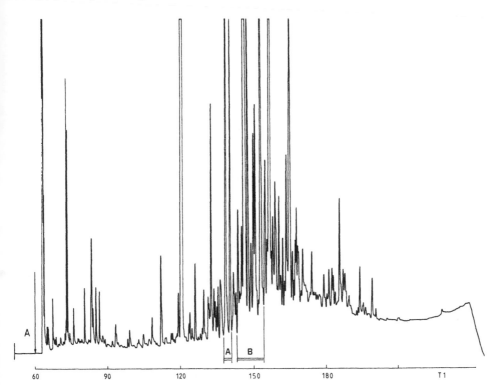

FIGURE 26 Multidimensional analysis of the essential of the Magnolia plant. Chromatographic conditions: see Figure 24. (A) Precolumn separation on apolar column. (B) Heartcut of fraction A. (C) Analysis of transferred fraction on Superox 20M column. No intermediate trapping. Oven 2: 60°C to 200°C at 3°C/min. (D) Analysis of transferred fraction on Superox 20M column. No intermediate trapping. Oven 2: 150°C; 20 min isothermal; then to 200°C at 3°C/min.

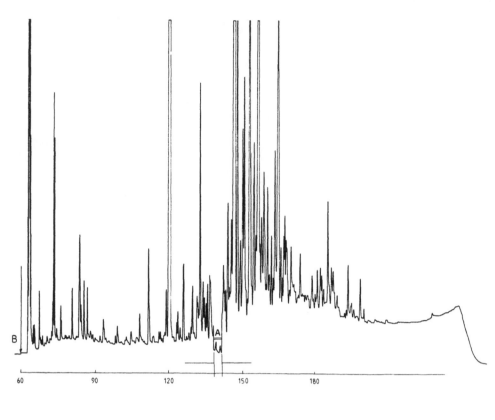

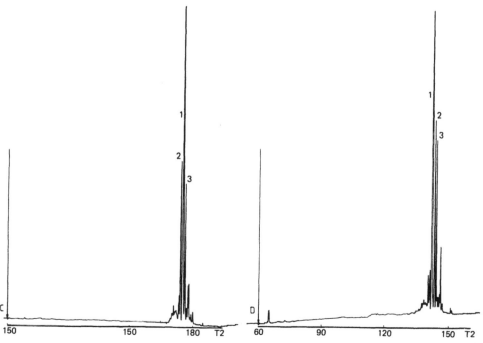

FIGURE 26

REFERENCES

1. M. J. E. Golay, Anal. Chem. 29: 928 (1957).
2. M. J. E. Golay, in Gas Chromatography, D. Desty (ed.), Butterworths, London, 1958, p. 36.
3. M. J. E. Golay, in Gas Chromatography, V. J. Coates, H. J. Noebels and I. S. Fagerson (eds.), Academic Press, New York, 1958, p. 1.
4. R. Dandeneau and E. Zerenner, HRC & CC, 2: 351 (1979).
5. M. Proot, F. David, P. Sandra, and M. Verzele, HRC & CC, 8: 426 (1985).
6. P. Sandra, LC-GC, 5: 236 (1987).
7. K. Grob and G. Grob, HRC & CC, 6: 133 (1983).
8. P. Sandra, I. Temmerman and M. Verstappe, HRC & CC, 6: 501 (1983).
9. L. S. Ettre, Chromatographia, 17: 553 (1983).
10. L. S. Ettre, in The Science of Chromatography, F. Bruner (ed.), J. Chromatogr. Library, Vol. 32, Elsevier, 1985, p. 87.
11. P. Sandra, in The Science of Chromatography, F. Burner (ed,), J. Chromatogr. Library, Vol. 32, Elsevier, 1985, p. 381.
12. F. David, M. Proot, and P. Sandra, HRC & CC, 8: 552 (1985).
13. K. Grob, Chromatographia, 10: 625 (1977).
14. T. J. Stark, P. A. Larson and R. D. Dandeneau, J. Chromatogr., 279: 31 (1983).
15. J. Buijten, L. Blomberg, K. Markides, and T. Wannman, J. Chromatogr. 237: 465 (1982).
16. J. Buijten, L. Blomberg, K. Markides, and T. Wannman, Chromatographia 16: 183 (1982).
17. P. A. Peaden, B. W. Wright, and M. L. Lee, Chromatographia 15: 335 (1982).
18. B. E. Richter, J. C. Kuei, J. I. Shelton, L. W. Castle, J. S. Bradshaw, and M. L. Lee, J. Chromatogr. 279: 570 (1983).
19. M. Verzele, F. David, M. Van Roelenbosch, G. Diricks and P. Sandra, J. Chromatogr. 279: 99 (1983).
20. M. Verzele, F. David and P. Sandra in Analysis of Organic Micropollutants in Water, (G. Angeletti and A. Bjorseth, eds.) Proc. Third Europ. Symp., D. Reidel Publ. Comp., 1983, p. 169.
21. F. Berthou, Y. Dreano, and P. Sandra, HRC & CC 7: 679 (1984).
22. E. Vanluchene, D. Vandekerckhove, P. Sandra, and F. David, HRC & CC 7: 644 (1984).
23. E. Geeraert and P. Sandra, HRC & CC 8: 415 (1985).
24. B. E. Richter, J. C. Kuei, L. W. Castle, B. A. Jones, J. S. Bradshaw, and M. L. Lee, Chromatographia 17: 570 (1983).

25. B. A. Jones, J. C. Kuei, J. S. Bradshaw, and M. L. Lee,
     J. Chromatogr. 298: 389 (1984).
26. K. Markides, L. Blomberg, J. Buijten, and T. Wannman, J.
     Chromatogr. 254: 53 (1983).
27. K. Markides, L. Blomberg, J. Buijten, and T. Wannman, J.
     Chromatogr. 267: 29 (1983).
28. K. Markides, L. Blomberg, S. Hofmann, J. Buijten, and T.
     Wannman, J. Chromatogr. 302: 319 (1984).
29. S. R. Lipsky and W. J. Mc Murray, J. Chromatogr. 279: 59
     (1983).
30. S. R. Lipsky and W. J. Mc Murray, J. Chromatogr. 289: 129
     (1984).
31. F. David, P. Sandra and G. Diricks, Proc. Eighth Int. Symp.
     on Cap. Chrom., Riva del Garda, 1987, P. Sandra (ed.),
     Huthig Verlag, Heidelberg, p. 264.
32. I. Temmerman, Ph.D. thesis, University of Gent, 1984.
33. G. Diricks, Ph.D. thesis, University of Gent, 1986.
34. P. Sandra, F. David, M. Proot, G. Diricks, M. Verstappe,
     and M. Verzele, HRC & CC 8: 782 (1985).
35. P. Sandra, M. Van Roelenbosch, I. Temmerman and M. Verzele,
     Chromatographia 16: 63 (1982).
36. K. Grob Jr., G. Grob, and K. Grob, J. Chromatogr. 156: 1
     (1976).
37. H. J. Maier and O. C. Karpathy, J. Chromatogr. 8: 308 (1962).
38. G. P. Hildebrand and C. M. Reilley, Anal. Chem. 36: 47 (1964).
39. R. J. Laub and J. H. Purnell, J. Chromatogr. 112: 71 (1975).
40. R. J. Laub and J. H. Purnell, Anal. Chem. 48: 799 (1976).
41. D. F. Ingraham, C. F. Shoemaker, and W. Jennings, J. Chro-
     matogr. 239: 39 (1982).
42. M. F. Mehran, W. J. Cooper and W. Jennings, HRC & CC, 7:
     215 (1984).
43. P. Sandra and M. Van Roelenbosch, Chromatographia 14: 345
     (1981).
44. J. H. Purnell, P. S. Williams, and G. A. Zabierek in Proc.
     Fourth Int. Symp. on Cap. Chrom., 1981, Hindelang (R. E.
     Kaiser ed.), Huthig Verlag, Heidelberg, p. 573.
45. J. Krupcik, G. Guiochon and J. M. Schnitter, J. Chromatogr.
     213: 189 (1981).
46. J. Krupcik, J. Mocak, A. Simova, J. Garaj, and G. Guiochon,
     J. Chromatogr. 238: 1 (1982).
47. G. Takeoka, H. M. Richard, M. Mehran and W. Jennings,
     HRC & CC, 6: 145 (1983).
48. J. H. Purnell and P. S. Williams, HRC & CC, 6: 569 (1983).
49. J. H. Purnell and P. S. Williams, J. Chromatogr. 292: 197
     (1984).
50. J. H. Purnell and P. S. Williams, J. Chromatogr. 321: 249
     (1985).

51. J. H. Purnell, M. Rodriguez, and P. S. Williams, J. Chromatogr. 323: 402 (1985).
52. J. H. Purnell and P. S. Williams, J. Chromatogr. 325: 1 (1985).
53. J. H. Purnell, M. Rodriguez, and P. S. Williams, J. Chromatogr. 358: 39 (1986).
54. D. R. Deans and I. Scott, Anal. Chem. 45: 1137 (1973).
55. R. E. Kaiser, L. Leming, L. Blomberg, and R. I. Rieder, HRC & CC, 8: 92 (1985).
56. R. E. Kaiser, R. I. Rieder, L. Leming, L. Blomberg, and P. Kusz, HRC & CC, 8: 580 (1985).
57. T. Toth, H. Van Cruchten, and J. Rijks in Proc. Sixth Int. Symp. on Cap. Chrom., 1985, Riva del Garda (P. Sandra, ed.), Huthig Verlag, Heidelberg, p. 769.
58. V. Pretorius, T. W. Smuts, and J. Moncrieff, HRC & CC, 1: 200 (1978).
59. R. E. Kaiser and R. I. Rieder, HRC & CC, 1: 201 (1978).
60. R. E. Kaiser and R. I. Rieder, HRC & CC, 2: 416 (1979).
61. P. R. Boshof and T. W. Smuts, J. Chromatogr. Sci. 18: 315 (1980).
62. P. R. Boshof, J. Chromatogr. Sci. 19: 238 (1981).
63. T. W. Smuts, K. de Clerk, T. G. du Toit and T. S. Buys, HRC & CC, 3: 124 (1980).
64. T. S. Buys and T. W. Smuts, HRC & CC, 3: 461 (1980).
65. T. S. Buys and T. W. Smuts, HRC & CC, 4: 102 (1981).
66. T. S. Buys and T. W. Smuts, HRC & CC, 4: 317 (1981).
67. T. S. Buys and J. B. Wagener, HRC & CC, 5: 662 (1982).
68. H. T. Mayfield and S. N. Chesler, HRC & CC, 8: 595 (1985).
69. J. V. Hinshaw Jr. and L. S. Ettre, Chromatographia 21: 561 (1986).
70. J. V. Hinshaw Jr. and L. S. Ettre, Chromatographia 21: 669 (1986).
71. P. Sandra, unpublished results.
72. P. Sandra, M. Proot, G. Diricks and F. David, in Capillary Gas Chromatography in Essential Oil Analysis (P. Sandra and C. Bicchi, eds.), Huthig Verlag, 1986, p. 29.
73. P. Sandra, M. Schelfaut, and M. Verzele, HRC & CC, 5: 51 (1982).
74. G. Schomburg, F. Weeke, F. Muller, and M. Oreans, Chromatographia 16: 87 (1982).
75. F. David and P. Sandra, in Capillary Gas Chromatography in Essential Oil Analysis (P. Sandra and C. Bicchi, eds.), Huthig Verlag, 1986, p. 387.

5

Process Multidimensional Gas Chromatography

ULRICH K. GOEKELER* and FRIEDHELM MUELLER   *Siemens AG, Karlsruhe, Federal Republic of Germany*

I. INTRODUCTION

Process gas chromatography is one of the most important analytical techniques used for process control, not only for quality control and to monitor process performance, but also for the determination of hazardous or explosive concentrations of individual components. When used as a routine instrument for around the clock operation, the system must be quick and precise, without need for attendance and also must be able to operate in a very reliable manner (1-3). The sample streams analyzed with this technology usually consist of more or less the same sample matrix with changeable measuring component concentrations, alleviating some of the more stringent requirements for multidimensional gas chromatography.

An automatic process gas chromatograph system consists principally of the sample handling system, the process gas chromatograph (analyzer), controller and data handling or data registration system. A system extracts liquid or gaseous sample in sequence from one or several sample points and adapts the flow, temperature and pressure to the gas chromatograph's need. The analyzer then takes a precise sample fraction for quantitative or qualitative analysis of components in a certain measuring range. The controller controls and monitors the operating parameters of the entire configuration and evaluates the acquired data.

The most important part of a process gas chromatograph is the separation technique. Only components which are interference free can be determined quantitatively. This chapter reviews the use of multidimensional gas chromatography in process applications. The

---

*Current affiliation: ES Industries, Voorhees, New Jersey

techniques and instrumental requirements as well as some practical
applications will be described

II. MULTIDIMENSIONAL GAS CHROMATOGRAPHY

A. Advantages of Multidimensional Gas Chromatography

Multidimensional gas chromatography has significant advantages over
that of a single column system, since two or more columns are
coupled in such a way that individual peaks or peak groups can be
transferred from one column into another for increased resolution.

Accordingly, some of the analytically important chromatographic
parameters can be manipulated, such as the column inside diameter,
column length, polarity of the stationary phase, phase ratio of the
coupled columns, temperature and temperature program for each
column (only with a dual oven system).

Due to the need for high reliability of a process gas chromatographic system, and the exact retention time repeatability, the
analyses are usually performed isothermally. Therefore, the solubility of the components with different boiling points and polarities
can not be optimized in many cases by means of a temperature program. With this restriction in mind, the above mentioned variation
of parameters becomes essential.

The results of the optimization of the working conditions can
be an increase in resolution (better separation), shorter analyzing
time (faster results), avoidance of column and detector contamination, (increase of column lifetime and reliability), and increase in
sensitivity (improved detection by removal of overlapped peaks).

1. Increase in Resolution

The resolution within a particular group of overlapping peaks
can be improved by transferring a part (cut) containing the group
of peaks from the first column to a second, more efficient column
(changing length and diameter) or more effectively, by changing the
polarity of the stationary phase. By transferring a series of narrow cuts into a second column of different polarity, many more components of a complex mixture can be resolved and identified, and
overlapping which may occur during the second (main) separation,
can be avoided.

2. Shorter Analysis Time

When a partial analysis of complex mixtures is to be performed,
it is usually necessary to remove the components that are not of interest in order to prematurely terminate the separation and condition

the column for the next analysis. By reversing the flow (backflush-ing) within the precolumn, it is possible to achieve fast removal of unwanted components from the column. The use of backflush is therefore an essential technique for process gas chromatography, because it insures a definite analysis end.

3. Extended Column Life

Preseparations are performed in the first column, and compo-nents that are not of interest (either of higher or lower volatility) are separated from those of interest without the need to achieve optimal resolution of all components. The entire sample comes into contact only with the precolumn, and less volatile constituents of the sample that may contaminate the second column can be backflushed. The highly efficient second column is thus protected against such contamination, and need not be operated under extreme conditions for reconditioning. Therefore, even when less stable stationary phases are used, the column life can be prolonged.

4. Decrease in Detection Limits

Under optimal operating conditions for the main column, column bleed may be kept to a minimum; overlapping by long tailing solvent peaks or major peaks can be avoided by "heartcutting." With the very efficient main column, narrow peak profiles can be achieved at high resolution. Narrow peak profiles result in higher signal-to-noise ratios.

5. Preventing Detector Contamination

By selective transfer of only the analytically significant com-ponents into first the main column and subsequently into the detector, contamination of the detector by other matrix components such as solvents, derivatization reagents, and major components is avoided. A typical example of this type of contamination is by chlorinated sol-vents in an electron capture detector (ECD). Contamination of se-lective detectors may decrease the sensitivity, change the specificity, and distort the linearity.

B. Requirements for a Process Gas Chromatographic System

There are some basic requirements of a process gas chromatographic system(4):
Rapid Analysis. When analyzing samples with a wide boiling point range, it is necessary to backflush all components eluting from the first column after the components of interest have been trans-ferred. This ensures an exact analysis end as well as a clean ana-lyzing path for the next analysis.

Precision. Measured components have to be separated entirely
from any interfering components. By means of coupling columns and
using the heartcut technique, specific components can be separated
qualitatively and determined quantitatively.

Reliability. By means of preseparation with the first column
(cleanup) and by transferring only the peaks of interest into the
second column (main analytical column) main column and detector contamination is prevented that might otherwise interrupt the analysis.
In addition, the reliability of the other parts of the system such as
injector, detector, solenoid valves, and pressure regulators must be
also ensured.

Wide Range of Analytes. The analysis of components of very different characteristics such as boiling point, polarity, or solubility
using the same analytical system can be made with quick and precise
results when differing column lengths and polarities are selected for
optimum separation.

Process gas chromatography has always used a column switching system based on the requirements mentioned above. A multidimensional
gas chromatographic column switching system is essential.

III. INJECTION SYSTEMS

A. Gas Injector

The gas injector system of a process chromatograph is installed in the
thermostatically controlled analyzer chamber. It consists, for example,
of a 6-way diaphragm valve and an injector loop (Figure 1). In
switching position A (clear path) of the valve, the sample flows continuously over the connections 5,6,3,4 through the injector loop,
while the carrier gas supplies the separation column via the path
1, 2. In the switching position B (dotted path) the sample is short-
ed via 5, 4: the carrier gas flushes the sample which was measured
in the injector loop, to the separation column via 1, 6, 3, 2. After
completion of the injection (approx. 1 to 10 seconds) switching back
to switching position A occurs. For gas injections, volumes between
0.5 and 3 mL are used, depending on the analytical needs.

B. Liquid Injector

Liquid samples are introduced in liquid form. The measured volume
of liquid is then vaporized and supplied to the separation column as
a gas. A diagram of a liquid injection system is presented in Figure 2. This task is performed by a liquid injector valve which consists of three sections, the pneumatic drive, the sample throughput section, and the vaporization system.

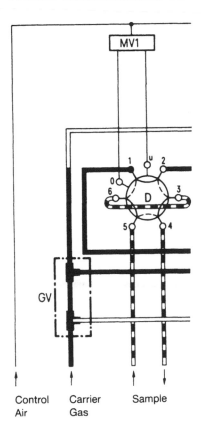

FIGURE 1   Gas injection system for processing gas chromatography.
D = gas injection valve, GV = gas supply, MV = solenoid.

A pneumatically driven injector push-rod transports the injection
volume (which is in the cross drilling of the push-rod) from the
sample throughput section, and the vaporization system.
A pneumatically driven injector push-rod transports the injection
volume (which is in the cross drilling of the push-rod) from the sample throughput section to the vaporization chamber. There, the
liquid is vaporized rapidly and completely, and flushed out by the
carrier gas as vapor. After the vaporization, the push-rod is returned to its starting position and the injector volume is again filled
with a sample. The vaporization chamber can be heated to temperatures up to 400°C. For liquid injection, injector volumes between 0.5
and 10 microliters are used, depending on the analysis needs.

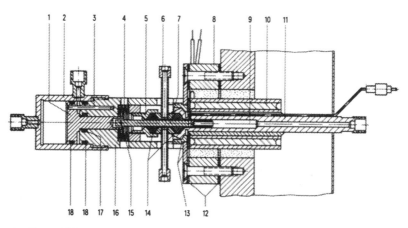

1	Control Piston	10	Heater
2	Control Cylinder	11	Evaporator
3	Guide Pin	12	Gasket (Seal)
4	Cup Spring	13	Pressure Screw w. Flange
5	Adjustable Intermediate Piece	14	Teflon Ring Seals
6	Sample Throughput Section	15	Pin
7	Injector Push Rod	16	Valve
8	Distancing Ring	17	O-Ring
9	Evaporation Chamber	18	O-Ring

FIGURE 2   Liquid injection system for process gas chromatography.

IV. FLOW SWITCHING SYSTEMS

Flow switching in coupled column systems can be achieved in different ways:

A. Multidimensional Gas Chromatography Using Valves (Packed Columns)

Commonly, a combination of packed columns with different column lengths, stationary phases, and stationary phase polarities are coupled by means of valves in process gas chromatography. This technique has been in use successfully for 30 years. A typical switching valve diagram is presented in Figure 3.

The advantages of using this configuration with packed columns are as follows: high amount of stationary phase translating to long column lifetime, short dead time, and rapid analysis times; high carrier gas flowrates are used—therefore, any small dead volume along

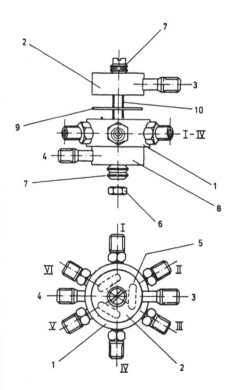

Component parts

1   valve plate (central section)
2   upper control head
3   upper control air
4   lower control air
5   position of the control recesses
    (broken lines)
6   nut M4
7   four plate springs
8   lower control head
9   diaphragm
10 cheese-head screw M4 x 40

I - VI   gas connections
I   sample in
II   sample out
III,VI   injection volume
IV   carrier gas to column
V   carrier gas in

FIGURE 3   Six-way diaphragm valve used for packed column switching.

the separation path is not of great importance; large sample amount can be injected, which means that injection amount repeatability and precision is easily attainable; as many different column configurations as necessary can be built.

However, there are also some disadvantages when using packed columns and valves, for example: unfavorable geometry and volumes of the various switching lines that become of importance for columns with smaller inside diameter and lower operating flow rates; temperature limitations due to the sealing material within the valve; adsorptive surfaces inside the valve, especially disadvantageous for trace component analyses; limited resolution with the packed column.

Utilizing the described column switching configurations, different procedures are possible, which are described in the following sections.

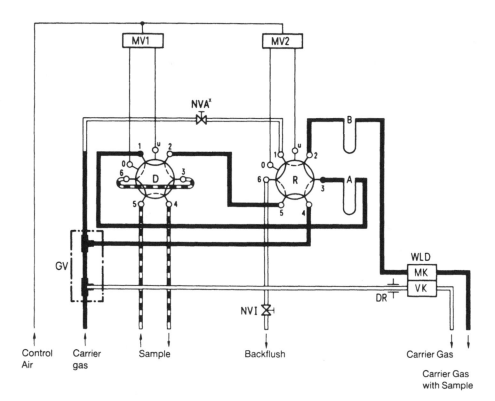

FIGURE 4   Column switching valves, backflush mode. D = gas injection valve, GV = gas supply, MV = solenoid valve, NV = needle valve, R = backflush valve, WLD = thermal conductivity detector, MK = measuring chamber, VK = reference chamber, DR = restrictor.

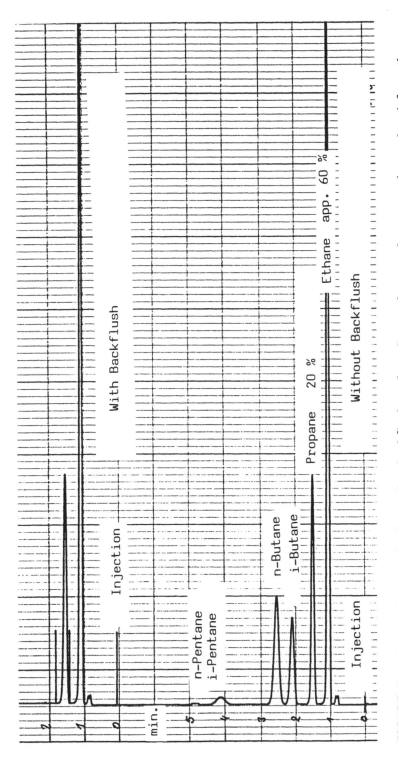

FIGURE 5 Shortened analysis time using backflush mode. Once ethane and propane have eluted from the first column into the second column for further separation, the flow to the first column is reversed. In this manner, the first column is cleaned while separation continues on the second column. This ensures a definite analysis end and prevents contamination.

1. Backflush

One of the most important column switching steps is the backflush.
As alluded to earlier, backflush is used to shorten analysis time and
to obtain a definite analysis end. Figure 4 shows a typical configuration of the backflush function. The valve R (backflush) connects
the columns A and B. In the switching condition A (clean paths)
the carrier gas stream flows through the path 4—5 of the R valve
and 2—1 of the injector valve over columnA and path 3—2 of the R
valve to column B. In switching conditions B (dotted paths) of the
valve, however, the carrier gas is lead to column A via the path
4—3 in the reverse direction. Thus, column A is backflushed via
the path 1—2 of the injector valve and 5—6 of the R valve. This
means that all components that have not left the column A at the time
of the switching are flushed out from the separation system via
needle valve NV 1.

The carrier gas supply of column B occurs via path 1—2 of the
R valve. All components which have been transferred into this column are carried to the detector. If the R valve is switched then
the measuring component with the longest retention time has transferred from column A to column B. A definite analysis end is reached
and thus a shortened analysis time. As soon as these components
are registered by the detector, a new sample can be injected into the
column system. An example of a shortened analysis using the backflush mode is presented in Figure 5.

2. Backflush Sum

For certain applications it is required that all components be detected,
including high boilers (heavy end). To have all the late eluting components pass through the separation system prolongs the analysis
time unnecessarily, and in many cases, higher boiling materials do
not elute at all, building up on the column and eventually deteriora-
ting the performance of the system. In these cases a backflush sum
is used (Figure 6).

The same switching configuration as for backflush is applicable,
but in this case the backflushed effluent is not flushed out of the
system but bypassed to the detector. All components in the first
column, which remained there after the first eluting components are
transferred into the second column, are flushed back, and the already achieved separation for these components is eliminated during
flow reversal. In this manner, all components are eluted simultaneously, and are detected as one peak. An example of this technique is presented in Figure 7.

3. Cut

The cut is mostly used when small contaminants in a pure product
are to be determined, and it is always used in conjuntion with back-

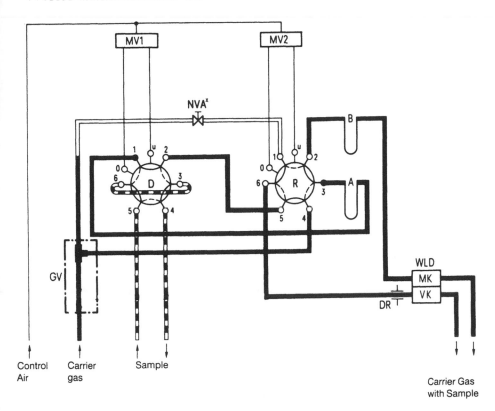

Control   Carrier   Sample
Air   gas

                                                              Carrier Gas
                                                              with Sample

FIGURE 6   Column switching with valves, backflush sum mode. D =
gas injection valve, GV = gas supply, MV = solenoid valve, NV =
needle valve, R = backflush valve, WLD = thermal conductivity detector, MK = measuring chamber, VK = reference chamber, DR = restrictor.

flush. In this case, the only concern is with the contaminants but
not with the main components. Due to the high concentration of the
main components, the separation of the components that have higher
retention time than the main component is difficult, but can be significantly improved if the main component is flushed out of the system after passing through one or two separation column sections.
Only the trace components with a reduced concentration relative to
the main component are supplied to the last separation column section. This switching configuration is shown in Figure 8. In switching position B (dotted path) of the valve (cut valve) the separation
columns B and C are connected, clear carrier gas flows out of the
system via the path 3—4 and 5-6. In switching position A (clear
path) the column C is supplied with carrier gas via the path 3—2,

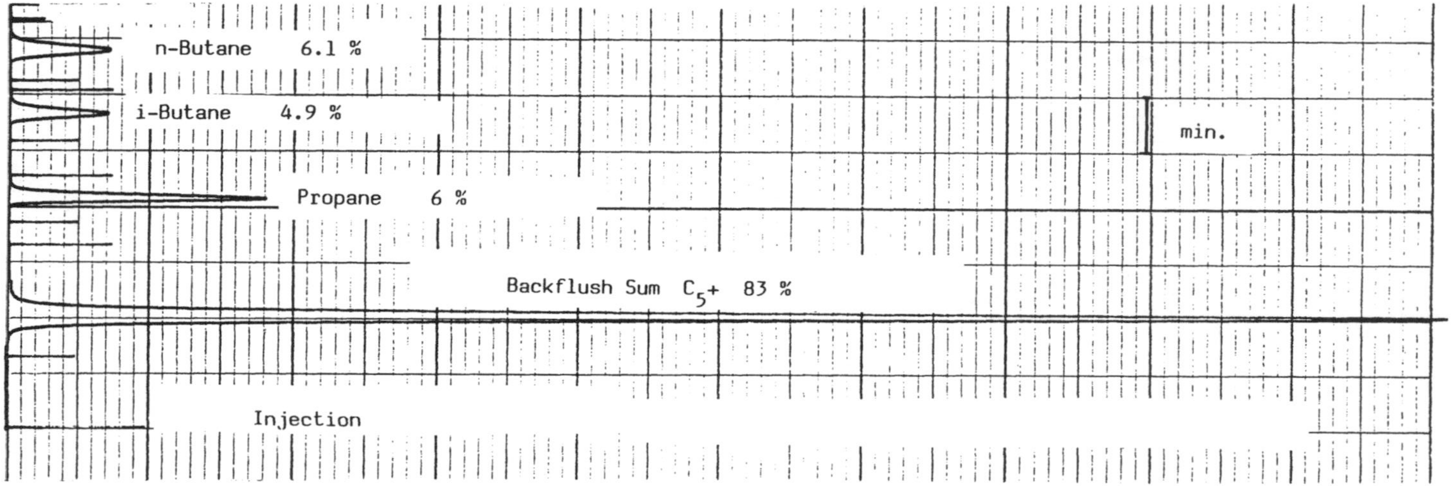

FIGURE 7   Example of backflush sum. Once propane, iso and n-butane are transferred to the second column, the flow is reversed in the first column, causing the remaining components to elute out of the first column into the detector, appearing as one peak.

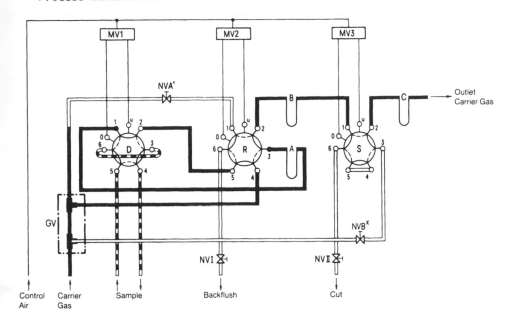

FIGURE 8   Column switching with valves, heartcut mode. A. backflush column, B. precolumn, C. main column, D. gas injection valve,
GV = gas supply, MV = solenoid valve, NV = needle valve, R = backflush valve, S = heartcut valve.

while the outlet of the separation column B leads out of the system
via path 1—6 and needle valve NV 2. All components leaving column
B in this switching position are not transferred to column C, but
flushed out. When switching to position B, another component transfer to column C occurs. Since the concentration ratio of the remaining main component to the trace components on Column C is now more
favorable, a complete separation can be attained (Figure 9).

4. Distribution

The column switching configuration called distribution is used in
order to separate different groups of components on two different columns simultaneously. Groups of components can be separated under optimal column conditions and analysis time can be shortened significantly
(Figure 10). In the initial position (clean path), eluate from column
B is flushed via 1—6 to column CII, or if no column is used there,
directly to the detector. In the switching position (dotted path),
eluate from column B is flushed to column CI via 1—2. Using this configuration, special care must be taken that the components eluting
from both columns to the same detector do not elute together, but sequentially (Figure 11).

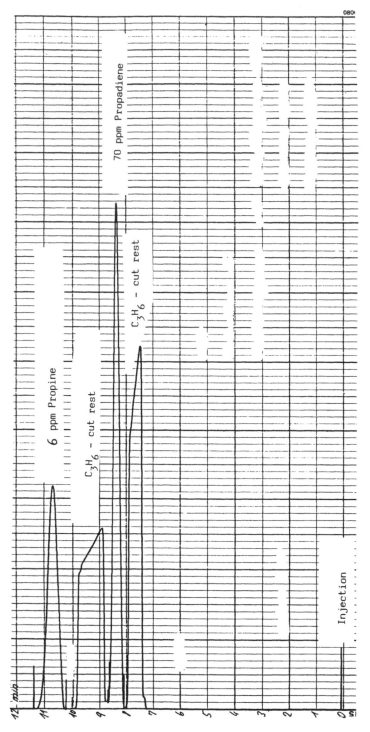

FIGURE 9   Example of heartcut.   Propine and propadiene are overlapped by propane at the large concentration ratio.   To obtain better separation, once propine elutes from the first column, backflush is triggered.   A slight preseparation is obtained, and the components of interest are transferred into the main separation column with a greatly reduced concentration of propane.   This results in interference free detection of the components of interest.

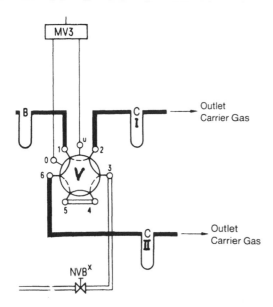

FIGURE 10   Column switching with valves, distribution mode. MV = solenoid valve, NVB = needle valve, V = distribution valve, CI = first column, CII = second column.

5. Stop and Go

When a sample mixture is very complex and the components to be measured are numerous, sometimes it is not possible to detect all components present without interference. A solution to this problem can be the "Stop and Go" configuration (Figure 12). In this case, a part of the preseparated peak groups is locked inside a column by shutting off the carrier gas flow to this column for a set period of time. The particular group is locked while other groups are separated. The column is connected to carrier gas again and the separation components have reached the detector (Figure 13).

B. Valveless Multidimensional Gas Chromatography

1. Valveless Column Switching Using DEANS System

The first valveless column switching system was introduced in the sixties by Deans(5). Originally developed for packed columns, it was applied in process gas chromatography successfully. Two columns are coupled together, and one pressure regulator before the first column represents the head pressure of the second column only.

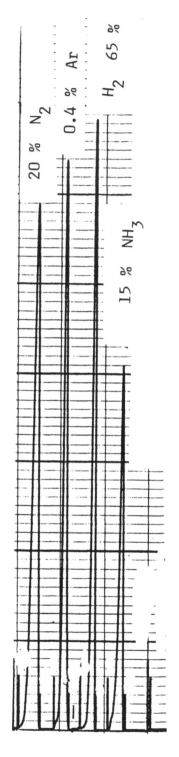

FIGURE 11   Example of distribution. Nitrogen, argon and helium elute together into the second column, where they are resolved. Ammonia, which would be retained considerably in the second column, is flushed directly to the detector after eluting from the first column.

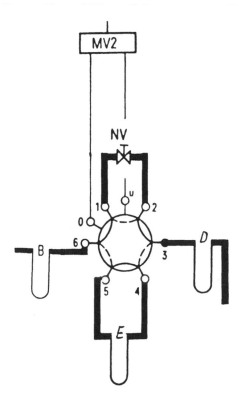

FIGURE 12   Column switching with valves, stop-and-go mode. MV = solenoid valve, NV = needle valve, B = first column, D = second column, E = stop-and-go column.

The cut exit is on the end of the first column and separated by means of a flow restrictor from the entrance of the second column and the second pressure regulator

Both pressure regulators and the cut exit flow are balanced in such a way that the entire sample is flushed into the second column as long as the cut exit is closed.

By opening the cut exit, the flow coming from the first column is vented out of the cut exit; in addition, a small flow from the second pressure regulator through the restrictor to the cut exit ensures that a clean cut is obtained. The second column is supplied with carrier gas through the second pressure regulator PM.

Utilizing this configuration, it is possible to transfer a fraction of the preseparation from the first column into the second column for

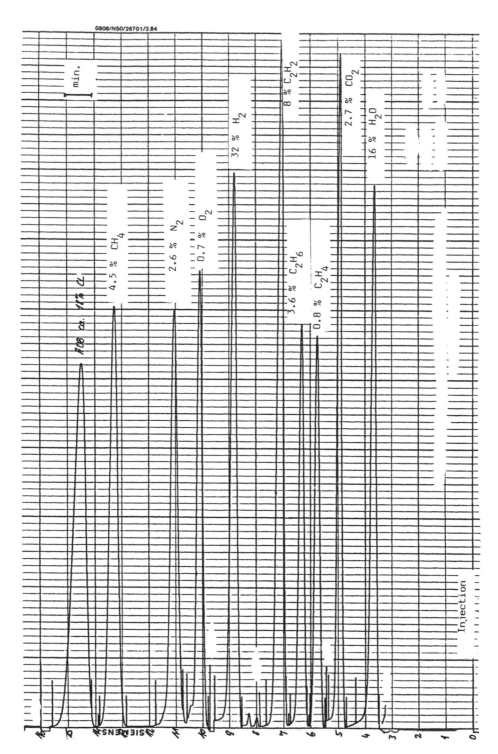

FIGURE 13 Example of stop-and-go. A complex sample with very different component characteristics is separated by holding the group containing hydrogen oxygen, nitrogen, methane and carbon monoxide in the stop-and-go column. After the remaining components are separated on the second column, the stop-and-go column is supplied with carrier gas and the components are transferred to the second column where the separation is completed.

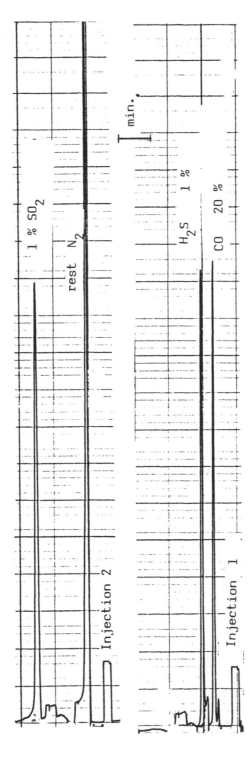

FIGURE 14 Chromatogram obtained using Deans switching technique. In a Claus process, hydrogen sulfide and sulfur dioxide must be monitored to ensure appropriate ratios; however utilizing valve switching, the system reliability is limited. Utilizing Deans switching, the required separation is obtained and longer system life time is insured; however, the flowrate variations experienced using this system create baseline instabilities, as can be observed in the peak shape deteriorations experienced.

further separation without interference of the other components, which are vented. The real advantage of the system is that no movable part is in the separation system.

However, the use of valveless column switching is limited by the following: only a few column configurations are possible (backflush and heartcut); the system is mainly applicable to packed columns usable in process gas chromatography (the use of capillary columns requires a focusing trap for the second column, which can't be used in process situations); high pressure differences in the coupling piece (depending on cut or transfer mode) results in retention time shifting on the first column (which depends on the cut time period); the cut valve is still in the sample path (therefore, limited lifetime of the valve); furthermore, the flow through the cut exit depends on the switching position and thus a constant flow is not available at that point, where detection is not possible (Figure 14).

2.   Valveless Column Switching System Using Live System

The Live Column Switching System (6) is based on the principle of the system developed by Deans. While the Deans system is based on significant flow differences between the two columns to be coupled, the Live system is based on small pressure differences which result in very small flow differences. With the Live system, a constant flow is insured through all the columns and restrictors, and these exits can be connected to the detector. An important difference in the Live system is a special coupling piece which is part of a "pneumatic bridge" configuration (Figure 15).

In a manner similar to the Weatstone bridge, the system consists of four flow resistors, needle valve NV2, NV3, and restrictors Dr1, Dr2, connected to pressure source. The differential pressure (ΔP) in the bridge diagonal is adjustable in the negative direction with NV2 and instantly switchable with solenoid valve MV2.

One difference of the "Live" switching system when compared to the Deans switching system is a flow channel with only a small resistance which is located in one diagonal, which either directs the total flow from the first column into the second column or diverts the flow from the first column via the restrictor DR.1 into the cut exit, due to a reverse flow dependent on the differential pressure applied.

The coupling piece is designed so that all gaps—especially the gap between the column and the coupling tube—are small and can therefore be flushed out well. Back diffusion of the sample into the gaps is thereby eliminated and peak deformations are prevented (Figure 16).

The advantages of such a system are easily recognizable: very fast switching time (heartcuts smaller than 1 second can be done); highly inert; no unflushed dead volume; no movable parts in the separation path; used for packed as well as for capillary columns.

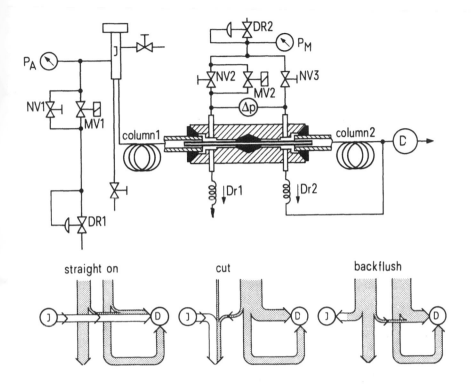

FIGURE 15   Valveless column switching using "Live" system. I =
injector, NV = needle valve, MV = solenoid, DR = pressure regulator,
DV = restrictor, PA = head pressure, D = detector.

    a. Column Switching Configurations. The principal column con-
figurations are as described earlier: backflush, heartcut, distribution, and backflush sum. Figure 17 is a schematic drawing of
the "valveless" separation column switching. The pressure regulator A is set to the head pressure, which affects the optimum carrier
gas flow for the separation of the column A and B. Between the
column A and B, coupled by means of the coupling-T-Piece, an additional pressure regulator B is used for carrier gas supply. The
pressure there results from the head pressure minus the pressure
drop along the column A.
    b. Heartcut Using Live System. The T-Piece contains a tube or
a capillary (depending on the separation column used) which reaches
into the end of both columns and the coupling tube. By means of
this gap, the auxiliary gas flow from pressure regulator B is connected to the end of column A and the head of column B.

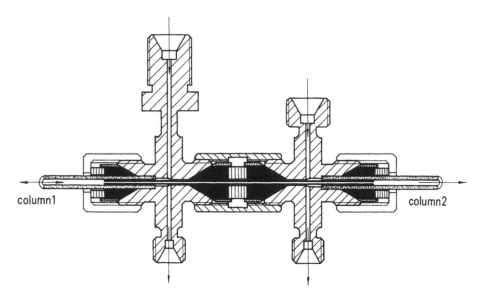

column1   column2

FIGURE 16   Diagram of live T-Piece for packed column operation.

    The column switching occurs by generating pressure differences
in the T-piece.   The auxiliary gas stream is directly connected via
the solenoid valve to the end of column A and the flow to the head
of column B is restricted by means of NV 3, thus, a pressure drop
occurs inside the T-piece which creates a flow in the direction of
column B, initiating sample transfer.   The entire carrier gas supply
of the column system occurs with pressure regulator A.   Pressure
regulator B contributes little to the supply of column B because of
the low overpressure.   Due to the small gap width, the flow reaches
a high speed which prevents diffusion of the sample in the gap.
    In the cut position the pressure regulator B is connected by
means of the solenoid valve to the head of column B.   The flow to
the end of column A is slightly reduced by means of NV 2.   In this
case, the effluent from column A cannot enter column B but is flushed
out via restrictor 1.
    c. *Backflush Using Live System.* Backflushing is triggered with
solenoid valve SV 1 and needle valve NV 1 by reducing the carrier
gas pressure at the head of the precolumn.   The pressure here falls
below the middle pressure and a reversal of flow is initiated in the
first column.   The components are flushed out via the needle valve
NV 4.   The injector is flushed further with carrier gas.   Pressure
regulator B handles the entire carrier gas supply for column B and
backwards for column A.

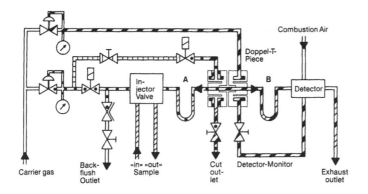

A

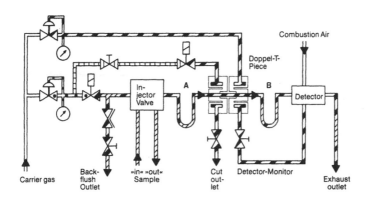

B

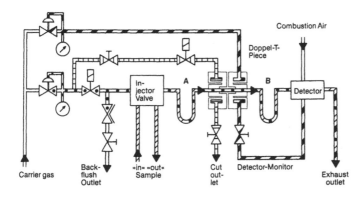

C

FIGURE 17   Flow path arrangements for the different switching positions using the live system.

   d.   Distribution Using Live System. If the cut exit is connected
to the detector either directly or by means of a capillary column, a
distribution can be achieved, bypassing column B.

V.   MULTIDIMENSIONAL GAS CHROMATOGRAPHY
   WITH CAPILLARY COLUMNS

Using capillary columns in gas chromatography has some significant
advantages over that of a packed column system, such as increase of
resolution, lower analyzing temperatures, highly inert separation
phase, and a shorter analyzing time (7). This results in: better separation, possibility of separating high boiling mixtures, analyses of
very polar or sensitive components, and usually quick results. However, the demands on a capillary switching system are much higher
since they require a dead volume free column coupling, absolute constant oven temperature, very small oven temperature gradients, precise carrier gas pressure regulators, uninterrupted carrier gas supply and small, precise and repeatable sample amounts for injections.

FIGURE 18   Connection of capillary columns using live system.

A. Capillary Column Coupling

The valveless Live column switching was originally developed for coupling capillary columns; therefore the system was adapted to the necessary requirements, that is, unflushed dead volumes and active surface areas are minimized.   Furthermore, the high temperature limit for the coupling piece makes this configuration ideal, as already described in Section IV.   Connection of two capillary columns using the live coupling is presented in Figure 18.

B. Oven Temperature

For capillary columns a small change in temperature has a greater influence than for packed columns.   Therefore an absolute constant oven temperature is required with a stability better that ± 0.1°C.

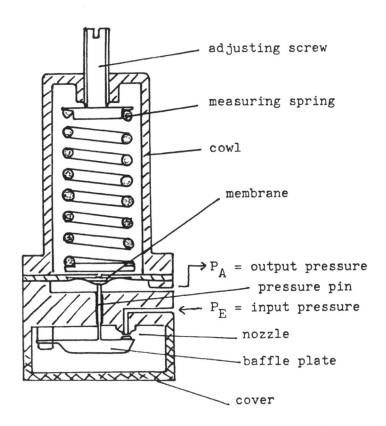

FIGURE 19   High precision pressure regulator for carrier gas supply.

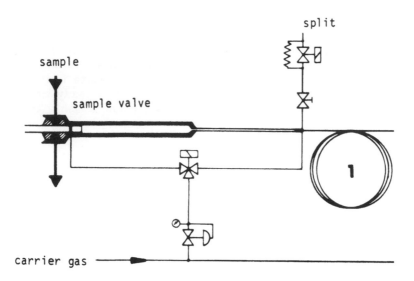

FIGURE 20.   Two-step injection system.   A liquid sample is injected
into the vaporization chamber (usually 0.5—2 microliter).   The vaporization chamber is bypassed with carrier gas, by means of an external
three—way solenoid valve.   The first capillary column is supplied with
carrier gas directly and the sample has time to vaporize smoothly.
After a certain period of time (5 to 30 seconds), when the sample is
in the gas phase, the carrier gas flow is switched to the vaporization
chamber and the sample is flushed to the column via a split.   Utilizing this technique, standard deviations smaller than 0.1% are achievable.

Temperature gradients across the capillary columns, which result in
decreased resolution, must be prevented.

C.   Carrier Gas Supply

A constant high purity carrier gas supply must be insured.   For
capillary columns a low flowrate as well as constant pressure regulation is necessary.   In addition, the pressure regulator must not allow
any diffusion into the carrier gas.   A pressure regulator based on
the principle of a nozzle/baffle plate system is shown in Figure 19.
Because a diaphragm is not used and a mechanical configuration is
built that eliminates ambient temperature change influence, it is a
preferred system for capillary columns.

D. Injection System for Capillary Columns

To have all the advantages of a capillary column separation system available, the quality of the injection system is of essential importance. Only a small sample amount can be injected into the capillary column without overloading (up to 100 ng of gas for example) and a split injector is typically used to have a small, precise and reproducible sample amount. For capillary columns a very small starting peak width is also required, which is easy to achieve in gaseous samples. However, vaporization of liquid samples automatically and rapidly requires a different injection procedure. One solution is a two step injection (8) illustrated in Figure 20.

VI. CONCLUSIONS

The use of multidimensional chromatography is essential for rapid reliable and precise analyses in process gas chromatography. The continuing need to investigate increasingly complex systems has generated advances in technology relating to producible sample transfer and analyses, including sample introduction, avoidance of contamination and the more widespread utilization of columns of capillary dimensions. As process technology progresses toward more automation, the use of multidimensional chromatography in process applications is expected to continue to grow at a fast pace.

REFERENCES

1. H. Oster, Prozess Chromatographie. Akademische Verlagsgesellschaft. Frankfurt am Main, 1976.
2. D. J. Huskins, Gas Chromatographs as Industrial Process Analyzers. Adam Hilger Ltd., Briston, 1975.
3. H. Ball, Maschinenmarkt, 82, 1976.
4. F. Muller, M. Oreans, Chromatographia 8: 27 (1977).
5. D. R. Deans, Chromatographia 1: 18 (1968).
6. Deutsche Patentschrift DE-PS 2955-387.
7. R. E. Kaiser, Chromtographie in der Gasphase, Band II: Kapillar-Chromatographie. Bibliographisches Institut Mannheim, 1975. Trenncassetten in der Gas-Chromtographie publ. by Institut for Chromtographie, Bad Durkheim, 1975.
8. Deutsche Offenlegungsschrift DE-05-2806-123.

6

Multidimensional High-Performance Liquid Chromatography

HERNAN J. CORTES and L. DAVID ROTHMAN *The Dow Chemical Company, Midland, Michigan*

I. INTRODUCTION

Liquid chromatography has been widely employed in multidimensional separations. The use of a packed column, gravity-flow liquid chromatographic separation as a means of preparing samples for subsequent analysis by another chromatographic separation is historically common in the fields of environmental, residue, toxicological and clinical trace analysis, where typical samples are complex matrices with a large number of components that would potentially interfere with the determination of the analytes or that would cause some harm to the analytical column or instrument if injected into a chromatograph. The goal of such "cleanup" separations is to reduce the complexity of the original sample matrix by separating a fraction of that matrix from the analyte(s). The desired result is a solution containing an amount of the analyte identical to the amount in the aliquot of sample applied to the cleanup column, ready for analysis and free of substances which might cause undesirable changes in performance in the analytical chromatography column or interfere with the determination of the analytes. Often, the separation scheme used in such cases is truly multidimensional, since the mode of separation in the gravity flow liquid chromatographic column cleanup step is frequently different than that of the analytical separation which follows. Typical examples of such multidimensional separations schemes would include sample preparation with a normal phase separation on a glass column packed with silica gel and eluted with organic solvent mixtures, followed by evaporation of the solvent from the collected fraction(s) containing the analyte(s), redissolution of the fraction(s) in an appropriate solvent and analysis by reversed-phase high-performance liquid

chromatography (HPLC) or gas chromatography. Variations on the
separation technique for sample cleanup include small packed beds
of relatively large diameter particles (50–200 μ diameter) of either
silica or silica with bonded functional groups or other packings contained in plastic syringe bodies or tubes which may be connected to
a source of vacuum or pressure (for example, syringe) to speed the
sample cleanup portion of the analysis. Techniques involving these
various sample preparation schemes prior to analysis by chromatographic separation can be grouped under the general heading of off-
line multicolumn separations, since the sample preparation and analytical columns are not physically coupled and there is some manual
manipulation of the sample between columns. As mentioned earlier,
many of these schemes can also be described as off-line multidimensional separations. It is not the goal of this chapter to discuss the
subject of off-line multidimensional liquid chromatography, but rather
to cover the area of on-line multidimensional HPLC (MDHPLC), where
two relatively high efficiency columns are coupled in an instrument
via the use of valves, traps, and other means. On-line multidimensional HPLC is practiced far less frequently than off-line techniques,
in the authors' experience. The general concepts of off-line and on-
line multidimensional separations are very similar, but the on-line
approach can offer the advantages of increased analytical throughput
and reduced operator effort due to the relative ease with which on-
line separations may be automated. The penalty of on-line MDHPLC
is that more thought must be given to design of experiments and
instrumentation to develop a successful application, compared to the
design of successful off-line approaches. This chapter will discuss
many of the considerations involved in designing and applying on-
line MDHPLC to practical problems, concluding with a series of examples of such applications drawn from the literature.

     It is further pointed out that this chapter will only deal with
HPLC separations which are truly multidimensional in nature. A
body of literature may be found in which liquid chromatographic
separations involving multiple columns connected by valving are applied to sample cleanup and analysis, but the modes of separation in
the two columns are not truly different. Examples would include the
use of a short column packed with a reversed phase material used
for the first separation, followed by a separation on a longer column
packed with the same phase and using the same or a similar mobile
phase. In some cases, a third separation on an off-line low resolution column may precede the on-line separations with all three columns
containing similar bonded stationary phases (1). The first column
can clean up or preconcentrate the sample, which is very useful in
extending the lifetime of the analytical column or increasing the sensitivity of the analysis to traces of analyte, but substances which tend
to co-elute with the analyte on the first column are not unlikely to
do so in subsequent separations. The resulting separation does not

offer the peak capacity of a truly multidimensional separation (2)
and can lead to interferences in analyses. However, it is clear that
such single dimensional, multiple column separations do have their
place, since a literature search on the subject of column switching
or multiple column separations in liquid chromatography leads to the
discovery of far more single dimensional separations than multidimensional ones. Perhaps this is because single dimensional multiple column separations almost always achieve the goal of removing
sample components via the cleanup column which would otherwise
irreversibly bind to the analytical column and change its performance.
The mechanical considerations in setting up such a "column-switching"
application are similar to those for multidimensional separations and
much of what is said in this chapter about MDHPLC has some application to column switching, as well. On-line multidimensional chromatographic separations have been performed by both gas and liquid chromatography, but, between the two, liquid chromatography appears to
offer a much richer number of possibilities for separation modes of
truly different selectivity and, therefore, more possible dimensions.
In gas chromatography, a wide variety of stationary phases are available, offering clearly different selectivities, but relative retention
times in gas chromatography are strongly correlated with analyte
boiling point, regardless of the stationary phases involved. Because
of this, separations with truly orthogonal selectivity in gas chromatography are rarely found. Consider, for example, that regardless
of the type of gas chromatographic stationary phase selected, a homologous series of compounds will always elute in the order of increasing boiling point (the authors would be interested in learning of any
exceptions). In addition, the mobile phase (carrier gas) employed in
most gas chromatography applications usually plays little or no role
in determining the selectivity (α) of the separation. Although the
modification of mobile phases by addition of water vapor or organic
solvents to produce changes in column selectivity are well-known, no
examples of such modified mobile phases applied to multidimensional
gas chromatography were found in our literature searches. Consider
some of the common liquid chromatographic separation modes: normal
phase, reversed phase, ion exchange, size exclusion chromatography,
and affinity chromatography. Selectivities of these various modes are
very different, due to the very different solute/stationary phase interactions involved. Complete reversal of component elution order is
frequently observed when normal and reversed-phase separations
of nonhomologous components are compared. Consider the previous
example of a gas chromatographic separation of a homologous series
in which compounds elute in order of increasing boiling point or molecular weight. Similar elution behavior is frequently observed for homologous series in normal or reversed phase HPLC, but in size exclusion chromatography, the same homologous series would elute in order
of decreasing molecular weight, or with complete reversal of relative

retention as compared to the other separation modes. Consider also that the mobile phase in liquid chromatography plays a major role in determining the partition coefficients of solutes. Reversals in solute elution order upon substitution of one solvent for another in a mobile phase are well-known, even though the mode of the separation is unchanged. Substantial effects on solute retention may be achieved by the use of ion-pairing agents, surfactants, buffers and other mobile phase modifiers. Overall, it would appear that liquid chromatography offers great opportunities for multidimensional separations. The experimental difficulties of the liquid chromatographic multidimensional experiment, which will be discussed later, as compared to either multidimensional gas chromatography or simple column switching liquid chromatography, may account for the relatively greater number of the latter two applications that are found in the literature.

The very different selectivities of the various modes of liquid chromatography permit relatively complex mixtures to be analyzed with minimal sample preparation. In fact, many of the examples of multidimensional liquid chromatography in the literature involve applications to trace analysis in difficult matrices. Historically, the use of noncoupled column multidimensional chromatography has been common in such cases, typically involving the initial "cleanup" of the sample on a silica or ion-exchange column followed by reconcentration of the sample (often by solvent evaporation) and injection of a fraction of the sample into an analytical HPLC system. Such a procedure can be very effective in removing potentially interfering components from the sample, as well as removing substances which would reduce the useful lifetime of the analytical column. The disadvantages of the noncoupled column approach are typically the time involved (solvent evaporation is often the slowest step) and potential problems with contamination and quantitative recovery of the analyte from the first separation. These disadvantages may be eliminated by application of a well-designed coupled-column separation scheme, whether it is multidimensional or not. An example of a multidimensional approach is the use of size exclusion chromatography with an aqueous mobile phase to perform an initial separation, followed by an ion-exchange separation (3). In this case, the effluent from the size exclusion column was directed onto the ion-exchange column during elution of the analyte(s) of interest. At all other times, the effluent is sent directly to waste. When the analyte-containing portion of the aqueous mobile phase entered the ion-exchange column, the analyte concentrated at the head of the column and remained there until the ion exchange column was eluted with a mobile phase of sufficiently high ionic strength. In such a case, sample preparation often is limited to dissolving and filtering the sample, then injecting it onto the first column. The total analysis time is greatly reduced by comparison to classical off-line sample preparation and recovery of the analyte is

generally very good with minimal chance for contamination. The reduction in analysis time may in large part be attributed to the on-
line reconcentration of the analytes by refocusing of the zones eluting from the size exclusion column on the inlet of the ion exchange
column. Refocusing on-line is often far faster and more convenient
than solvent evaporation off-line.

There are a number of challenges to the application of the multidimensional liquid chromatography. The equipment is significantly
more complex than that of the conventional single-column experiment.
Typically, additional HPLC pumps, valves and detectors are required.
There are events which must take place during the experiment, including valve switching, mobile phase changes, etc. The additional
complexity can often be easily dealt with by automation of the equipment. Another set of difficulties arises when all the possible combinations of liquid chromatography separations modes are considered.
There are some modes which have incompatible mobile phases, for
example, normal phase and ion-exchange separations. Potential problems arise with liquid phase immiscibility, precipitation of buffer salts
and incompatibilities between the mobile phase from one column and
the stationary phase of another (e.g., the swelling of some polymeric
stationary phase supports by changes in solvents or the deactivation
of silica by small amounts of water). These difficulties can reduce
the number of choices available to someone wishing to apply multidimensional HPLC to a separations problem, although inventive approaches have been found to "compatibilize" seemingly incompatible
separation modes (4).

II. INSTRUMENTATION OF MULTIDIMENSIONAL
LIQUID CHROMATOGRAPHY

Much of the instrumentation for multidimensional liquid chromatography is the same as that used in conventional "one-dimensional" experiments. Commercial pumps, injectors and detectors are generally
useful for either type of experiment. Certain multidimensional experiments, however, may place additional demands on equipment.
Consider the various scenarios which follow:

A. Trace Analysis Performed in a Multidimensional HPLC System

The detector chosen for such work should be one which shows minimal response to changes in mobile phase flow rate. Otherwise, the
switching of valves or other events which change or interrupt the
analytical mobile phase flow rate must be performed at times that do
not coincide with elution of the analyte peaks from the analytical
column. Simple testing of detectors will reveal any problems with
either short-term or long-term changes in baseline signal level with

either impulse or long-term changes in flow rate. Absolute baseline
levels can be measured at different flow rates as an indication of
detector sensitivity to flow rate changes. During these tests, it is
important to replicate the conditions of detector cell pressure which
will be seen in actual use. Responses to flowrate changes seen in
some detectors would make trace analysis impossible, were these responses to occur as a result of a switching event during elution of
the analyte.

B. Column-Switching Times Determined by Monitoring the
    Separation on the First Column and Switching Valves
    at Times which Are Dependent on the Observed Signal

The simplest approach to valve switching in multidimensional chromatography is to assume that the separation on the first column is
sufficiently reproducible that the analyte will always elute from that
column within a fixed time window. There are cases in which this
assumption is not valid. Since the first column serves, among other
things, as a filter to remove sample matrix components which might
foul the analytical column, it is not unlikely that the selectivity of
the first column will change from injection-to-injection and that,
therefore, the retention time window within which the analytes elute
will also drift. In such a case, valve switching times are best selected by monitoring the separation with a detector and actuating
valves at times determined by monitoring the retention time of a reference peak. In such a case, it is possible that the system design
will, at some time, place the detector flow cell in-line between two
HPLC columns. When this occurs, the flow cell of the detector must
be designed to withstand the pressures to which it will be exposed.
This may rule out certain types of detection, such as refractive
index, fluorescence or electrochemical, where high pressure flow
cells designed into commercial instruments are rare or nonexistent.
It is also important to choose a detector which causes minimal band
broadening if analyte bands must pass through the detector before
being introduced onto the second column. Fittings and internal
tubing between the detector cell and the solvent outlet port on some
detectors are not always designed with this in mind. Techniques for
refocusing analyte bands on the second column will correct for this
flaw, at least partially, but it is a good practice to minimize band
broadening wherever possible. Again, simple testing of the detector
will reveal such a flaw. The column outlet in a single-dimensional
HPLC experiment can be connected to the outlet of the detector,
thus operating the flow path of the detector in reverse. If observed
peak shapes are unaffected (or even improved) as compared to the
same detector used in the normal flow direction mode, the entire
flow path of the detector is probably suitable for multidimensional
work as described here. Alternatively, a second detector may be

attached to the outlet of the first to see how much the peak profiles
are distorted by the first detector.

C. A Single Pump Is Used for Mobile Phase Delivery with
Solvent Switching on the Pump Inlet Used to Change
Phases for the Different Stages of the Separation

In this case, it is best to use a pump that has a low internal volume
and has an effective means of bypassing the HPLC columns during
the solvent changeover phase of operation. The latter may be handled via external valving. It is also important to use a pump which
readily tolerates the changeover from one solvent to another. This
is sometimes a problem when the two solvent systems either have
very different densities, are immiscible, or when one of the solvents
poorly wets the pump head internal parts. A pump that allows remote
control of flow rate is desirable if the system is to be automated.
This will allow programming of the system to stop mobile phase flow
if desired or select a high flowrate when the system is being purged
out, as during a solvent changeover operation. Pump systems which
have quaternary solvent gradient capability may be most easily
adapted to an application involving solvent changeover, since they
allow the use of one or two "compatibilizing" solvents, miscible with
each mobile phase, to be directly plumbed to the pump in addition
to the desired mobile phases to ease the changeover of solvents in
the pump.

D. The First Dimension of the Separation Is a Very Simple One
with Relatively Easy Separation of the Analyte(s) from the
Matrix

In this case, the first column may be a small cartridge column placed
in the sample loop of the injection valve. It is important to use a
valve which allows either use of a second, external pump for the
first separation on the cartridge or else makes it easy to supply the
mobile phase for this separation via a syringe inserted in the injection port. Valves make on-line multidimensional HPLC possible and it
is here that the biggest differences exist between these and conventional HPLC instruments. Some of the factors to be kept in mind
when considering valving arrangements are the number of components
of interest, their capacity factors, the different chromatographic
modes employed, the number of solvents required for analysis and
the sample type and complexity. In general, transfer of components
is accomplished using rotary type valves which at present appear to
be the most convenient means for switching operations, as such valves
can easily be automated using pneumatic or electric actuators and can
routinely withstand the typical operating pressures of modern liquid
chromatography. It is required, however, that such valves be inert

towards mobile phases and sample so that components of interest can
be transfered without loss or memory effects. Valve body construction materials can be stainless steel, Hastelloy C, or tantalum, the
choice of which depends on the requirements of the analysis. Rotors
are usually constructed of Vespel (polyimide) or fluorocarbon polymers.
A wide variety of valving schemes have been used for multidimensional
experiments and some will be discussed here. Vendors of HPLC valve
systems generally can provide application literature which covers the
most common schemes (5).

There are two general types of multidimensional chromatography
separation schemes: those in which the effluent from one column flows
directly onto a second column at some time during the experiment
and those in which some type of trap exists between the two columns
to decouple them. The purpose of a trap is often to allow collection
of a fixed eluate volume to reconcentrate the analyte zone prior to
the second separation step or to allow a changeover from one solvent
system to another. In either mode, either all or only a portion of
the analyte eluting from one column may be collected and switched
onto the next column. Techniques involving partial transfer of analyte
between columns have such names as "heartcutting" or peak sampling.
These are useful when the analyte is present at relatively high concentration and there is a need to minimize the amount of material transferred between columns or when introducing large amounts of eluent
from one column to another would result in unacceptable band broadening due to solvent-column incompatibilities. Such might be the case
if, for example, the effluent from a normal phase separation is introduced onto a reversed-phase column. Qualitative analysis by this approach is fairly simple, but quantitative analysis demands very rigorous
control of a variety of factors, including flow rates, temperatures
and switching times. An interesting application of partial transfer
MDLC is in the examination of chromatographic peaks for coeluting
interferences. Portions of the leading and trailing edges and the center of an analyte peak on the first column can be sampled and separated on the second column to look for components other than the
analyte which may also be present under the peak. The extent of
interference can be estimated from such a study, which is useful in
validating the accuracy of an analytical method, even if the eventual
method is intended to be a simple, single dimensional experiment. Two
examples of systems suitable for partial transfer MDLC are shown in
Figure 1. In Figure 1A, an experimental setup is shown using two
6-port valves, two HPLC columns and two pump systems. If the two
columns have different modes of selectivity, this is a simple MDLC
setup. The valves used for this purpose are standard high-pressure
6-port switching valves (commonly available from such companies as
VALCO and Rheodyne). In this setup, a sample may be loaded into
the loop of the first valve while pump 1 pumps mobile phase 1 through

column 1 (hereafter referred to as the cleanup column) and pump 2
pumps mobile phase 2 through column 2 (hereafter referred to as the
analytical column). The sample injection valve is switched to initiate
the experiment. A separation occurs on the cleanup column. As the
analyte zone elutes from the cleanup column, some or all of that zone
can be diverted from the cleanup column waste stream to the analytical column by switching the second 6-port switching valve. When
the desired portion of the analyte zone has been transferred to the
analytical column, the 6-port switching valve is again switched and
the cleanup column effluent is redirected to waste. A detector on
the cleanup column waste stream could be used to monitor the progress of the separation in that column, if necessary. Elution of the
analytical column proceeds to complete the analysis while the cleanup
column is re-equilibrated with mobile phase 1 prior to the next injection. With attention to separation times on the two columns, it is
possible to inject a second sample onto the cleanup column while the
analysis of the first sample occurs on the analytical column. Such a
system makes very efficient use of instrument time.

In Figure 1B, a 10-port valve is used to control flows within the
system. The general operation of this system is like the two-valve
system just discussed, but fewer moving parts are involved. This
may present some advantages in automated systems. When the valve
is in position A, sample may be loaded into the sample loop. The
columns are connected in series and pump 1 pumps mobile phase
through them while the output from pump 2 goes directly to waste.
After the sample loop is loaded, the valve is switched to position B,
where pump 2 washes the sample onto the inlet of the cleanup column
and the analytical column continues to be pumped by pump 1. At
some time during the elution of the solute zone of interest, the valve
is switched back to position A and the eluting zone, instead of passing out of the valve to waste, is directed onto the inlet of the analytical column. When the zone has been transferred from the cleanup
column to the analytical column, the valve is switched back to position
B and the elution of the analytical column continues with the effluent
being monitored by the detector.

The requirements of a system such as this one are that both
columns must be compatible with both mobile phases, since in position
A, the same phase is being pumped through both columns by pump 1,
but at the moment of switching to position A, the cleanup column is
filled with the mobile phase from pump 2. If the proper combination
of columns and mobile phase is chosen, no problems should arise.
Another consideration is that the separation in the cleanup column
must not be seriously interfered with by the presence of pump 1
mobile phase at the time of injection, even though the mobile phase
from pump 2 is the actual eluting solvent. Similarly, the presence of
a slug of solvent from pump 2 at the time of transfer of analyte from

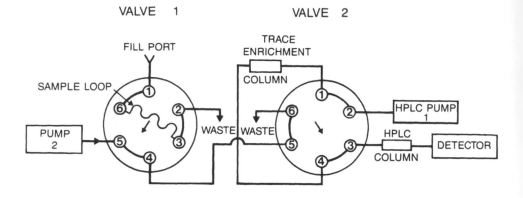

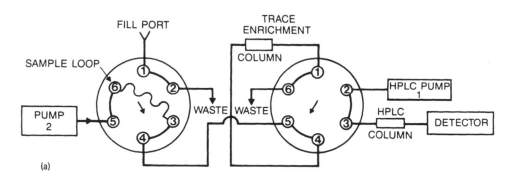

(a)

FIGURE 1 Valving schemes for multidimensional liquid chromatography. (a) Trace (or sample) enrichment using two six-port valves. (b) Trace (or sample) enrichment using a 10-port valve.

the cleanup column must not seriously affect the separation in the analytical column.

These constraints require careful thought in setting up a method of analysis. Some potentially compatible combinations might include a system in which the cleanup column is a size exclusion chromatography column compatible with organic-containing mobile phases and water, and the analytical column is a reversed-phase HPLC column. If the solvent for pump 1 is an organic/aqueous mixed solvent and the eluent for pump 2 is either water or a water/organic solvent mixture like phase 1 with equal or lesser amounts of organic solvent,

POSITION A

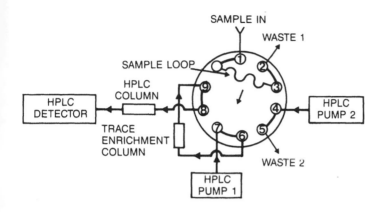

POSITION B

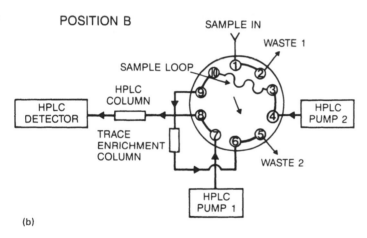

(b)

FIGURE 1 (Continued)

a successful multidimensional separation can probably be developed.
Other potentially compatible combinations would be of ion-exchange
and reversed-phase columns or ion-exchange and SEC columns. In
any of these cases, the compatibilities of each column with each
mobile phase must be carefully considered and the user must be
aware of any potential long-term problems which may arise from repeated step changes in mobile phase character, such as swelling and
shrinking of some polymer-based column packings (6).

Trapping systems in MDLC usually involve the quantitative transfer of an analyte from one column to another with an opportunity for additional analyte manipulation between columns. Some manipulations may include trapping the analyte zone in a sample loop of known volume, reconcentration of the analyte zone or a change in the nature of the solvent system. The use of a loop of fixed volume for trapping does not allow quantitative transfer of analyte between columns unless the volume of the loop exceeds the volume occupied by the analyte upon elution from the cleanup column. This technique may be more useful when a known volume is to be transferred between columns, as in a "heartcutting" application (7). Without reconcentration, the analyte zone applied to the second column by this approach may result in very broad peaks in the final analytical result, but it is likely that proper design of the second separation will allow reconcentration of the analyte zone on the inlet of the second column prior to the actual analytical separation, particularly if the second separation is performed by gradient elution. Reconcentration of the analyte zone by another means is important if the conditions of the second separation will not accomplish this on a large volume of analyte-containing mobile phase from the first column. This can be done in several different ways, including the use of a third, more retentive, column as a trap between the cleanup and analytical columns or use of a column similar to the cleanup column as the trap with modification of the mobile phase to increase k' of the analyte. In the first case, an example chosen from the literature (8) involves the use of a C_8 reversed-phase column for the initial separation, a C_{18} reversed-phase column for trapping the analyte and, finally, a strong anion exchange column for the analytical separation. In this case, the trap column was of the same general phase type (bonded alkyl hydrocarbon) as the cleanup column, but more retentive due to the longer alkyl chain length bonded phase. Another example of the use of a more retentive column as a trap involves an initial anion exchange separation with the analyte being trapped on a short reversed-phase column followed by elution of the analyte onto a longer reversed-phase column for the analytical separation (9). Post-column modification of the mobile phase to increase k' of the solutes on a trap column similar to the cleanup column has been performed by one of the authors (10). In that case, the effluent from a reversed-phase cleanup column passed through a tee, where it was diluted with water before passing through a second short reversed-phase column. A dilution ratio was chosen which resulted in effective trapping of the solutes on the inlet of the second column, which was then eluted with a mobile phase compatable with a subsequent ion-exchange separation, resulting in quantitative trapping and transfer of the analytes with minimal band broadening. Effective trapping techniques are important for trace analysis, since quantitatively collecting the analyte eluting

from a cleanup column and refocusing it into a narrow band which
gives the most narrow (i.e., tall) peaks in the analytical separation
maximizes the sensitivity of the analysis.

As mentioned earlier, there is sometimes a problem of incompatibility of solvents between two steps of an MDHPLC scheme. An example of the use of a trap to achieve solvent compatibility in such a
case is shown in Ref. 4, where a short "dummy" column was interposed between the initial normal phase separation and the following
reversed-phase separation. The dummy column provided a convenient
volume to contain the eluated fraction collected and a surface for adsorption of the solutes when the trapped solvent was evaporated from
the dummy column by a stream of gas. The mobile phase from the
second separation was then used to elute the solutes from the dummy
column and carry them to the analytical column. By eliminating the
mobile phase from the cleanup separation before proceeding to the
analytical separation, any mobile phase incompatability problems which
existed between the two columns were circumvented. This general
approach, while somewhat involved, is practical.

III. DIFFICULTIES OF THE MDHPLC EXPERIMENT

The additional complexity of MDLC experiments leads to greater difficulties than those found in conventional HPLC. First, two different
HPLC separations must be developed. If the application is to a sample
with a complicated matrix which would otherwise require off-line
liquid chromatography for sample preparation, devlopment of two LC
separations is required anyway and the additional time to develop the
second HPLC step for MDHPLC is usually not great. If the first
HPLC step is size exclusion chromatography, development of that separation step should be quite short unless problems are encountered
with sample/column compatability. A greater challenge is coupling
the two separations on-line without seriously compromising the performance of either. Careful attention must be paid to compatability of both
columns with all the mobile phases which will pass through them. The
mobile phases must be compatible with each other (e.g., a high ionic
strength aqueous buffer should not be mixed with most nonaqueous
mobile phases, nor should immiscible phases be used). The alternative,
if phase incompatibility is unavoidable, is to find some way to isolate
the desired eluate fraction from one column and effect a suitable solvent changeover before the solutes enter the next column. In addition
to the chemical considerations, there are typically more pieces of equipment involved in the experiment than in conventional HPLC, leading
to longer startup time for experiments, increased maintenance and
greater possibility of equipment failure. In addition, some operations
must take place in each separation which do not normally occur in
conventional HPLC, such as valve switching events. The timing of

these events can have a great effect on the reproducibility of the
experiments. As discussed elsewhere in this volume, automation of
as many system functions as possible results in a mechanically more
complex system, but one which is easier to operate and which is
likely to produce more reliable results. Finally, in the authors' experience, there is some psychological resistance to the use of on-
line multidimensional chromatography by those accustomed to practicing off-line separations. The resistance appears to be due to recognition that the results obtained by the on-line system may differ
little from those achieved with similar off-line methods, but requires
additional capital investment in equipment, more complex instrumentation and a somewhat different approach to separation development.
When off-line techniques have successfully solved the available problems, this resistance is at its highest (such resistance to change is
known elsewhere where less-conventional techniques are suggested
as replacements for existing conventional methodology or instrumentation (11)). In such cases, it is wise to question whether an on-
line multidimensional separation is necessary. Frequently it can be
demonstrated that, although the results of off-line and on-line techniques applied to the same problem appear similar, the on-line technique offers higher analytical throughput per hour of the analyst's
time and that the results can be more reproducible and perhaps
interference-free than the off-line approach. If the application calls
for a large number of routine analyses, the additional time to develop
and set up an on-line MDHPLC separation scheme can pay off through
these advantages.

IV. SELECTED APPLICATIONS

The number of applications which utilize the coupled column approach
in modern liquid chromatography is extensive when one considers the
various combinations which can be used to resolve a complex mixture,
such as for example, heartcut (sampling of a section of the chromatogram for further resolution), on-column concentration or trace enrichment (the use of a weaker eluent to reconcentrate components of
interest at the head of a precolumn prior to separation on an analytical column), or the use of columns having stationary phases with
similar characteristics but differing phase ratios. Based on the
criteria imposed for multidimensional separations, that is, that different separation mechanisms must be employed to achieve the required resolution in the sequential steps, and that components resolved in the initial separation stage remain resolved (1), this section
will concentrate on selected applications which can be considered multidimensional by meeting the above requirements. In addition, examination of the literature indicates that multidimensional HPLC techniques
have been applied to a broad range of problems relating to the en-

vironmental, toxicological, pharmaceutical, fuel, and biomedical fields. The applications described in this section have not been grouped by category, but rather on a chronological basis, while Table 1 summarizes the analytes and separation modes used by matrix type.

Erni and Frei (12) utilized size exclusion chromatography (SEC) in an aqueous mode coupled to a reversed-phase column (Nucleosil C-18) to separate senna glucoside extracts. The use of an aqueous buffered mobile phase in the SEC separation allowed transfer of the components of interest into the reversed-phase column without detrimental band broadening effects, and elution of the transfered section was obtained in the second system using a 7-step gradient of varying concentrations of acetonitrile/water.

Johnson et al. (13) coupled SEC in a nonaqueous mode (Micropak TSK gels eluted with tetrahydrofuran) to a gradient LC in the reversed-phase mode using acetonitrile/water gradients to determine malathion in tomato plants, lemonin in grapefruit peel and to study butadiene-acrylonitrile and styrene-butadiene copolymers.

Barnett and co-workers (14) developed a method for the simultaneous determination of vitamin A acetate, vitamin D_2 and vitamin E acetate in multivitamin tablets using a column with a phenyl stationary phase coupled to a C-18 column and eluted with gradients of methanol/water. Detection was by UV and using an internal standard (cholesterol benzoate) relative standard deviation (RSD) of 0.06 to 0.03 were obtained. Analysis time was 50 minutes.

Aqueous SEC coupled to a reversed-phase column was used (7) to determine caffeine and theophylline in plasma and urine, DOPA in urine, sugars in molasses and candy, and B vitamins in food protein supplements. The SEC system consisted of a Micropak TSK-2000 SW column and various mobile phases, such as methanol/water (10:90) with 0.1 M KH_2PO_4 and 0.01 M heptanesulfonic acid. The reversed-phase system consisted of a C-18 column eluted with methanol/water gradients, or isocratically using aqueous buffered mobile phases. An example of the separation of DOPA in urine using this system is presented in Figure 2. Amino glycoside antibiotics in blood serum (15) were determined using a cation exchange column coupled to a C-18 column and a mobile phase consisting of 50 mM EDTA in both systems. Postcolumn derivatization with o-phthaldehyde was used to enable fluoroescence detection, and the method described allowed direct injection of deproteinized serum. Alfredson (16) applied a cyanopropyl bonded column coupled to a highly activated silica gel column for hydrocarbon group-type separations of gasoline, light and heavy oils, and solvent refined coals. Saturates, olefins, aromatics and polars were separated using a microprocessor controlled system and pneumatically actuated valves. Terbutaline in human plasma (8) was determined in the range of 5–50 pmole/mL using a coupled system consisting of a C-8, C-18 and a strongly acidic ion exchanger. Electrochemical detection was used and precision of 2.2% to 3.5% was obtained.

TABLE 1 Selected Applications by Matrix Type

Reference	Component(s) of interest	Sample matrix	Separation systems
FOODSTUFFS			
[12]	Senna glucosides	Plant extract	Aqueous SEC/C-18
[13]	Malathion	Tomato plants	Nonaqueous SEC/C-18
[13]	Lemonin	Grapefruit peel	Nonaqueous SEC/C-18
[7]	Sugars	Molasses, candy	Aqueous SEC/C-18
FUELS			
[16]	Hydrocarbon groups	Gasoline, light/heavy oils solvent ref. coal	Cyanopropyl/silica
[4]	PAHs	Shale oil	Amino/C-18
[32]	Cycloparaffins	Gasoline, kerosene	Cation exchange/PONA
TISSUE			
[19]	Peptides	Rat pituitary and brain	Gel protein
[3]	Metallothionein	Various	Gel filtration/ion exchange
[26]	Salinomycin	Chicken skin and fat	Silica/C-18
[30]	Catecholamines	Brain	Affinity/ion pair
[37]	4-Hydroxy debriso- quinone	Liver	Silica/PRP1/silica
BIOLOGICAL FLUIDS			
[7]	Caffeine	Plasma	Aqueous SEC/C-18

[7]	Theophylidine	Urine	Aqueous SEC/C-18
[7]	DOPA	Urine	Aqueous SEC/C-18
[15]	Antibiotics	Serum	Cation exchange/C-18
[9]	Terbutaline	Plasma	C8/C18/anion exchange
[17]	Metropolol	Plasma	C-2/cyanopropyl/C-18
[18]	Riboxamide	Human and canine plasma	Cation exchange/C-18
[20]	Catecholamines	Plasma, urine	Affinity/C-18
[9]	Caphalosporin	Serum, urine	Anion exchange/C-18
[21]	Prostaglandins	Plasma	Polar amino-cyano/silica
[22]	Glycoproteins	Serum	Affinity/size exclusion
[6]	Amoxillin	Various	C-18/ion pair
[23]	Felodipine	Urine	Silica/C-18
[24]	4-Hydroxyindole-3-acetic acid	Cerebrospinal fluid	Ion exchange/C-18
[25]	Biphenyl metabolites	Plasma, urine	β-cyclodextrin/C-18
[28]	Digoxin	Serum	Aqueous SEC/C-18 or diol
[29]	Cyclosporin A	Whole blood	Cyanopropyl/TMS
[30]	DOPA	plasma	Affinity/ion pair
[33]	Amitriptiline and metabolites	Plasma	C-18/cyanopropyl
[34]	THC metabolite	Urine	Cyanopropyl/C-18
[35]	5-S-cysteiyldopa	Urine	Affinity/C-18
[36]	Leucovorin	Serum	Phenyl/bovine serum albumin

TABLE 1 (Continued)

Reference	Component(s) of interest	Sample matrix	Separation systems
MISCELLANEOUS			
[13]	Additives	Copolymers	Size exclusion/C-18
[14]	Vitamins	Multivitamin tablets	Phenyl/C-18
[31]	Heroin	Opium	Alumina/C-18

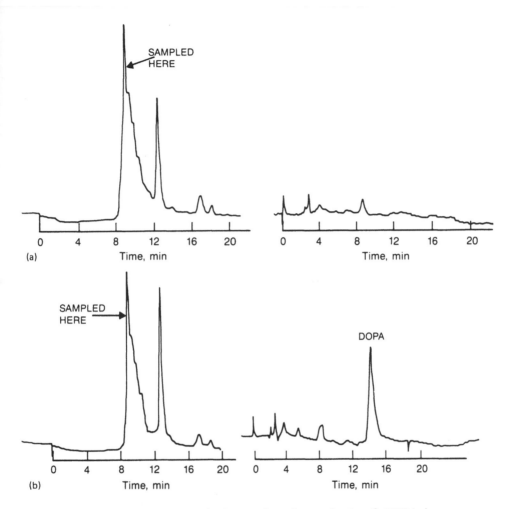

FIGURE 2   Heartcutting techniques for the analysis of DOPA in urine.   (a) Normal urine.   (b) Abnormal urine.   SEC analysis (left) column, MicroPak TSK 2000SW (30 cm × 7.5 mm i.d.); water at 1.0 ml/min; injection volume, 10 μl raw urine; detection at 230 nm; 2.0 a.u.f.s.   RPC analysis (right): column, MicroPak MCH-10 (30 cm × 4 mm i.d.) water with 20 mM camphorsulfonic acid, 100 mM NaH_2PO_4 and 0.1 mM NaEDTA at 1 ml/min; detection, electrochemical detector with glassy carbon electrode operated at 0.720 mV vs. Ag-AgCl; attenuation, 2 nA/V; sample, 100 μl.   (Reproduced with permission from Ref. 7.   Copyright Elsevier Science publishers.)

Metropolol and its α-hydroxy metabolite were determined in human plasma (17) using a system consisting of an RP-2 precolumn coupled to a cyanopropyl analytical column and a mobile phase of acetonitrile/0.22 M acetic acid buffer (75:25). The components of interest were transfered to a RP-8 column using a mobile phase of acetonitrile/0.088 M acetic acid (75:25). Fluorescence detection was used. Polynuclear aromatic hydrocarbons (PAHs) in shale oil fractions were studied (4) by coupling an amino bonded column using methylene chloride/n-pentane (5:95) to a reversed-phase column (Vidac TP201 ODS) using acetonitrile/water (35;65) followed by a linear gradient to pure acetonitrile. The coupling of the normal phase system to the reversed-phase system was accomplished via a concentrator column having affinity for the components of interest, and after collection of the appropriate fraction, nitrogen and heat were used to remove the normal phase eluent. Following a cooling period, the concentrator column was switched to the reversed-phase system and the aqueous eluent was used to transfer the components to the analytical column for further resolution. A diagram of the multidimensional system used is presented in Figure 3, and the resulting chromatograms obtained on the shale oil sample are presented in Figure 4.

Riley et al. (18) developed a coupled column system for the determination of riboxamide in human and canine plasma. A solvent generated anion exchanger on silica gel was used for the preliminary separation step and a solvent generated ion exchanger on an ODS support for the final separation step. Detection limits of 40 ng/mL were reported. The applicability of the method was demonstrated by following plasma levels of riboxamide after intravenous administration in the dog.

The characterization of β-endorphin related peptides in rat pituitary and brain (19) was accomplished using a series of gel protein analysis columns (2 × I125 and 2 × I60) and a mobile phase of 6 M guanidine-HCl/0.2M triethylamine phosphate coupled to a reversed phased column using 0.13% heptafluorobutyric acid/35% acetonitrile with a gradient to 40% acetonitrile. The procedure demonstrated tissue specific differences in the processing of β-endorphin in pituitary and brain regions and suggested the general applicability to the analysis of neuropeptide heterogeneity.

Edlund and Westerlund (20) utilized the multidimensional approach for the determination of adrenaline, noradrenaline and dopamine in plasma and urine. The catecholamines were selectively absorbed on a boronic acid gel column using a citrate buffer containing 2 mM decylsulfate and 0.03 mM EDTA during conditioning, and 25% methanol during elution. Compounds of interest were eluted from the main column using a phosphate (pH 6.65), citrate (pH 6.65), methanol mobile phase (37.5:37.5:5.25) containing 2 mM decylsulfate and 0.03 mM EDTA. Detection of dopamine was by coulometry while the other catecholamines were detected fluorometrically after postcolumn reaction as the trihy-

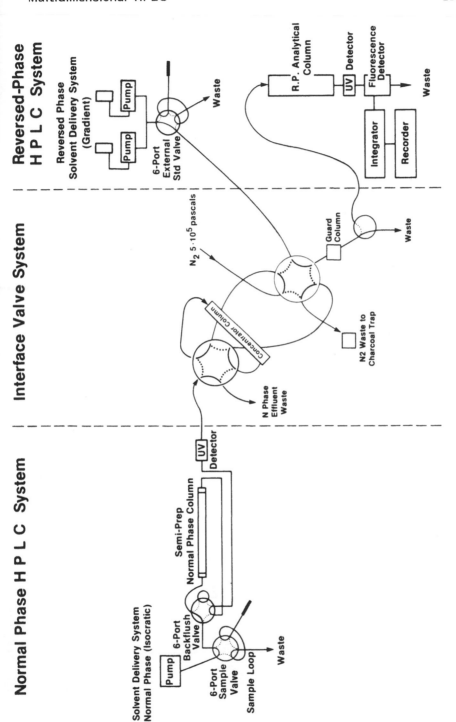

FIGURE 3 Schematic of on-line multidimensional system and interface valve system using standard HPLC six-port valves. (Reproduced with permission from Ref. 4. Copyright American Chemical Society.)

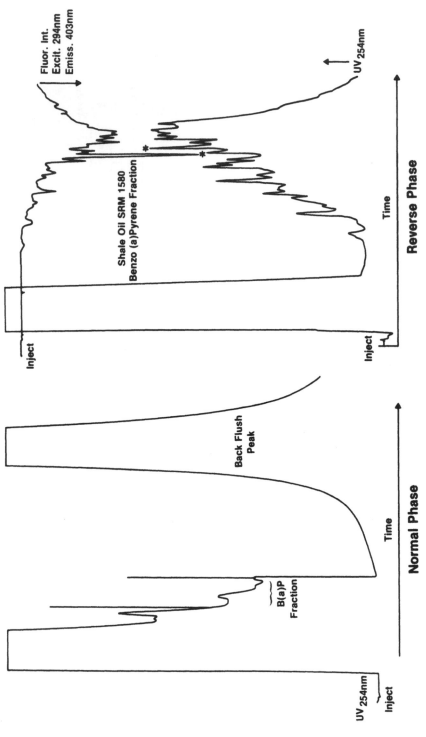

FIGURE 4 (Left) Normal-phase chromatogram of SRM shale oil. Conditions: 5% Ch_2Cl_2/m-pentane at 5mL/min on µBondapak NH_2 semi-pre column. (Right) Reversed-phase chromatogram of benzo[a]pyrene fraction after using on-line multidimensional system. Conditions: 35% CH_3CN/H_2O for 10 min followed by linear gradient to 100% CH_3CN over 25 min. Column: 5 µm Vidac TP201 ODS. (Reproduced with permission from Ref. 4. Copyright American Chemical Society.)

droxy indoles. Detection limits obtained range between 0.04 and
1.6 pmol. RSD reported were 2—4%.

Cephalosporin in human serum and urine (9) was determined
using an anion exchange column coupled to a reversed-phase column.
Detection limits of 0.05 µg/mL in serum and 0.5 µg/mL in urine were
reported using UV detection.

Suzuki et al. (3) applied gel filtration coupled to ion exchange
using a Tris-HCl buffer eluent for the determination of metallothio-
nein in tissue.

A method for the determination of E prostaglandins to be used
for the analysis of human plasma samples (21) was developed for the
panacyl ester derivatives using a system consisting of a combination
of a polar amino-cyano precolumn coupled to a cyanopropyl and a
silica analytical column with fluorescence detection. Mobile phases
used were hexane/methylene chloride/isopropanol (70:30:1) and
hexane/methylene chloride/tetrahydrofuran/isopropanol (60:20:20:1).
Quantitation limits were estimated as 10 pg injected.

Human serum glycoproteins (22) were separated according to
lectin affinity and molecular size using a coupled system consisting
of lectin phyto hemagglutinin on porous silica as the adsorbent in
the affinity separation and a TSK-G-3000 SW size exclusion column.
Serum samples were resolved into peaks representing glycoproteins
exhibiting differing molecular sizes but with carbohydrate contents
compatible with the specificity of the affinity system.

Amoxillin, a polar amino penicillin, was determined in biological
fluids (6) using a reversed phase (C-18) column eluted with methanol/
phosphate buffer (pH 7.4) (10:90), containing hexyl sulfate. The
component of interest was transfered to a second C-18 column using
methanol/phosphate buffer (30:70) containing 0.003 M tetrahexyl ammonium hydrogen sulphate, and retained as the ion pair. Postcolumn
derivatization with fluorescamine allowed fluorescence detection which
yielded detection limits of 10 ng/mL in plasma and 25 ng/mL in urine.
Precision obtained was in the range of 0.5%.

Wiedolf (23) determined urinary metabolites of C-14 labeled felo-
dipine using an underivatized silica stationary phase and an eluent
consisting of 0.01 M tetrapropyl ammonium in 5% methanol/0.05 M
phosphate buffer (pH 5.0) to separate the metabolites into carbox-
ylic acid and hydroxylated analogs. Each group was transfered to
an octyl bonded column and separated using methanol gradient.
Compounds were detected using an on-line radioactivity detector.
The method was applied to urine samples collected from rats after
oral dosing of the parent compound.

Edlund (24) utilized a coupled system to determine homovanillic
acid and 5-hydroxyindole-3-acetic acid in plasma, urine and cerebro-
spinal fluids. Acidic compounds were isolated by ion exchange and
separated on a reversed-phase column. Components of interest were
detected by coulometry and amperometry using two working electrodes

in series.   Detection limits in plasma of 4–5 mM were reported.   RSD obtained was 2–4%.

The hydroxylated metabolites of biphenyl following in vitro incu-bation with hepatic 9000 g supernatant (25) were studied using a system consisting of a β-cyclodextrin bonded phase to a reversed-phase column with fluorescence detection.   The method was used successfully on samples from rats and mice treated with Araclor, β-naphthoflavone or phenobarbitol and in monkeys dosed with Araclor,

The analyis of chicken skin/fat for salinomycin was accomplished on a coupled system (26) consisting of a silica gel column and a C-18 column.   Samples were extracted with methanol and partitioned with carbon tetrachloride prior to analysis.   The purified solutions were derivatized with pyridinium dichromate and detected using UV, which yielded detection limits in the order of 100 ppb.

Lang and Rizzi (27) separated nucleosides, nucleotides and purine bases using gradients to provide separation of bases and nucleotides on a C-18 column, and nucleosides on an anion exchange system.   The method was used for the determination of adenosine metabolites in human erythrocytes.

A diagram of the column switching arrangement and valve program used is presented in Figure 5A.   Chromatograms of a synthetic mixture containing the components of interest, and of a real sample are pre-sented in figures 5B and 5C.

Serum samples from human patients receiving digoxin therapy (28) were analyzed using an aqueous size exclusion system coupled to a C-18 or a diol bonded column.   0.1 M phosphate buffer at pH 6.6 was used as the mobile phase with amperometric detection of NADH, the digoxin homogeneous enzyme immunoassay product.   Correlation with immunoassay results in the ranges of 0.6 to 4.1 ng/mL was ob-tained.

A multidimensional system utilizing a cyanopropyl stationary phase coupled to a trimethylsilicone bonded column was used for the determi-nation of cyclosporin A in whole blood (29).   UV detection yielded minimum detectable concentrations of 5 ng/mL.

The determination of adrenaline, noradrenaline, DOPA in brain tissue, and DOPAC in plasma (30) was accomplished by injection of brain homogenates or plasma into a coupled system consisting of a boronic acid gel and two reversed-phase columns.   The catecholamines and DOPA were selectively absorbed on the boronic acid gel column and separated by ion-pair chromatography on the reversed-phase columns.   Using coulometry and amperometry limits of detection ob-tained were 0.05 nmol/g in brain tissue and 2 nmol/L in plasma for DOPA and DOPAC respectively.

Billet et al. (31) analyzed heroin and opium samples on a coupled alumina/reversed-phase system using 0.01 M tetramethyl ammonium hydroxide-citric acid buffer (pH 6.0) and acetonitrile/methanol/buffer (17:28:55).   Both switching and the use of the coupled system in series were described. Peak purity was confirmed using photodiode array detection.

The volume fraction of total cycloparaffins in gasolines and kerosines with distillation end points >300°C (32) was accurately quantified using a series of columns comprising an unsaturated selective column (strong cation exchanger in the silver form bonded to silica), 2 naphthenic selective columns (Micropak PONA), Freon 123 as the mobile phase, and a dielectric constant detector. The absolute error reported for each structural group type was within 2% on the basis of hydrocarbon standards, while limits of detection for naphthenes was found to be 4.0 vol. %.

Amitryptiline and its metabolites, nortriptyline, 10-hydroxynortryptiline and 10-hydroxyaminotryptiline (33) were determined in plasma using a coupled system consisting of a reversed phase concentrating column and a cyanopropyl bonded column. An acetonitrile/acetate buffer (60:40) mobile phase was used and detection limits for direct plasma injections were 5-10 ng/mL for each of the four drugs. The procedure was applied to samples from patients receiving oral doses of amitryptyline.

Karlsson (34) developed a procedure consisting of a cyanopropyl column coupled to a C-8 column for the determination of Δ^9-tetrahydrocannabinol-11-oic acid in hydrolyzed urine from humans. Dual detection consisting of UV and electrochemical yielded detection limits on the order of 5 ng/mL, and precision obtained was 2.8% at 85 ng/mL and 13.4% at 6 ng/mL.

5-S-cysteinyldopa in urine (35) was separated using a column of 3-aminobenzene boronic acid covalently bonded to silica coupled to a reversed-phase column. Electrochemical detection was used, and RSD of 6.4% at the μmol/L range was obtained. The eluent conditions used as well as representative chromatograms obtained are presented in Figure 6.

Wainer and Stiffin (36) determined diastereoisomers of leucovorin and 5-methyltetrahydrofolate in blood serum by coupling a phenyl bonded column used to isolate fractions of interest from interfering serum peaks to a bovine serum albumin chiral stationary phase for the determination of stereoisomeric composition. Mobile phases used consisted of a gradient of 0.25 M phosphate buffer (pH 5.0) to methanol/buffer (50:50) in the preliminary separation step and the buffer solution in the chiral separation. The system was used for the determination of stereoisomers in human subjects after administration of racemic leucovorin. Chromatograms obtained are presented in Figures 7 and 8.

Mellstroem (37) used a system consisting of two silica gel precolumns, two polymer based (PRP-1) concentrating columns and two silica analytical columns for the determination of 4-hydroxydebrisoquine in human liver microsomal incubates. The mobile phase used consisted of acetonitrile/5 mM sodium octanesulfonate in phosphate

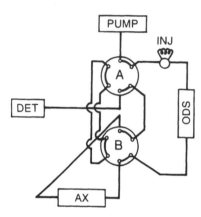
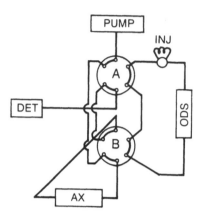

A and B in position LOAD   A and B in position INJECT

(a)

FIGURE 5 (a) The column switching arrangement. AX = Anion-exchange column. (b) Two column HPLC with ODS column followed by anion-exchange column. Chromatogram of a synthetic mixture containing purine bases, nucleosides and nucleotides. Peaks: 1 = hypoxanthine; 2 = xanthine; 3 = inosine; 4 = guanosine; 5 = 2'-deoxyinosine; 6 = 2'-deoxyguanosine; 7 = adenosine; 8 = 2'-deoxyadenosine; 9 = uric acid; 10 = IMP; 11 = AMP; 12 = guanosine 5-phosphate; 13 = xanthosine 5'-phosphate; 14 = inosine 5'-phosphate; 15 = adenosine 5'-diphosphate; 16 = guanosine 5'-diphosphate; 17 = xanthosine 5'-diphosphate; 18 = inosine 5'-triphosphate; 19 = ATP; 20 = GTP; 21 = XTP. A small eluent fraction of about 2.5 ml was switched from the first to the second column. The most important switching events are indicated by "SW" arrows. Temperature 30°C. (c) Two column HPLC of ultracentrifuged lyzed human erythrocytes grown in a medium spiked with adenosine (final concentration 100 μM) and incubated there for 5 min. Chromatographic conditions and symbols as in (b). (Reproduced with permission from Ref. 27. Copyright Elsevier Science Publishers.)

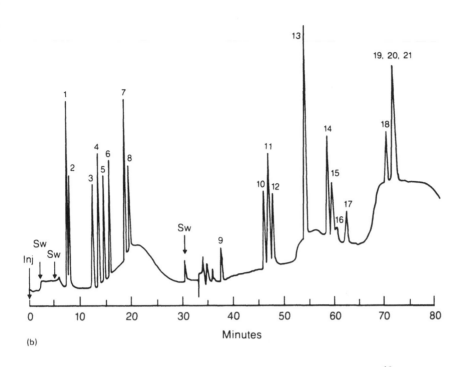

(b)

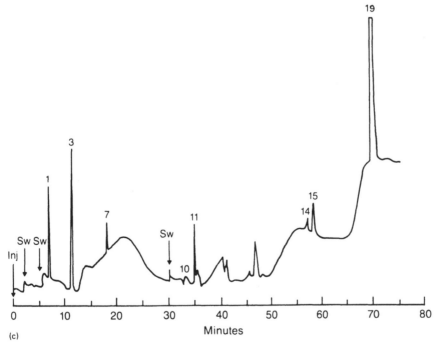

(c)

FIGURE 5 (Continued)

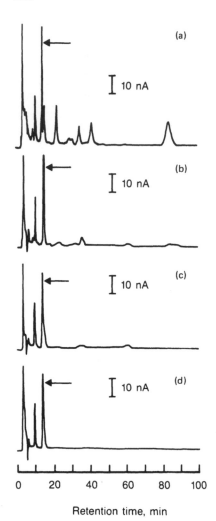

FIGURE 6. Urine chromatograms obtained with the coupled column system. The arrow indicates the 5-S-cysteinyldopa peak. The boronate column was loaded with 100 μl of urine and then washed with a 0.1 M phosphate buffer containing 0.2 mM disodium EDTA at a flowrate of 2 ml/min. Buffer pH and washing times were: (a) pH 7.0 for 5 min. (b) pH 6.0 for 5 min. (c) pH 6.0 for 10 min. (d) pH 6.0 with 10% methanol for 10 min. Chromatography on the analytical column was performed with 0.1 M formate containing 0.2 mM disodium EDTA (pH 3.0) at a flowrate of 1 ml/min. (Reproduced with permission from Ref. 35. Copyright Elsevier Science Publishers.)

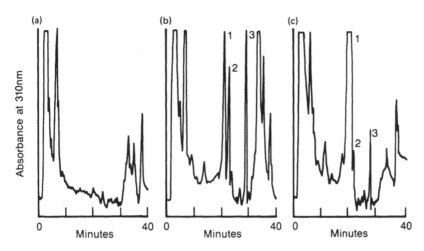

FIGURE 7 Representative chromatograms on the achiral phenyl support of extracted blank serum (a), serum with added leucovorin (1250 ng/ml), 5-methyltetrahydrofolate (1250 ng/ml) and internal standard (300 ng/ml) (b), serum sample from a patient after 15 min postinfusion (c). Peaks: 1=leucovorin; 2=5-methyltetrahydrofolate; 3=internal standard.

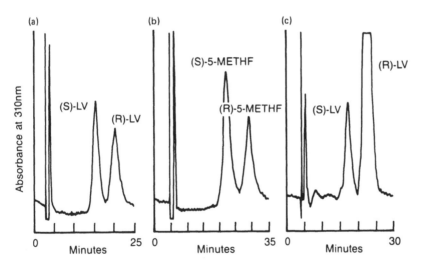

FIGURE 8 Representative chromatograms of the BSA-CSP of a leucovorin serum standard (419 ng/ml per isomer) after switching from the achiral system (a), a 5-methyltetrahydrofolate serum standard (419) ng/ml per isomer) after switching from the achiral system (b), and leucovorin from a patient serum sample acquired 4.5 h postinfusion (c). (Reproduced with permission from Ref. 36. Copyright Elsevier Science Publishers.)

247

buffer of pH 2.7 (12:88). Detection limits of 200 µM were obtained using UV detection.

V. CONCLUSIONS

Multidimensional HPLC offers very high separating power to those who make use of it. The challenges in applying the technique are in efficiently coupling the columns for transfer of analytes and maintaining mobile phase/column compatibility. Applications in which liquid phases on both columns are miscible and compatible are frequently reported. Perhaps the greatest area for development is in the coupling of columns with incompatible mobile phases. It is clear that this poses a challenge, based on the relatively few examples available in the literature. The use of off-line multidimensional techniques (conventional sample cleanup) with incompatible mobile phases is common in the literature, and replacing these procedures with automated on-line multidimensional separations will require continued work to develop, improve and simplify sample trapping and transfer techniques. Additional efforts to improve sensitivity will probably involve the use of relatively large diameter fractionating columns followed by small diameter (microbore (1-2 mm), and packed capillary (< 1 mm)) analytical columns. While some work is being done in this area, there is clearly room for additional effort here, with analyte trapping and transfer representing part of the challenge. Applying on-line multidimensional HPLC to routine analyses requires a better understanding of both instrumentation and the separation processes than do conventional off-line techniques. The advantage that offsets this requirement is the promise of greater efficiency in preparing and analyzing samples of high degree of complexity. As more practicing chromatographers recognize the value of said efficiency improvement, applications of on-line multidimensional HPLC will appear with greater frequency.

REFERENCES

1. H. Omori, K. Okabe, T. Nakashizuka, and S. Yamasaki. HRC & CC 9: 477 (1986).
2. J. C. Giddings. Anal. Chem. 53: 945A (1981).
3. K. T. Suzuki, H. Sunaga, and T. Yajima. J. Chromatogr. 303: 131 (1984).
4. W. J. Sonnefeld, W. H. Zoller, W. E. May and S. A., Wise. Anal. Chem. 54: 723 (1982).
5. Valco Instruments, Houston, Texas, and Rheodyne Inc., Cotati, California.

6. J. Carlqvist and D. Westerlund. J. Chromatog. 344:285 (1985).
7. J. A. Apfel. T. V. Alfredson, and R. E. Majors. J. Chromatogr. 206: 43 (1981).
8. L. E. Edholm, B. M. Kennedy, and S. Bergquist. Chromatographia 16: 341 (1982).
9. Y. Tokuma, Y. Shiozaki, and H. Noguchi. J. Chromatogr. 311: 339 (1984).
10. L. D. Rothman, Unpublished work, 1984.
11. M. Verzele, C. Dewaele, and M. De Weert. LC-GC, 6: 966 (1988).
12. F. Erni and R. W. Frei. J. Chromatogr. 149: 501 (1978).
13. E. L. Johnson, R. Gloor, and R. E., Majors. J. Chromatogr. 149: 571 (1978).
14. S. A. Barnett, and L. W. Frick. Anal. Chem. 51: 641 (1979).
15. G. L. Smith and W. Slavin. Chromatogr. Newsl. 9:21 (1981).
16. T. V. Alfredson. J. Chromatogr. 218: 715 (1981).
17. J. B. Lecallion, C. Souppart, and F. Abadie. Chromatographia 16: 158 (1982).
18. C. M. Rilesy, L. A. Sternson, and A. J. Repta. J. Chromatogr. 276: 93 (1983)
19. M. Dennis, C. Lazure, N. G. Seidah, and M. Cretien. J. Chromatogr. 266: 163 (1983).
20. P. O. Edlund and D. Westerlund. J. Pharm. Biomed. Anal. 2: 315 (1984).
21. J. W. Cox and R. H. Pullen. Anal. Chem. 56: 1866 (1984).
22. C. A. Borrebaeck, J. Soares, and B. Mattiasson. J. Chromatogr. 284: 187 (1984).
23. L. Weidolf. J. Chromatogr. 343: 85 (1985).
24. P. O. Edlund. J. Pharm. Biomed. Anal. 4: 641 (1986).
25. D. E. Weaver and R. B. Van Lier. Anal. Biochem. 154: 590 (1986).
26. G. P. Dimenna, J. A. Creegan, L. B. Turnbull, and G. J. Wright. J. Agric. Food Chem. 34: 805 (1986).
27. H. R. Lang and A. Rizzi. J. Chromatogr. 356: 115 (1986).
28. D. Wright, B. H. Halsall, and W. R. Heineman. Anal. Chem. 58: 2995 (1986).
29. H. Hosotsubo, J. Takezawa, N. Taenaka, K. Hosotsubo, and I. Yoshiya. J. Chromatogr. 383: 349 (1986).
30. P. O. Edlund. J. Pharm. Biomed. Anal. 4: 625 (1986).
31. H. Billet, R. Wolters, L. DeGalan, and H. Huitzer. J. Chromatogr. 368: 351 (1986).
32. P. C. Hayes Jr. and S. D. Anderson. J. Chromatog. 387: 333 (1987).
33. D. Dadgar and A. Power. J. Chromatog. 416: 99 (1987).
34. L. Karlsson. J. Chromatogr. 417: 309 (1987).

35. C. Hansson, B. Kaagedal, and M. Kaellberg. J. Chromatogr.
     420: 146 (1987).
36. I. W. Wainer and R. M. Stiffin. J. Chromatogr. 424: 158 (1988).
37. B. Mellstroem. J. Chromatogr. 424: 435 (1988).

7

Multidimensional Chromatography Using On-Line Coupled High-Performance Liquid Chromatography and Capillary-Gas Chromatography

HERNAN J. CORTES *The Dow Chemical Company, Midland, Michigan*

I. INTRODUCTION

The analysis of complex matrices, such as those encountered in environmental, medical, agricultural, fossil fuel and biotechnology areas has become a very important area of separation science, motivating advances in theory and technique to the point where modern high resolution chromatography operates at levels which are close to the theoretical limits.

By utilizing specific detectors such as a mass spectrometer for gas chromatography (GC) or an electrochemical detector in liquid chromatography (LC), the selectivity and specificity of a chromatographic system can be increased in order to improve resolution or to increase detection limits.

The increases in efficiency of a chromatographic system, brought about by the use of open tubular columns in gas chromatography (1) as well as smaller particle packings (2) packed capillary columns (3–5) and open tubular systems (6–7) for LC have further increased the resolving power necessary to study such complex matrices.

Challenges still remain, however, with the continuing need to increase limits of detection and increase the information content of an analysis, since even with state of the art technology, there are limitations to the resolving power attainable with a single chromatographic system (8).

An estimate of the separating power of a chromatographic system can be obtained by considering the maximum number of components that can be placed into the available separation space with a given resolution that satisfies the analytical goals (9). This value, termed the peak capacity of the system (10), is in the range of 100 to 300 for modern high resolution chromatographic systems, which would

appear adequate for separations in which the number of components
in a mixture is less than the peak capacity of the system; however,
components in complex samples are seldom evenly separated and appear randomly, in most cases overlapping each other. Utilizing the
model of statistical overlap developed by Giddings and Davis (11),
the seriousness of said component overlap becomes apparent when one
estimates the number of visible components in a chromatogram, (assum-
ing that the total number of components in a mixture can be estimated)
yielding a value of 37% of the system's peak capacity.

The reader is referred to Chapter 1 for a more rigorous examina-
of the mathematical treatment and theoretical advantages of multidimensional separations.

Various strategies can be used to overcome the problem described,
for example, overdesigning the system to generate peak capacities far
in excess of the number of components in the sample. Although this
is a common approach, as the sample complexity increases so does the
analysis time and the technical difficulty of the experiment. Another
approach used to obtain greater resolution is the technique of multidimensional chromatography.

In general, multidimensional systems are designed to separate
selected components in a primary system that are then transfered in
some manner to a secondary separating system where further resolution is attainable. Perhaps the most common use of multidimensional
separations is the pretreatment of a complex matrix in an off-line
mode.

Typically, a sample is introduced into a separating system, eluting
fractions are collected, either manually or via a fraction collector,
and are concentrated, derivatized or evaporated to remove incompatible
solvents, prior to introduction into a second separating system. This
type of sample treatment is widely used due to its simplicity; however,
it is recognized that off-line techniques can be time consuming, operator intensive (expensive), and difficult to automate and reproduce,
although the latter problems have been somewhat alleviated with the
use of robotics. Another problem of off-line technology is that the
probability of affecting the integrity of a sample detrimentally, increases with the number of manipulations the sample is subjected to,
particularly when operating at trace levels.

The coupling of two or more chromatographic systems in an on-
line mode (that is, the introduction of selected fractions eluting from
one separating system into a second one by means of valves or pressure controls) offers the advantages of ease of automation and greater
reproducibility usually in a shortened analysis time. Also, the loss
of sample material during preparation and analysis is significantly reduced, due to the decreased sample handling steps.

A comparison of the practical advantages and constraints of a
single-column separation system and a multidimensional system points

out the strengths and weaknesses of each mode, and can serve as a guideline to select the appropriate technique depending on the analysis needs. The main source of separation on a single-column system depends predominantly on the efficiency of the system, and to some extent on the selectivity. A multidimensional system generates its resolving power mainly from the selectivity differences of the separating modes used. Therefore, the resolution of a single column system can be limited by the peak capacity of the system while in a multidimensional system selectivity differences can be fully exploited.

The ability to quantitate components of interest in a single column system is excellent if said components are fully resolved from interferences; for a multidimensional system, quantitation is highly dependent on the technique used to transfer components from one separating system to another. In terms of quantitation, transfer of only fractions of a peak are not advantageous; the ability to quantitate components in a multidimensional system is superior to a linear system when adequate resolution is not attainable on a single column, and the total peak(s) of interest can be transferred reproducibly from one separation system to another. For qualitative analysis, where the complete peak volume may not need to be introduced, multidimensional systems offer more than one set of independent retention data, yielding more information and therefore can be superior to single column systems. In terms of cost, operator expertise necessary, and availability, single column systems are prefered, since multidimensional systems can be expensive, are limited by the few commercial sources and are more complicated to set up and use.

With these consideration in mind, the major practical advantages of a single column system in summary are: relative simplicity, low cost, wide availability, and ease of operation; the disadvantages can be the limited resolving power and information attainable.

For a multidimensional system, the major advantages are the greater resolution obtained when selectivity differences between the separation modes are exploited fully, the availability of more than one set of retention data, and the high probability that the conditions chosen for the primary separation step will be applicable to a wide range of different matrices, translating into less method development. The disadvantages are that multidimensional systems are more difficult to operate, the initial setup can be time consuming and expensive, interfaces are relatively complicated, method development is less straightforward as two or more separations need to be developed, and few commercial sources exist.

Examples of on-line coupling of chromatographic systems in the gas phase and in the liquid phase have existed for some time, and are reviewed in other chapters of this volume. A brief review of some of the main practical considerations of the more widely used multidimensional systems can also serve as a guideline for selection of the appropriate multidimensional technique.

Multidimensional gas chromatography is perhaps the most widely used system (other than thin-layer methodology) due to the mobile phase compatibility between the primary and secondary separating systems, which allows relatively simple coupling with less complicated interfaces. In addition, the wide availability of sensitive and selective detectors, commercial instruments or add-on accessories and conversion kits to carry out switching operations, makes multidimensional GC very attractive. When using columns of capillary dimensions, very high peak capacities are obtained. It is necessary, however, that the components of interest be sufficiently volatile to be transported in the gas phase (although derivatization techniques alleviate some of these problems), and that the samples be relatively clean so as not to deteriorate the performance of the primary column by contamination with nonvolatile or highly polar components. Also, multidimensional GC lacks the selectivity variation dependence on mobile phase, and is somewhat limited by the selectivity differences which are obtained using common stationary phases. For example, separation of components using stationary phases with very different characteristics, such as methylsilicone and cyanopropyl are still highly correlated by boiling point.

Multidimensional liquid chromatography offers other advantages, as the mobile phase polarity can be adjusted in order to obtain adequate resolution, and a wide range of selectivity differences can be employed when using the various available separation modes, such as adsorption, partition, size exclusion, ion exchange and affinity. Highly polar and nonvolatile compounds are routinely separated using any of the above modes, while relatively complex samples can be introduced without severe deterioration of the performance of the system. Total peak capacities however, tend to be lower, detection systems are generally not as sensitive or universal, and mobile phase incompatibilities can limit the applicability of multidimensional LC.

The coupling of a liquid chromatograph to a gas chromatograph in an on-line mode offers a different perspective of multidimensional separations, since selectivities that are difficult to obtain in multidimensional systems using either gas or liquid phases alone, are possible using the wide range of variables available, such as mobile and stationary phases, temperature profile combinations, and detector systems of the two techniques. A system of this type combines the selectivity of liquid chromatography with the efficiency and sensitivity of gas chromatography, yielding relatively high peak capacities. It should be pointed out, however, that matching the two technologies is not a trivial matter, since the two separation techniques operate in phases which are in two different physical states. Ways to make the transfered fraction from the LC system compatible with the gas chromatograph have only recently begun to be explored. The basic approaches used are: to introduce a sufficiently small volume of the peak of interest from the LC so that the components of interest are not distorted by the large volumes of solvent; to develop introduction techniques which allow large volumes of effluent to be introduced into the GC, such as the use of uncoated inlets; to reduce the LC column diameter in order to keep the components of interest diluted in a smaller volume, and to combine this approach with the use of uncoated inlets.

The number of applications of multidimensional LC-GC is experiencing rapid growth, as evidenced by the growing number of publications appearing in the literature.

This chapter will review the development of the on-line coupling of a liquid chromatograph with a gas chromatograph, and describe the approaches used to introduce LC effluents into capillary GC. In addition, we will summarize the applications of this powerful technology.

II. TRANSFER TECHNIQUES

The techniques used for introduction of effluents from a liquid chromatograph into a capillary gas chromatograph can be divided into three major categories: the use of autosamplers or autoinjectors (direct introduction), sampling valves without external loops which allow introduction of the LC effluent to take place directly by the LC pump (eluent feed), and sampling valves with external loops of sufficient volume to contain the complete volume of the component(s) of interest (gas transfer interface). In keeping with the historical terminology, the following sections will include this nomenclature. However, a concept which we call Stop-Flow Introduction is additionally introduced in order to clarify some of the confusion which exists concerning interface classification in the field at the present time.

A. Autosamplers

The first reported on-line coupling of liquid chromatography and gas chromatography was by Majors et al. (12,13) in which a conventional LC was connected to a GC autosampler. In this application, the effluent from a liquid chromatographic detector is directed to a flow-through syringe and goes to a waste container, the syringe being continuously flushed with liquid chromatographic solvent. A signal from the electronics module initiates the injection cycle; the syringe is lifted from the waste container and lowered into the GC injector port where the syringe plunger is depressed to make the injection. The connection between the LC detector and the side arm syringe is made via a 0.009 in. i.d. Teflon[R] tube. (Figure 1). The described system was used successfully for the determination of atrazine in sorghum fodder, as illustrated in Figure 2.

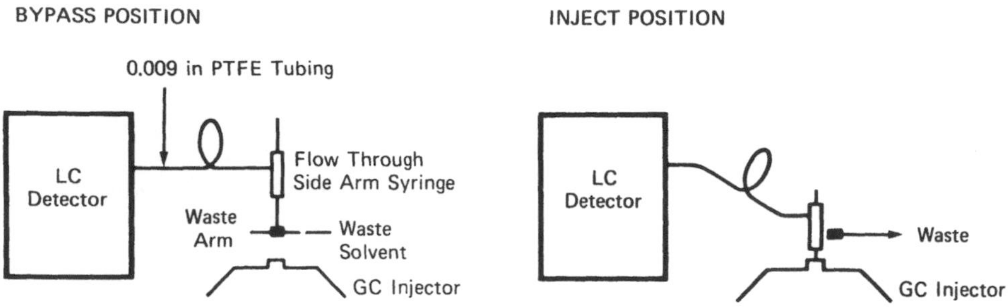

BYPASS POSITION   INJECT POSITION

0.009 in PTFE Tubing

FIGURE 1   Schematic diagram of the interface for an automated on-
line LC-GC multidimensional system.   (Reproduced with permission
from Ref. 13.   Copyright Preston Publications, Inc.)

Polynuclear aromatics in solvent refined coal and shale oil (14)
were determined by separation according to ring size on a cyanopropyl
column eluted with hexane/chloroform (95:5).   Sections of the separated fractions were transferred to a 30 m SE-54 capillary column
using the interface described.

The analysis of plant extracts for the study of terpenoids as
sources of energy (15) was accomplished by normal phase separation
(silica column eluted with hexane/isopropanol (98:2)) and introduction
of the terpenoid fraction into the GC using a split injection.

A similar approach was used for the fractionation of gasoline and
diesel fuel (16), where saturates, olefins, and aromatics were separated from each other by LC and the peak maximum of each family of
compounds was introduced into a capillary GC.

Utilizing a system of switching valves in a modification of the
above technique (17), coal liquids were separated by size exclusion
chromatography and transfered to a conventional GC injector in the
split mode.   In this manner, fractions containing alkanes, aromatics
and alkylated phenols were examined qualitatively.

Folpet in hop samples (18) was determined by extraction of 20
grams of sample with 250 ml of acetone; a 2 ml aliquot was then filtered and diluted with 2 ml of water and 1 ml of brine.   The resulting
solution was passed through a solid phase extraction column and partitioned into hexane.   This solution was concentrated to 2 ml and injected into a liquid chromatographic system consisting of a diol followed
by a cyanopropyl column, eluted with hexane on the first and hexane/
ethanol on the second.   The section of interest was transferred to the
GC and separated on a 15 meter, 0.53 mm methyl silicone column.
A typical chromatogram is presented in Figure 3.

An eluent splitter constructed of a bundle of capillaries (isotachic
splitter) was incorporated between the outlet of the LC system and

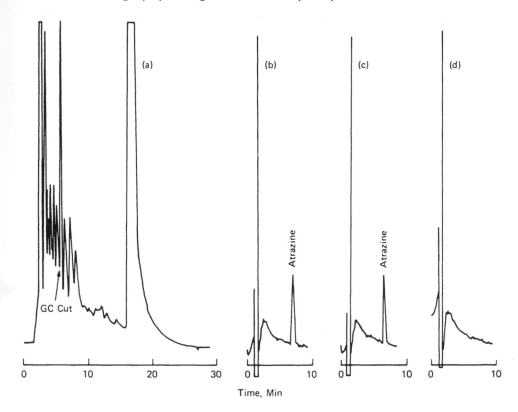

FIGURE 2 Chromatograms of atrazine in sorghum: (a) HPLC chromatogram of sorghum extract containing atrazine (15 cm × 4 mm column containing cyano bonded phase, 2 cm^3 min^{-1} flowrate, 2% isopropanol
in hexane mobile phase, column was flushed for 1 min with 100% isopropanol at 18 min mark to remove strongly retained components);
(b) GC chromatogram of HPLC-GC cut from (a) (25 m OV-101 glass
capillary column, 8 μL splitless injection, 200°C isothermal column temperature, nitrogen selective thermionic detector; (c) GC chromatogram
of atrazine standard carried through entire HPLC-GC procedure (0.2
ppm atrazine, conditions same as for (b); and (d) GC chromatogram
of sorghum control carried through entire HPLC-GC procedure (conditions same as for (b). (Reproduced with permission from Ref. 13.
Copyright Preston Publications, Inc.)

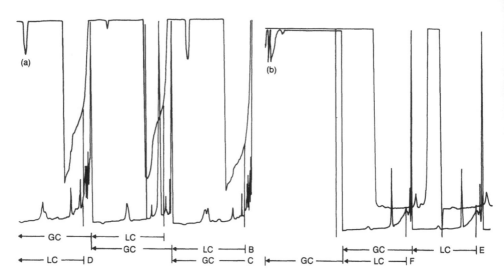

FIGURE 3 (a) Gas chromatogram of hop extract after on-line HPLC clean up. Sample injected into the LC column: 250 μl. Sample transfered to the GC column: 2 μl. Top: UV trace (LC). Bottom: ECD trace (GC). A = Hop sample spiked at 0.1 mg/kg. B = standard LC injection, 2 ng folpet; C, D = previous sample injection for GC and next sample injection for LC. The injection of the next sample for LC was made about 2 min after the GC injection. The UV trace after the second LC column is off-scale. (b) Gas chromatogram of a hop extract without LC column switching (sample directly injected into the second LC column). E = standard injection, 10 ng folpet; F = hop sample, spiked at 0.5 mg/kg; no detection possible. (Reproduced with permission from Ref. 18. Copyright Elsevier Scientific Publishers.)

the autoinjector interface (19), the effluent volume being reduced while providing a representative sample to the GC. The system was used to analyze solvent refined coal samples separated on a 25 cm × 1.0 mm i.d. ODS column using a methanol/water (80:20) mobile phase; the sections of interest were transfered to a 16 m capillary column coated with SE-30 and detected using an FID. The system was later coupled on-line to a mass spectrometer for determination of polycyclic aromatics (20,21). A representation of the multichannel isotachic splitter is included in Figure 4.

Although the use of the autoinjector interface for introduction of effluent from a liquid chromatograph to a gas chromatograph can be applied in some cases, especially when qualitative information is sufficient, the major limitation still remains: the liquid volume that can

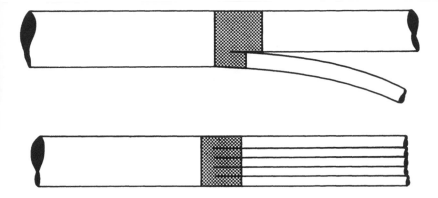

FIGURE 4   Typical two channel splitter design is shown on top.
A chromatographic zone is represented by the shaded area.   Note the
change in linear velocity of the zone as it enters the two split legs.
A multichannel isotachic splitter design is shown on the bottom.
(Reproduced with permission from Ref. 19.   Copyright Elsevier Scien-
tific Publishers.)

be introduced directly into a capillary GC column, an obvious incompatibility when using conventional LC columns that are operated at
flow rates of 1–3 ml/min, resulting in peak volumes of hundreds of
microliters.   Due to this limitation, only a small fraction of a selected
peak can be sampled for introduction into the GC column, and the injection made represents only a portion of the LC peak, making quantitative analysis difficult.

The use of multichannel isotachic splitters allows a representative
fraction to be introduced into the GC and can in principle be used
for quantitative analysis, however, the sensitivity of the analysis can
be limited, as the major portion of the component of interest is bypassed to waste.

B.   Uncoated Inlets

The introduction of relatively large volumes of sample into a capillary GC column has been studied for some time.   For this technique
to be successful, various criteria must be met: no overloading (solvent
or concentration) should take place, and the sample components of
interest should not be distorted or broadened.

When the solutes of interest are not effectively focused at the head
of the capillary GC column, a split-peak pattern is observed, which is
created by the larger amount of condensed solvent spreading out the
solute at the column inlet.   The described broadening and peak splitting effect is illustrated in Figure 5.

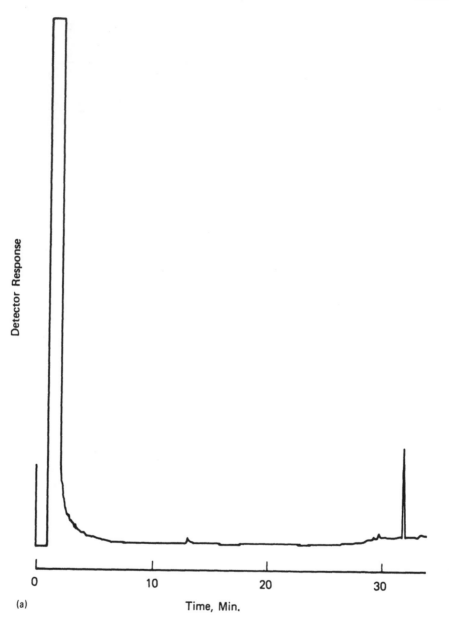

(a)

Detector Response

Time, Min.

0   10   20   30

FIGURE 5 Effect of injection volume on peak shape for on-column injection. (a) 1.0 μl of 10 μg/mL 3,5-dichlorobiphenyl. Conditions:
column, 30 m × .25 mm i.d. Supelcowax 10. 0.25-μm film; program,
115°C for 7 min, 5°C/min to 240°C; carrier, helium at 68 cm/sec;
detector, flame ionization; attenuation, ×4. (b) 10.0 μl injection.
Attenuation, 16×; other conditions as in (a).

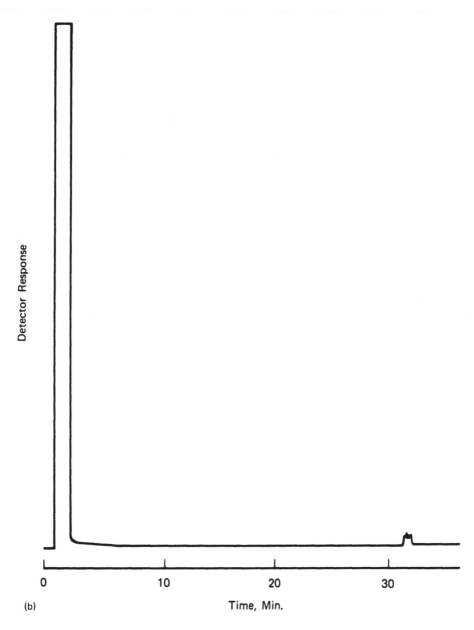

(b)

Time, Min.

FIGURE 5   (Continued).

One process in which relatively large amounts of liquid volume can be introduced into a capillary GC was described (22) where, by maintaining the column temperature at 100°C or more below the elution temperature of the most volatile solute, the stationary phase acts to focus the components of interest at the head of the column.

A second process utilized for large volume introduction is called the solvent effect (23-25). In this technique, a liquid sample forms a film at the inlet of an injection system which subsequently evaporates, transferring the focused solutes to the column.

If the solutes are not effectively focused at the head of the column, the resulting bands will be broadened and split in a manner similar to the effect illustrated in Figure 5. This effect has been termed band broadening in space, and the concept of using an uncoated inlet to contain the large solvent volume (retention gap) was introduced to overcome this problem (26,27).

In the use of a retention gap, an inlet section of the column having negligible retention for the components of interest is placed in front of the separating column, allowing various reconcentration mechanisms to take place, and deposition of the solutes of interest at the front of the film of the stationary phase. The various reconcentration mechanisms which have been postulated to accomplish large injections into a capillary GC can be divided into different categories.

Stationary phase focusing (28) is the general term used to describe reconcentration occurring by phase ratio focusing and cold trapping. Phase ratio focusing (29) refers to the reconcentration obtained when the migration of components of interest in a column with a stationary phase is slower than the migration in an uncoated inlet. This difference in migration speed allows the rear edge of the band to catch up with the leading edge, and depends solely on the phase ratio of the uncoated inlet to the separation column.

Cold trapping (30) refers to the reconcentration mechanism obtained when solutes reach the stationary phase at a temperature which is too low for elution, and are retarded sufficiently so that the rear edge of the band catches up to the leading edge in a manner analogous to phase ratio focusing. Cold trapping and stationary phase focusing are sometimes used interchangeably.

Stationary phase swelling (31) refers to the reconcentration obtained due to the change in characteristics of the stationary phase induced when large volumes of solvent vapor saturate the carrier gas. The properties of the stationary phase can be altered by swelling (thicker apparent film), changed viscosity or changed polarity. This effect has also been called phase soaking (32).

The solvent effect (33) also called solvent trapping (34) refers to the reconcentration obtained within an uncoated inlet, and is based on the theory that the solvent forms a film in said inlet whose rear edge evaporates by passage of carrier gas and moves forward. This evaporation and forward movement forces solutes to reconcentrate within the liquid film prior to reaching the analytical column, as long as the partition coefficient of the solute between the gas and liquid phases is greater than that of the solvent. A pictorial representation of each reconcentration mechanism is presented in Figure 6.

When an uncoated inlet is used to introduce large volumes of solvent into a capillary GC, consideration must also be given to the surface characteristics of the inlet. Typically, fused silica or glass capillaries of lengths ranging from a couple of meters up to 50 meters are used. In our opinion, fused silica is preferred due to the ease of handling and wide availability. The diameter of the inlet should also be considered, in terms of the surface area necessary to contain the solvent volume and is typically chosen to be of the same internal diameter as the capillary GC column. It is critical that liquid volume does not reach the separating column, as this will cause peak splitting, and this is dependent both on the inlet lengths and operating temperatures. (The influence of operating temperatures will be discussed in the following sections.) Fused silica is a high energy surface, with critical surface tension values ranging between 50 and 70 dyne/cm (35). As such, special consideration must be placed on the characteristics of the surface, in order to avoid irreversible adsorption of the components of interest. Inertness towards the components of interest should be a primary consideration. In our experience, some general guidelines can be used to determine the need for surface deactivation of said inlets. When using nonpolar eluents (normal phase), deactivation of the uncoated inlet is necessary if the components of interest have some degree of polarity. If said components are nonpolar, deactivation of the surface may not be necessary, but the inertness should be investigated. Typical deactivation procedures (36) involve the use of sylation reagents such as hexamethyldisalazine, diphenyltetramethyldisalizene or polymethylhydrosiloxane. The thickness of the deactivation layer should be kept to a minimum in order to avoid retention of the components of interest on the inlet surface at the particular temperature used. Other deactivation materials have also been investigated, such as Carbowax or cyanopropyl, but some problems remain unsolved in terms of stability, specially when dealing with aqueous (reversed-phase) solvents. In our experience when using reversed-phase eluents, the need for deactivation of the inlet surface is minimized. Preliminary data (37) indicate that water forms a film on the surface of the undeactivated inlet, preventing adsorption of components that are less water soluble. In addition, the data obtained suggest that water-containing eluents wet the surface of an untreated fused silica inlet (37) which would argue towards using untreated inlets if adsorption of the components of interest does not occur. It must be stressed that when using reversed-phase eluents the need for deactivation must be investigated for each particular compound to be analyzed, but in general it appears that it is not absolutely necessary to use deactivated inlets to obtain adequate chromatography.

RECONCENTRATION MECHANISMS

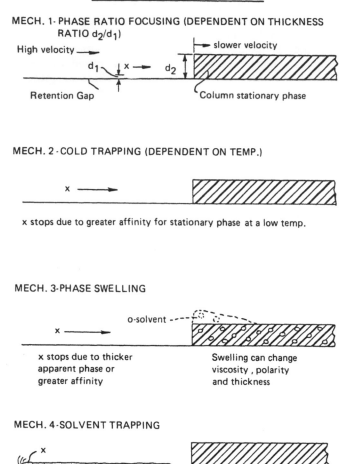

MECH. 1- PHASE RATIO FOCUSING (DEPENDENT ON THICKNESS
RATIO d_2/d_1)

High velocity

slower velocity

d_1 x d_2

Retention Gap

Column stationary phase

MECH. 2 - COLD TRAPPING (DEPENDENT ON TEMP.)

x

x stops due to greater affinity for stationary phase at a low temp.

MECH. 3-PHASE SWELLING

o-solvent

x

x stops due to thicker
apparent phase or
greater affinity

Swelling can change
viscosity , polarity
and thickness

MECH. 4-SOLVENT TRAPPING

x

FIGURE 6   Pictorial representation of the reconcentration mechanisms
occurring via an uncoated inlet.   x = analyte, o = solvent.

1. Solvent Introduction Below the Boiling Point

The solvent effect mechanism postulated for the introduction of large volumes of solvent into a capillary GC requires that said introduction be made at temperatures below the boiling point of the solvent, and in addition, the solvents used must wet the uncoated inlet.

The first requirement translates into the use of long retention gaps (up to 50 m) and long analysis times, since the solvent must be fully evaporated prior to elution of the components of interest by temperature programming; on the other hand, components with elution temperatures close to the solvent elution temperature appear as sharp peaks. The second requirement precludes the use of aqueous solvents generally used in reversed-phase liquid chromatography, particularly when using deactivated inlets whose critical surface tension values fall below the surface tension of the solvents used. The introduction of eluents below their boiling point has become less popular for these reasons, and most recent applications involve temperatures at which the eluent vaporizes to some extent during introduction.

2. Solvent Introduction Close to the Boiling Point

When solvents are introduced into an uncoated inlet at temperatures which are at or above the solvent boiling point, the evaporation of the solvent does not take place solely from the rear edge, but throughout the uncoated inlet, and the requirement that solvents wet the retention gap is decreased. The first reported use of this sample introduction process was made in 1985 (38), and later called concurrent solvent evaporation (39). In this process, the solutes of interest are reconcentrated at the head of the analytical column based on the various mechanisms described above, with the possible exception of solvent trapping. Solutes with elution temperatures close to the introduction temperature can be broadened, but judicious selection of solvents, inlet surface characteristics, and capillary column stationary phases can produce some sharp peaks eluting close to the introduction temperature (37). The main advantages of this technique is that short retention gaps are sufficient (3 to 20 m), the evaporation time required when compared to operation below the boiling point of the solvent is significantly reduced, and the importance of solvent wetting the uncoated inlet is minimized.

3. Transfer Control: Stop-Flow Introduction

In addition to the use of the autosampler interface described previously (without the use of uncoated inlets), the mechanics of sample introduction to the GC can be accomplished by directing the effluent to a conventional GC injector or by the use of switching valves or a combination of both.

The concept of stop-flow introduction refers to the process of discontinuing carrier gas flow to the uncoated inlet/capillary GC col-

umn arrangement while the LC effluent is introduced. The time period
for which carrier gas flow is interrupted can be relatively long, as
when the effluent is introduced by the LC pump, or relatively short,
as when the effluent is trapped in an external valve loop and introduced via carrier gas flow pushing the contents of the loop into the
GC system. Although the use of a switching valve with external
loops has been considered to be different from the use of a valve
without external loops, in either case the important variables that affect the quality of the results are the introduction temperature of the
effluent and the introduction rate. Therefore, the eluent feed interface (40) and the gas transfer interface (40) can be considered to be
essentially equivalent. The introduction of effluent into the GC system under conditions in which carrier gas is introduced at the same
time (56) is not to be considered stop-flow introduction, and for
clarity purposes we will call it simultaneous introduction.

Effluent from the LC system can be directed to the GC system by
interposing a switching valve (4-, 6- or 10-port) between the LC
and the GC uncoated inlet/capillary column. The components of interest are bypassed to waste when the valve is in one position, and
transfered to the GC injector (56) (simultaneous introduction) or directly to the uncoated inlet (38) (stop-flow introduction) when the
valve is switched in the alternate position. Carrier gas is either
simultaneously introduced into the GC inlet (56) or is halted during
transfer (38). After the transfer is complete, the connecting tube
between the valve and GC injector may be backflushed to decrease the
probability of contamination of the next transferred section (56). If
the system does not involve the GC injector or simultaneous introduction, the transfer line is flushed by the eluting mobile phase and backflushing is not required (38). This type of transfer, which is controlled by the LC pump pushing the solvent through the valve and
into the GC injector or uncoated inlet, has been called "eluent feed
interface."

Direct transfer into the GC injector can also be accomplished by
temporarily interrupting the flow in the LC (41). When the section of
interest appears on the recorder, the LC injection valve is placed in
the intermediate position between the fill and inject positions. As is
well-known, this procedure does not affect the LC separation. The
LC detector outlet (equipped with a transfer tube) is then manually
inserted into a cold on-column injector. Once the component(s) of
interest have been transferred, the procedure is repeated, the transfer line is withdrawn, and the GC temperature program can be started.
As mentioned above, stop-flow introduction can also be accomplished
using a loop injector. A 6- or 10-port valve is connected to the LC
detector outlet, and the components are introduced into a fixed loop
of known volume, corresponding to the volume of the fraction of interest. When the valve is switched, the carrier gas flushes the sample
loop and forces the liquid plug into the GC column. If a 10-port

switching valve is used, a second loop can be added to either introduce other components into the GC column or to flush the sample loop if the second loop is filled with solvent, in order to decrease contamination. The use of a sampling valve with external loops has been termed "gas transfer interface" and should also be classed as stop-flow introduction. A diagram of a 10-port valve arrangement is presented in Figure 7.

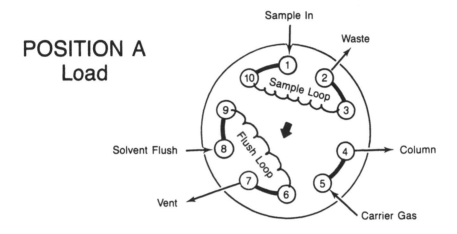

POSITION A
Load

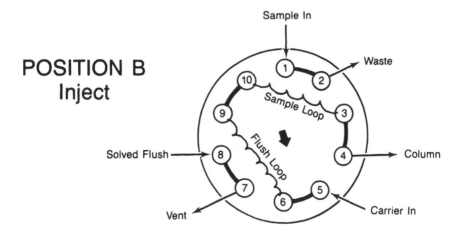

POSITION B
Inject

FIGURE 7   Ten-port valve configuration with fixed-volume sample loop.

III. MICROCOLUMN LIQUID CHROMATOGRAPHY

The reduction of the internal diameter of the LC column utilized for
multidimensional chromatographic applications to microcolumn dimensions (0.1–1.0 mm) introduces various advantages to the technique:
the volume of eluent used is significantly reduced, which means the
solutes of interest are diluted in much less eluent. For example, the
peak volume of a compound eluting from a microcolumn of 250 μm i.d.
at a k' of 3 is about 4 μl, a significant reduction when compared to
the peak volume given by a conventional column under the same conditions, approximately 500 μl. This reduction in peak volume is very
important, as it allows the introduction of aqueous (reversed-phase)
eluents into the capillary GC without the severe limitations experienced
using conventional size columns. In addition, much larger sections of
the LC chromatogram can be introduced into the capillary GC, allowing quantitative transfer of the components of interest, resulting in
greater reproducibility and better opportunity for quantitative analyses. Because microcolumns can be effectively prepared at lengths
greater than the conventional 25 cm, greater total column efficiencies
can be obtained.

   This emerging technology however, is somewhat more demanding
than the use of conventional column liquid chromatography, and the
advantages and constraints must be understood for the successful
use of microcolumn LC. This section will cover some details in the
preparation and use of this technology.

   In the use of microcolumns for LC, extra column effects must be
minimized to obtain in practice the theoretical performance (42). For
example, the maximum detector cell volume that does not significantly
contribute to system band broadening has been defined (43) as

$$V_d = 0.18(L \cdot d_p)^{\frac{1}{2}} d_c^{\ 2}$$

(1)

where L is the column length, d_p the particle size, and d_c the capillary diameter. A plot of plate height vs. detector cell volume is presented in Figure 8. It is evident that a compromise must be accepted
between efficiency attainable and sensitivity requirements, and although
the effect is more pronounced at smaller k', it should be considered.
The column diameter does not appear to have an effect on efficiency
on packed systems (44), but fluctuations in the packing density which
may occur across the column diameter (45) and temperature gradients
generated due to viscous friction (46) may be detrimental contributions
that can be minimized by reducing the column diameter. A plot of
plate height vs. column diameter obtained on fused silica columns of
various internal diameters is presented in Figure 9. A decrease in
plate height was observed when comparing columns of 320 μm and
250 μm. However, the 100 μm column did not appear to perform as

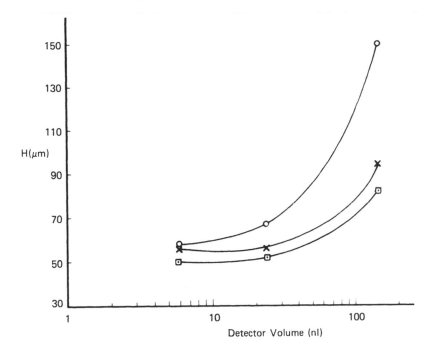

FIGURE 8 Effect of detector volume on plate height (H) for varying k' values in packed capillary system: o) phenol, k' = 0.70, x) aceto-phenone, k' = 1.14, ⊡ = methyl benzoate, k' = 1.49. Column: 75 × 250 μm i.d. fused silica packed with Zorbax ODS, dp = 7 μm. Eluent: 75% acetonitrile-water. Flow: 5.0 μl/min. (Reproduced with permission from Ref. 38. Copyright Dr. Alfred Huethig Publishers.)

well, which may be due to wall effects which are not completely under-stood at this time or to the possibility that the system used did not meet the dead-volume requirements. Similar results have been re-ported by others (3). The conditions for packing such columns re-producibly have been described (47,48). Briefly, it involves the intro-duction of a slurry, typically at a 4:1 ratio in a suitable solvent, and applying pressure in the range of 6,000 to 10,000 PSI.

Various techniques have been developed to hold the particles in the column and at the same time, minimize flow resistance, such as the use of glass wool (49), wires (50), porous polymer discs (51), and porous ceramic frits (52). Data obtained on hydrodynamic variables obtained on these columns are presented in Table 1.

The maximum volume that can be injected into a micro LC column has been estimated using the following equation, which assumes a 10% loss of efficiency (43).

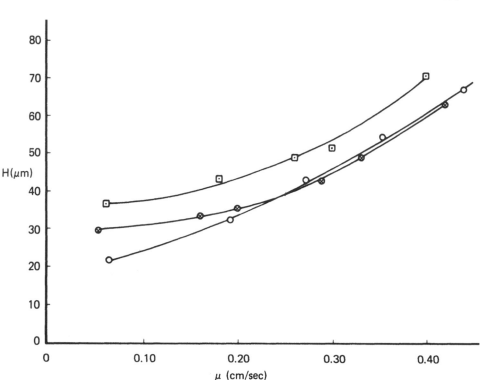

FIGURE 9 Effect of column diameter on plate height (H) for packed capillary system: ▣) 100 μm i.d., ⊛) 250 μm i.d., ○) 320 μm i.d., obtained at k' = 1.49, other conditions as in Figure 8. (Reproduced with permission from Ref. 38. Copyright Dr. Alfred Huethig Publishers.)

$$V_s = 0.36(L \cdot d_p)^{\frac{1}{2}} d_c^{2} \tag{2}$$

Although some questions exist as to the maximum volume that can be injected in to a microcolumn, the use of a weaker solvent than the mobile phase (peak compression) allows introduction of relatively large volumes without excessive loss in efficiency (3), as long as the equilibrium isotherm remains linear.

More detailed reviews on the advantages and utilization of this emerging technology have been presented recently (42,43,53), and it is worthwhile to mention that the equipment needed for this work is commercially available, while microcolumns for LC have recently been made available from various sources (54,55).

TABLE 1   Comparison of Columns Terminated with Various Bed Supports

Column bed support	ϕ	$K^\circ \times 10^{10}$ (cm^2)	ε_t
0.1 cm glass wool	451	5.54	0.50
0.01 cm ceramic	485	5.15	0.49
6 cm ceramic while packing cut to 0.1 cm prior to use	550	4.54	0.46

Flow resistance factor = ϕ = $dp^2 \Delta P / \mu L \eta$, where dp = particle diameter, ΔP = pressure drop across the column, μ = linear valocity, L = column length, = η eluent viscosity.   Specific column permeability = K° = ϕ / dp^2, total porosity = ε_t = $Fto/\pi r^2 L$, where F = volumetric flow rate, to = elution time of unretained component, r = column radius.
Source: From Reference 41, with permission copyright Dr. Alfred Huethig Publishers.

In the use of micro LC columns for multidimensional separations when coupled to a capillary GC, the microcolumn is used as a highly efficient preseparation (cleanup) step, or a chemical class fractionation.   Therefore, a decrease in efficiency (within limits) due to larger injection volumes can be tolerated.   In addition, samples which may contain higher molecular weight material, or components which may be irreversibly adsorbed can, as a last resort, be introduced into the system.   In our experience, when the front of the column becomes severely contaminated, removing a small section of the column front usually restores the column to its original operating conditions.

4.   Applications

The use of uncoated inlets for multidimensional LC-GC has gained widespread popularity, as evidenced by the varied applications found in the literature.   In this section, said applications will be reviewed. Table 2 summarizes the applications by matrix type (foodstuffs, fuel, environmental, biological fluids, and miscellaneous).   Unless otherwise noted, the internal diameters of the uncoated inlets used correspond to the internal diameters of the capillary GC column.   The use of an uncoated inlet to introduce LC effluent below its boiling point into a capillary GC was described (56).   In this application, a conventional LC column (100 × 3 mm ID) packed with Spherisorb[R] S-5W

TABLE 2 Applications of Multidimensional LG-GC by Matrix Type

Reference	Analyte	Matrix	Interface	LC mode	Volume transferred to GC (µL)
FOODSTUFFS					
(13)	Atrazine	Sorghum fodder	Autoinjector	Normal phase	2
(56)	Azulenes	Toothpaste	Eluent feed[a]	Normal phase	150
(58)	Chlorpyrifos pesticide	Rat feed	Eluent feed[b]	Normal phase	48
(61)	p-Hydroxy-phenyl-1-butanone	Raspberry sauce	Gas transfer[b]	Normal phase	450
(62)	Wax esters	Olive oil	Gas transfer[b]	Normal phase	450
(18)	Folpet	Hops	Autoinjector	Normal phase	2
(71)	Various	Butter	Gas transfer[b]	Normal phase	400
(72)	di-(2-ethyl) phthalate	Salad oil	Gas transfer[b]	Normal phase	1000
(37)	2-Chloro-6-trichloro-methylpyridine	Corn grain	Eluent feed[b]	Reversed-phase	7
FUELS					
(14)	Polynuclear aromatics	Solvent refined coal; shale oil	Autoinjector	Normal phase	2
(15)	Terpenoids	Plant extracts	Autoinjector	Normal phase	2

(16)	Group-type analysis	Gasoline, diesel fuel	Autoinjector	Normal phase	2
(38)	PCBs	Coal tar	Gas transfer[b]	Reversed-phase	40
(57)	Chlorinated benzenes	Fuel oil	Stop-flow[b]	Normal phase	22
(58)	Aliphatic and aromatic hydrocarbons	Gasoline	Eluent feed[a]	Normal phase	100
(17)	Various	Coal liquids	Autoinjector	Size exclusion	0.1
(19)	Polycyclic aromatics	Solvent refined coal	Autoinjector (isotachic splitter)	Reversed-phase	1.0
(63)	Polycyclic aromatics	Diesel exhaust	Gas transfer[b]	Normal phase	150
(70)	Hydrocarbons	Lignite tar, shale oil	Gas transfer[b]	Normal phase	200
(41)	PACs, group-type analysis	Gasoline	Eluent feed[a]	Normal phase	3
(76)	Methylated dibenzo thiophenes	Oil spills	Gas transfer[b]	Normal phase	500

ENVIRONMENTAL

(64)	PCBs	Fish	Gas transfer[b]	Normal phase	400
(66)	Chlorinated pesticides, PCBs	Water	Eluent feed[a]	Liquid-solid extraction (normal phase)	85
(67)	PCBs	River sediment	Eluent feed[a]	Normal phase	180

TABLE 2 (Continued)

Reference	Analyte	Matrix	Interface	LC mode	Volume transferred to GC (μL)
ENVIRONMENTAL (cont.)					
(68)	Chlorinated pesticides	Water	Eluent feed[b] (cold trap)	Liquid-solid extraction (normal phase)	70
(73)	Propachlor herbicide	Soil	Eluent feed[b]	Reversed-phase	8
BIOLOGICAL FLUIDS					
(60)	Diethyl-stilbesterol	Bovine urine	Eluent feed[a]	Normal phase	300
(65)	Broxaterol	Plasma	Gas transfer[b]	Normal phase	450
(68)	Heroin metabolites	Urine	Gas transfer[b]	Normal phase	500
(74)	Idaverine	Drug standards	Autoinjector[b] (uncoated inlet)	Reversed-phase	7
(75)	Diazepam	Urine	Eluent feed[b]	Reversed-phase	3
MISCELLANEOUS					
(77)	Triglycerides phospholipids	Staphilococcus aureus cells	Post column reaction, eluent feed	Normal phase	50
(78)	Polymer characterization	Styrene-acrylonitrile copolymer	Pyrolisis chamber	Size exclusion	6

[a]Simultaneous. [b]Stop-flow. [c]Autoinjector.

274

was operated using a nonpolar eluent (cyclohexane) at a flow rate of 0.4 mL/min. The LC effluent was introduced into a capillary GC to qualitatively identify the components in a methanol extract of tooth paste. The components of interest were anethol and guajazulene. No information was supplied as to the concentration of the components which were present in the tooth paste. The chromatograms obtained are depicted in Figure 10. The LC chromatogram in this application is fairly clean, and the components are well resolved, making the identification by GC retention times simple. A 50 m retention gap was used, implying that the effluent volume may have been too large to allow the introduction of both components in one run.

The first application of multidimensional LC-GC utilizing micro-columns for LC, reversed-phase eluents and transfer of the sections of interest above the boiling point of the solvent was described* (38). A block diagram of the system used is included in Figure 11.

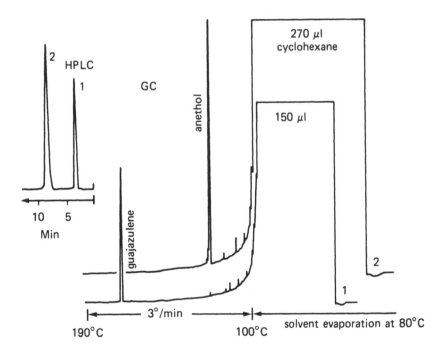

FIGURE 10 LC-GC of two solutes visible on the HPLC trace (left), run for the analysis of azulene dyestuffs in a tooth paste. Peak 1 was transferred with 150 μl of eluent and coeluted in the GC system with guajazulene. Peak 2 was identified as anethol by LC-GC-MS. (Reproduced with permission from Ref. 56. Copyright Elsevier Scientific Publishing Company.)

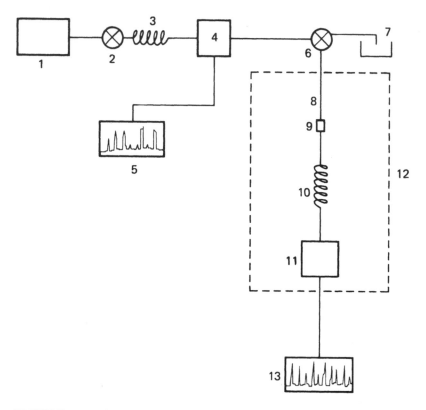

FIGURE 11 Block diagram of on-line packed capillary LC-capillary
GC system. 1. pump; 2. injection valve; 3. packed capillary LC
column; 4. detector; 5. recorder; 6. switching valve; 7. effluent
waste; 8. retention gap; 9. butt connector; 10. capillary GC column;
11. detector; 12. gas chromatograph oven; 13. recorder. (Reproduced
with permission from Ref. 38. Copyright Dr. Alfred Huethig Publishers.)

A complex hydrocarbon sample (coal tar extract) was analyzed for the presence of two polychlorinated biphenyls. The LC system consisted of a fused silica capillary (95 × 250 μm i.d.) packed with Zorbax[R] ODS of 7 μm particle diameter. Acetonitrile was used as the eluent. The length of the retention gap was 4 meters and the volume introduced into the capillary GC was 40 μL. The polychlorinated biphenyls of interest were resolved from all other components in the mixture, allowing a separation that was not possible using HPLC or GC alone.

In another application (57), the applicability of microcolumn LC-GC for quantitative analysis was demonstrated by the analysis of a sample of fuel oil for the presence of various chlorinated benzenes. In this case, a normal phase eluent was used in a capillary fused silica column packed with Zorbax silica of 7 μm particle diameter. The sample was introduced into the system without any prior sample preparation. The section of interest was introduced into a 30 m Carbowax column preceded by a 10 m undeactivated inlet. The liquid chromatograph effectively removed all interfering components from the sample, allowing quantitation of the components of interest. The detection limits obtained in this application ranged from 8 to 17 μg/g utilizing a flame ionization detector. The chromatograms obtained are included in Figure 12.

Aliphatic and aromatic hydrocarbons in gasoline were analyzed (58) by separating fractions of aromatic and aliphatic character using a normal phase LC system consisting of a 10 cm × 1 mm i.d. column packed with silica and eluted with pentane. A 10 m deactivated inlet was connected to a 30 m capillary GC column coated with OV-1 which separated the individual components within each class. Gasoline samples were diluted in pentane prior to analysis.

The determination of Chlorpyrifos, (O-O-diethyl-O-(3,5,6-trichloro-2-pyridyl)phosphothioate) in rodent feed used for toxicological studies was made (59) utilizing a normal phase system with an eluent consisting of heptane/methyl-t-butyl ether (95:5). Fifteen grams of rodent chow were extracted with 100 mL of benzene, the solvent was allowed to evaporate under a nitrogen stream and the residue was dissolved in 5.0 mL of eluent. The resulting solution was filtered and injected into the micro LC system using a 110 cm × 250 μm i.d. fused silica column packed with Zorbax[R] silica of 7 μm particle diameter. The section of the LC chromatogram known to contain the component of interest was introduced into the capillary GC consisting of a 10 m uncoated inlet and a 50 m 5% phenylmethylsilicone GC column. Detection limits obtained in this application were estimated as 20 ppb using an electron capture detector.

The determination of diethylstilbesterol in bovine urine (60) was accomplished by enzimatic hydrolysis of a 10 ml sample, followed by an off-line cleanup step through a Sep-pak and chemical treatment of the component of interest converting it to the dipentafluorobenzyl

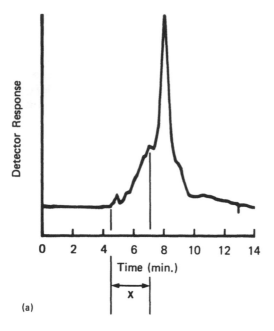

FIGURE 12 (a) μ-LC Chromatogram of fuel oil sample. Column: 105 cm × 250-μm i.d. packed with Zorbax Silica, dp = 7 μm. Mobile phase: heptane. Flow: 10.6 μl/min. Detector: Jasco Uvidec II at 214 nm. Pressure: 3800 psig. Sample prepared to give concentration of 1.00 g/10 ml in heptane. X = section introduced into the gas chromatograph. (b) Capillary gas chromatogram of fuel oil from LC. Column: 30 m × 0.25-mm i.d., Supelcowax 10 (df = 0.25 μm). Retention gap: 15 m × 0.25 mm fused silica. Oven temperature: 105°C for 9 minutes, program to 245°C at 5°C/min. Carrier: helium at 70 cm/sec. Make up gas: nitrogen at 30 ml/min. Detector: FID at 275°C. Injected volume: 21 μl. Retention times of chlorobenzenes of interest are indicated. 1. chlorobenzene; 2. 1,2-dichlorobenzene; 3. 1,2,4,5-tetrachlorobenzene; 4. 1,2,3,4-tetrachlorobenzene; 5. pentachlorobenzene; 6. hexachlorobenzene. (Reproduced with permission from Ref. 57. Copyright Elsevier Scientific Company.)

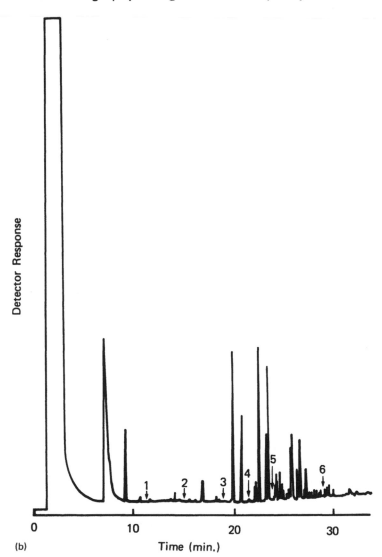

(b)

Time (min.)

Detector Response

FIGURE 12   (Continued)

ether derivative. An aliquot of the derivatized material was then injected into a 10 cm × 3.0 mm i.d. silica column using cyclohexane/ tetrahydrofuran (99:1) mobile phase. The component of interest was introduced into a 50 m deactivated inlet and separation was obtained on a 15 m × 0.32 mm i.d. capillary column coated with OV-73 of 0.25 μm film thickness. Part per billion sensitivity was reported using an electron capture detector.

The determination of raspberry ketone (p-hydroxy phenyl-1-butanone) in raspberry sauce (61) was accomplished by diluting 1 ml of sauce with 9 ml water, the solution extracted with 2 ml of diethyl ether, and a 10 μl volume of the extract was then injected on a silica column (10 cm × 2.0 mm i.d.) eluent with pentane/diethylether (80:20) mobile phase. The fraction of interest was transferred to the GC and resolution of the component of interest was obtained by using a 2 m deactivated inlet and a 15 m × 0.32 mm i.d. OV-1701 column of 0.3 μm film thickness. Wax esters in hexane into a silica column (10 cm × 2.0 mm i.d.) eluted with methylene chloride/hexane (10:90). The transferred cut was separated on a 15 m × 0.32 mm SE-54 of 0.15 μm film thickness and a 2 m deactivated inlet. Flame ionization detection yielded part per million detection limits (62).

Trace polycyclic aromatic compounds in diesel exhaust particulate extracts were determined (63). 100 liters of exhaust were collected on glass filters which were then Soxhelet extracted with benzene/ methanol (80:20) for 8 hours. The resulting extracts were evaporated to dryness and redissolved in pentane to a volume of 300 μl. Extracts were injected into a liquid chromatograph equipped with an aminosilane column coupled in series to a silica gel column (25 cm × 1.0 mm i.d.). Samples were eluted with pentane, and the sections of interest were transferred to a GC equipped with a 25 m × 0.33 mm i.d. BP-5 column of 0.5 μm thickness and a 25 m deactivated inlet. Chromatograms obtained in this manner are presented in Figure 13. This application described the first reported automated system for LC-GC.

Polychlorinated byphenyls in fish were determined (64) by sox-helet extraction of 40 g of fish with 100 ml of petroleum ether. The extracts were brought to dryness and redissolved in pentane. The resulting solutions were injected into a cyanopropyl bonded silica column (10 cm × 3.0 mm i.d.) using pentane as the eluent. The cuts of interest were transferred to the GC and separated using a 3 m uncoated, deactivated inlet connected to a 13 m × 0.32 mm i.d. methyl silicone column of 0.6 μm film thickness.

Concentrations of Broxaterol in plasma (65) were investigated after precipitation of proteins, addition of an internal standard, and extraction with diethylether. The residue after evaporation was derivatized with trifluoroacetic acid anhydride, evaporated and re-dissolved in 500 μl diethyl ether. 20 μl were injected into a 15 cm ×

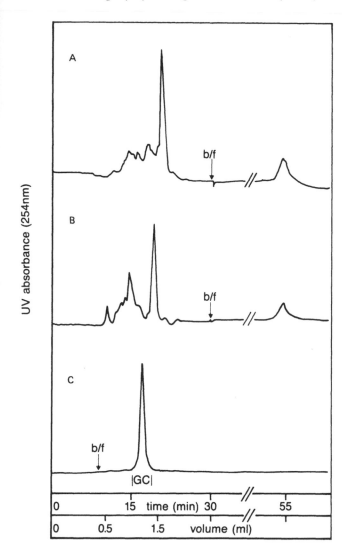

FIGURE 13(a)　　HPLC chromatograms of (A) diesel particulate extract and (B) kerosene particulate extract obtained with forward
eluent flow; both chromatograms gave chromatogram (C) when back
flushed at 8.5 min after aliphatic elution.　The 100 μl fraction
transfered on-line into the GC is indicated.　(Reproduced with permission from Ref. 63.　Copyright American Chemical Society.)

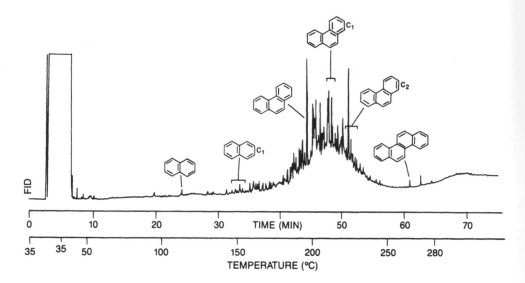

FIGURE 13(b)   GC Chromatogram of aromatic back flush fraction obtained by on-line HPLC-GC of the diesel particulate extract.

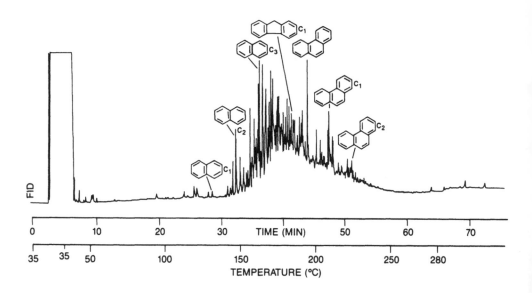

FIGURE 13(c)   GC chromatogram of aromatic back flush fraction obtained by on-line HPLC-GC of the kerosene particulate extract.

3.9 mm i.d. silica column and eluted with pentane/diethyl ether
(55:45). The transferred section was separated on a 25 m × 0.32 mm
i.d. column of 1.0 μm SE-54 film, using a 2 m uncoated inlet. Trace
enrichment of a series of chlorinated pesticides and chlorinated biphenyls (66) was accomplished prior to GC analysis by the use of a
micro LC precolumn, where a large volume of an aqueous sample was
passed through the precolumn, nitrogen flushing and vacuum was used
to remove the residual water, and the enriched components of interest
were then desorbed with hexane directly into the GC system which
consisted of a 5 m uncoated inlet and a 25 m × 0.32 mm i.d. CPSil-5B
column of 0.13 μm film thickness. A diagram of the LC injection
valve containing the micro LC injection valve containing the micro
LC precolumn is included in Figure 14.

The determination of polychlorinated biphenyls in hexane extracts
of river sediment (67), was made by separation on a 15 cm × 0.7 mm
i.d. silica column eluted with hexane, and introduction of the fractions of interest into a 5 m uncoated inlet connected to a 50 m × 0.32
mm CPSil-5B column of 0.13 μm film thickness.

Determination of trace chlorinated pesticides in aqueous samples
was accomplished using similar principles (68) with an important modification. In this application, a cold-trap/vaporizer arrangement was
implemented by using a vent following the uncoated inlet, in order to
allow rapid venting of the solvent without the need for elution through
the separating column. The use of a cold-trap/vaporizer can be limited, as the more volatile components are also vented with the solvent;
nevertheless, compounds which are not volatile at the introduction
temperature are collected in the trap and appear as sharp peaks when
the oven temperature is increased. A schematic diagram of the arrangement is included in Figure 15.

Study of polyaromatic hydrocarbons and separation of light gasoline
fraction by group type was accomplished using the microcolumn LC-
capillary GC approach (41). A 320 μm fused silica column was packed
with silica and components were separated using a hexane/methylene
chloride (75:25) mobile phase for the PAC separation and pentane for
the gasoline separation. Components were transfered to a 5 m uncoated inlet connected to 25 m × 0.25 mm RSL 300 column of 0.25 μm
film thickness. The metabolites of heroin in urine (69) were determined using a 2 mm × 10 cm column packed with silica and eluted with
diethylether/methanol/dimethylamine (91.5:8.0:0.5). The section containing the components of interest was introduced into 2.5 m × 0.53 mm
i.d. deactivated inlet followed by a 20 m × 0.32 mm GC capillary of
OV-1 of 0.5 μm film thickness.

Hydrocarbons in lignite tar and shale oil samples (70) were studied
using an automated LC-GC system, where a 1.0 mm silver loaded silica
column was combined with a silica column and using pentane as the
mobile phase, yielded separations of alkane, alkene, aromatic, and

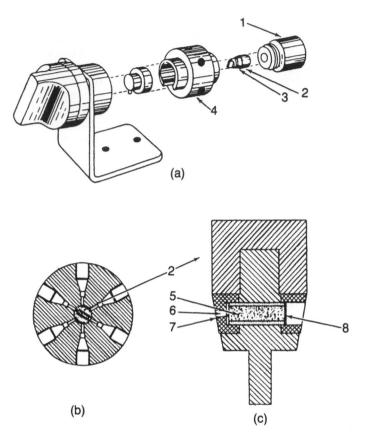

FIGURE 14 Switching valve with internal precolumn. (a) General
view, (b) top view of cross-section, (c) cross-section of rotor body.
5. precolumn, 6. outlet precolumn, and 7. and 8. screens. (Reproduced with permission from Ref. 66. Copyright Dr. Alfred Huethig
Publishers.)

polar families. The separated fractions were introduced into the
GC system using a gas transfer interface and a 20 m deactivated
inlet. The separation of the individual components within each
class was accomplished on a 25 m × 0.33 mm BP-1 column of 0.5 μm
film. For the shale oil sample, the LC separation used a combination
of aminopropyl and silica columns eluted with pentane/methylene chloride (90:10) and a 25 m BP-5 column of 0.33 mm i.d. and 0.5 μm film
thickness.

    The components in a butter sample (71) were separated on a
10 cm × 4.0 mm i.d. ODS column. The transferred section was se-

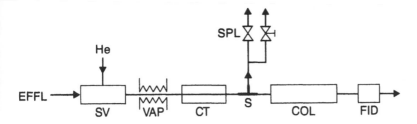

FIGURE 15 Schematic design of the LC-GC interface. EFFL = LC effluent, SV = sampling valve, VAP = vaporizer, S = all glass low dead volume T-splitting device, CT = cold trap, COL = GC column, SPL = splitter valves, FID = Flame ionization detector. (Reproduced with permission from Ref. 68. Copyright Dr. Alfred Huethig Publishers.)

parated on a 25 m × 0.32 mm i.d. SE-54 column. Eluent conditions and inlet characteristics were not described. A 10-port valve was used to collect more than one fraction per analysis in a secondary storage loop.

Di-(2-ethylhexyl)-phthalate plasticizer was determined in salad oil (72) by separation on a 4.6 mm i.d. silica column and a methylene chloride/cyclohexane/acetonitrile (50:50:0.05) eluent. Components of interest were transferred using a gas transfer interface into a 2 m precolumn and a 20 m × 0.32 mm GC column coated with SE-54 of 0.15 μm film thickness.

The determination of 2-chloro-N-isopropylacetanilide (PROPACHLOR* Herbicide) in soil, (73) was accomplished using a reversed phase system with an eluent of methanol/water (90:10). 50 g of soil were extracted with methylene chloride, filtered and subsequently evaporated. The residue was dissolved in eluent, filtered and injected into the LC system using a fused silica column (102 cm × 250-μm i.d.) packed with Spherisorb ^{R} ODS of 5 μm particle diameter. Detection limits obtained were estimated as 3.6 μg/g using an FID.

Studies of the drug idaverine (74) were made using a micro LC column (15 cm × 320 μm) packed with C-18 and eluted with acetonitrile/water/triethylamine (90:9.9:1.1). In this application, an autoinjector interface was used in conjunction with an uncoated inlet of 2 m × 0.53 mmm i.d. The GC column used was a 30 m × 0.32 mm coated with DB-1 of 0.25 μm film thickness. The effects of deactivation of the inlet with various stationary phases was studied in order to improve the wettability of the surface and at the same time retain its inertness. The determination of diazepam in urine was accomplished on a system consisting of a 15 cm × 320 μm microcolumn packed with reversed-phase (C-18) support. A mobile phase of methanol/water (80:20) containing 0.1% H_3PO_4 was used to separate

the component from the interferences in the urine matrix. An eluent
feed interface was used to transfer the LC fraction to 2 m × 0.25 mm
inlet deactivated with aminopropyl or cyanopropyl layers.

Methylated dibenzothiophene isomers in oil samples (76) were determined by the LC-GC approach. Samples of oil were diluted in
pentane (1% solutions) and injected into a 25 cm × 4.6 mm i.d. aminopropyl column using a pentane mobile phase. The components of
interest were transfered to a 5 m × 0.53 mm deactivated inlet preceding a 40 m × 0.32 mm i.d. RSL-300 of 0.25 μm film thickness.
Detection was made using a flame photometric detector operated in
the sulfur mode. The relative distributions of the methylated dibenzothiophene isomers allowed the qualitative characterization of the
oils.

The determination of N-SERVE nitrogen stabilizer (2-chloro-6-
trichloromethyl pyridine) in corn extract (37) was made using a
reversed-phase microcolumn system consisting of a 30 cm × 250 μm
i.d. fused silica column packed with Zorbax CN and a mobile phase
of acetonitrile/water (65:35). Transfer of the section of interest was
made via an eluent feed interface using a 15 m undeactivated inlet
and the component of interest was separated on a 30 m × 2.25 mm i.d.
DB-5 column. Electron capture detection yielded detection limits of
10 ng/g (ppb). This application demonstrated the advantages of
using microcolumns for LC by allowing introduction of relatively large
amounts of aqueous eluents into the GC system without detrimental
effects. The chromatograms obtained are presented in Figures 16
and 17.

IV. ADVANTAGES AND CONSTRAINTS

The multidimensional technique using on-line coupled LC-GC offers
various advantages over single column systems, and in some cases
over the more common multidimensional LC or GC techniques. These
include speed of analysis, minimal sample preparation, automation,
sensitivity, quantitation, and reproducibility. Some of the constraints
of the technique are the dependence on component volatility (although
approaches are being investigated to extend the range of applicability),
interface durability, the type of GC columns that can be used, and
some eluent restrictions, especially using LC columns of conventional
dimensions. These points are discussed in more detail in the following sections.

The LC section of the LC-GC system is used as a highly efficient
means of effectively removing undesirable interferences in the sample
matrix in a relatively short time. Manual extractions, preseparations
using LC columns and tedious collection of cuts can be eliminated,
especially when analyzing samples that are liquids in which a dilution
and filtration may be all the preparation necessary. Since pretreat-

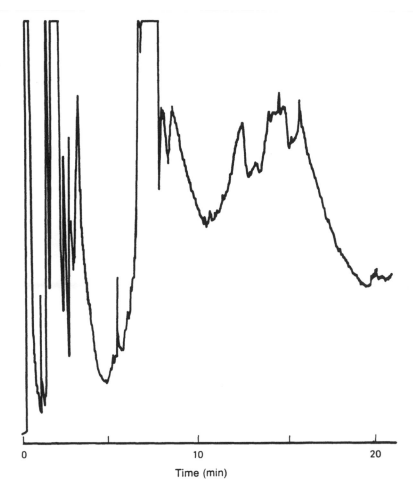

FIGURE 16 Chromatogram of corn extract. Column 30 m × 250 μm
DB-5, 0.25 μm film. Retention gap = 15 m × 0.25 mm fused silica.
Oven temperature: 115°C for 3 min, program to 280°C at 15°C/min.
Carrier: He at 40 cm/sec. Make-up: N_2 at 30 ml/min. Injection
size: 5 μl. (Reproduced with permission from Ref. 37. Copyright
Aster Publishing Co.)

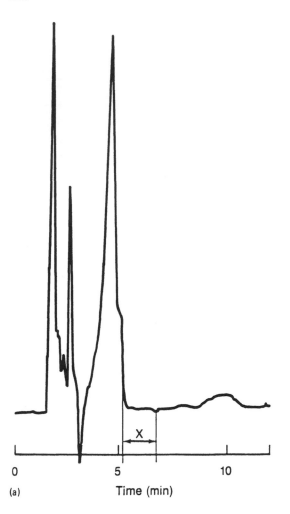

(a)   Time (min)

FIGURE 17   Chromatograms of corn extract. (a) Micro LC.   Column:
32 cm × 250 µm i.d. fused silica packed with Zorbax CN (d_p = 7 µm).
Mobile phase: Acetonitrile/water (65:35).   Flow rate: 4.2 µl/min.
Detection UV at 205 nm, 0.01 AUFS.   Injection size = 100 nl.   x =
section introduced into the capillary GC.   (b) Capillary GC chromatogram of transferred section.   1. 2-chloro-6-(trichloromethyl)pyridine
(100 ng/g).   Other conditions as in Figure 16.   (Reproduced with
permission from Ref. 37.   Copyright Aster Publishing Co.)

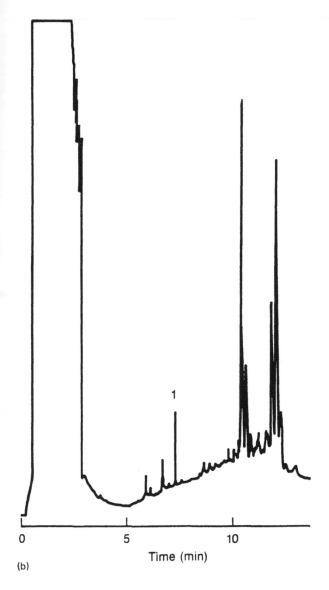

(b)

Time (min)

FIGURE 17   (Continued)

ment of the sample is often needed, the integrity of the sample is maintained, minimizing the possibilities of sample loss through handling or changing the components of interest in a detrimental manner.

A multidimensional system is easily automated once the appropriate chromatographic conditions are established. An autoinjector is needed to introduce the sample into the LC system. Utilizing time control, the eluting sections of interest can be introduced reproducibly into the GC. This can be easily accomplished with any integration device equipped with external relay controls. The GC can be modified to initiate the temperature program at a signal from the integrator.

The components of interest are retained at the head of the GC column by the utilization of an uncoated inlet and judicious initial temperature selection. It is possible to make multiple injections into the GC system to increase the amount of material to be detected, as long as the initial temperature is maintained below the elution temperature of the solutes of interest. By using inherently sensitive detectors, such as an electron capture, sensitivities in the low part per billion range are attainable.

The use of microcolumns for LC allows the quantitative transfer of the total component(s) of interest into the GC (without the volume restrictions experienced using conventional columns), in contrast to only sections of a peak. There is therefore less dependance on absolute reproducibility of retention times in the LC. This advantage allows better quantitation of the material since all the component eluted from the LC column is transfered to the GC. In a well designed micro LC system, the chromatographic reproducibility is comparable to that obtained in a conventional LC system. Retention time variability of 1% is attainable and adequate.

Validation data obtained in our laboratory yielded relative precisions at the 95% confidence level between 9.2% and 17.2% at concentrations of 0.9 parts per million. Recoveries of known quantities of the components of interest added to the matrix were found to range between 90% and 110%. (Table 3). These data are comparable to those obtained on conventional chromatographic systems. Utilization of conventional size columns is expected to yield similar results, if the complete section containing the components of interest can be reproducibly introduced into the GC.

Some consideration must be given to the areas in which the technique described may not be applicable and also to those experimental parameters that should be kept in mind when utilizing this technology. When using an uncoated inlet, the components of interest must be retained at the head of the GC column at the initial temperature selected. This means that in many cases, only compounds with boiling points above that of the solvent can be effectively trapped and quantitated. In addition, only compounds that can be chromatographed in the gas phase are amenable to the technique.

TABLE 3   Recovery and Precision Data for the Determination of Trace Impurities in a Proprietary Herbicide

Amount added μg	Amount found μg	% Recovery	Amount added μg	Amount found μg	% Recovery	Amount added μg	Amount found μg	% Recovery
1.09	0.99	90.8	1.18	1.19	101	0.88	0.83	94.3
1.20	1.08	90.0	1.00	1.00	108	0.90	0.81	90.0
3.10	2.70	87.1	3.50	3.50	100	2.40	2.20	91.7
3.64	3.21	88.9	3.92	4.28	110	2.92	2.56	89.7
6.37	5.94	93.2	6.86	7.06	102	5.11	4.68	90.8
7.78	7.89	101	7.84	6.81	86.7	5.84	5.98	102
18.2	16.6	90.6	–	–	–	–	–	–
1.09	0.98	88.0	1.18	1.29	109	0.88	0.89	101
1.37	1.28	93.4	1.47	1.14	77.6	1.10	0.96	87.3
0.91	1.14	–	0.98	1.14	117	0.73	0.82	–
0.56	0.54	96.4	0.59	0.56	03.2	0.45	0.43	95.6
	$\bar{x} =$	91.9			103			93.6
	$\sigma =$	4.23			9.23			5.11
	$2\sigma =$	8.45			18.4			10.2
	$(2\sigma/x)100 =$	9.2%			17.7%			10.9%

Two recent approaches have been used for the analysis of substances which are normally nonvolatile. In one case, the separation of triglycerides and phospholipids in extracts of Staphilococcus aureus cells (77) was made on a 25 cm × 2.0 mm i.d. silica column eluted with heptane/isopropanol (70:30). Prior to introduction into the GC system, the components of interest were submitted to on-line post-column derivatization using a fixed-bed reactor (15 cm × 4.6 mm i.d. packed with a strong cation exchanger). The derivatized components of interest were transfered to a 50 m deactivated inlet connected to the separation column, which was a 25 m × 0.32 mm i.d. coated with 5% phenylmethyl silicone.

The characterization of a styrene-acrylonitrile copolymer in terms of composition vs. molecular size (78) was made by separating solutions of the polymer on a micro size exclusion column (50 cm × 250 μm i.d. packed with Zorbax PSM-1000) using tetrahydrofuran as the mobile phase. The particular fractions of the SEC chromatogram were switched onto an on-line pyrolysis chamber where the nonvolatile polymer fractions were pyrolyzed and the products separated on a capillary GC column of 50 m × 0.25 mm i.d. coated with 5% phenylmethylsilicone. By measuring the ratios of styrene and acrylonitrile formed in each molecular size fraction, the variability in composition was studied. A diagram of the interface is presented in Figure 18, while the micro SEC chromatogram obtained on the polymer and a typical pyrolysis gas chromatogram of a fraction are represented in Figures 19 and 20. These two approaches demonstrate that the possibilities of using LC-GC for nonvolatile compounds are a reality, and further growth is foreseen in these areas.

For effective chromatography to take place, the uncoated inlet used must have minimal retention for the components being determined. The introduction of components that are nonvolatile may not be avoided in some cases, and these components can collect in the retention gap to a point where they can act as a stationary phase, broadening the components of interest. By monitoring changes in peak shape and retention times, the time when the uncoated inlet must be cleaned or replaced can be determined.

Since relatively large volumes of solvent vapor pass through the GC column, the stationary phase of said column must be stable so as not to be stripped off by the eluent. With the increasing commercial availability of "bonded" or cross-linked phases, which routinely withstand severe conditions such as those encountered when using supercritical fluids, this limitation becomes less significant but should still be considered.

The eluents used in the LC section of the technique have included acetonitrile, methanol, heptane, water, and others. It is expected that most other common mobile phases will behave in a satisfactory manner. However, nonvolatile buffer salts, which may be required in

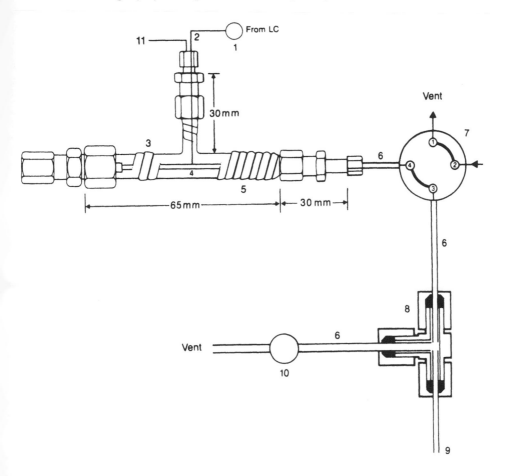

FIGURE 18 Diagram of microcolumn liquid chromatography-pyrolysis gas chromatography interface. 1. 10-port switching valve; 2. transfer capillary; 3. glass chamber; 4. pyrolysis ribbon; 5. heating tape; 6. transfer capillaries; 7. 4-port switching valve; 8. split tee; 9. capillary GC column; 10. micrometering valve; 11. auxilliary carrier gas. All components following the glass chamber were positioned within the GC oven. (Reproduced with permission from Ref. 78. Copyright American Chemical Society.)

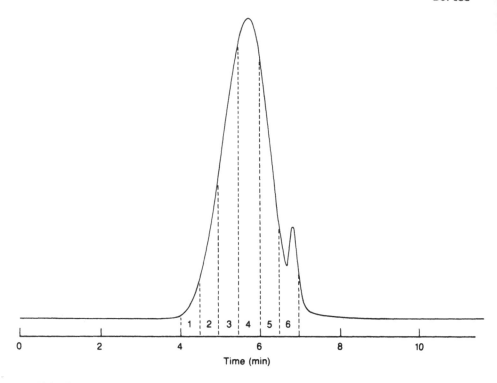

FIGURE 19 Microcolumn size exclusion chromatogram of styrene-
acrylonitrile copolymer. Fractins transfered to other pyrolysis systems are indicated. Column: 50 cm × 250 μm i.d. fused silica packed
with Zorbax PSM-1000 (d_p = 7 μm). Eluent THF. Flow rate: 2.0
μL/min. detector. Jasco Uvidec V at 220 nm. Injection size 200 nL.
(Reproduced with permission form Ref. 78. Copyright American
Chemical Society.)

some applications, can create problems by depositing in the inlets,
and in severe cases, plugging can occur.

V. CONCLUSIONS

Multidimensional chromatography, especially utilizing the combination
of LC and capillary GC, is a technique which is becoming more important in the analysis of complex matrices when a high degree of
resolution is required. The use of a microcolumn for LC offers a desirable approach, helping to overcome some of the problems associated
with the introduction of large volumes into a capillary GC.

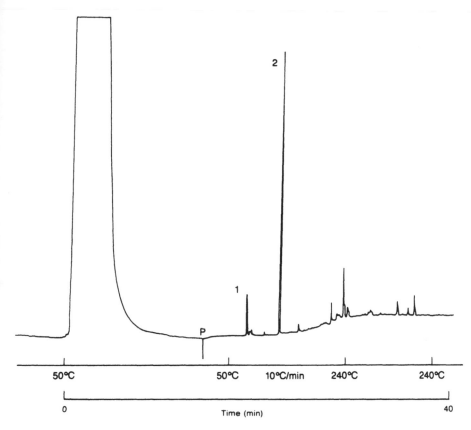

FIGURE 20   Typical chromatogram of fraction from Micro SEC system.
Column: 50 m × 0.2 mm i.d. 5% phenyl methyl silicone (d$_f$ = 0.33 μm).
Oven temperature 50 to 240°C at 10°C/min.   Carrier: helium at 60
cm/sec.   Detector: FID at 320°C.   Make-up: nitrogen at 20 mL/min.
P, point at which pyrolysis was made.   1. acrylonitrile; 2. styrene.
(Reproduced with permission form Ref. 78.   Copyright American
Chemical Society.)

Recent advances in the preparation and use of microcolumns have made the multidimensional approach feasible, as demonstrated by the various examples in the text. Although limitations still exist, especially in the types of components that can be chromatographed in the gas phase, many unique analytical opportunities can easily be envisioned. Routine analyses where tedious manual cleanup is required can be automated and the time savings can be significant. Continued research in microcolumn technology and interface performance and design can be expected to produce further improvements in this interesting area of chemical analysis.

REFERENCES

1. M. L. Lee, F. J. Yang, and K. D. Bartle, Open Tubular Column Gas Chromatography, Theory and Practice. John Wiley & Sons, New York, 1984.
2. J. Dicesare, M. Dong, and L. Ettre, Chromatographia, 14: 257 (1981).
3. F. J. Yang, J. Chromatogr. 236: 265 (1982).
4. M. Novotny, V. McGuffin, A. Hirose, and J. Gluckman, Chromatographia 17: 303 (1983).
5. D. Ishii, and T. Takeuchi, J. Chromatogr. 285: 97 (1984).
6. J. Jorgenson , and E. Guthrie, J. Chromatogr. 225: 335 (1983).
7. J. Jorgenson, L. Knecht, and E. Guthrie, Anal. Chem. 56: 418 (1984).
8. G. Guiochon, M. Gonnord, M. Zakaria, L. Beaver, and A. Siouffi, Chromatographia 17: 121 (1983).
9. J. C. Giddings, Anal. Chem. 39: 1927 (1967).
10. J. C. Giddings, Anal. Chem. 53: 945A (1981).
11. J. C. Giddings and R. Davis, Anal. Chem. 55: 418 (1983).
12. R. E. Majors, E. L. Johnson, S. P. Cram and A. C. Brown, III, Pittsburgh Conference on Analytical Chemistry and Spectroscopy, Cleveland, Ohio, Paper 115, 1979.
13. R. E. Majors, J. Chromatogr. Sci. 18: 571 (1980).
14. T. M. Chen, J. A. Apffel, and H. McNair, Amer. Chem. Soc. Div. Fuel Chem. 26: 7 (1981).
15. D. J. Luzbetak and J. J. Hoffman, J. chromatogr. Sci. 20: 132 (1982).
16. J. A. Apffel and H. McNair, J. Chromatogr., 279: 139 (1983).
17. C. V. Philip and R. G. Anthony, J. Chromatogr. Sci. 24: 438 (1986).
18. K. Ramsteiner, J. Chromatogr. 393: 123 (1987).
19. T. Raglione and R. Hartwick, Anal. Chem. 58: 2680 (1986).
20. T. V. Raglione, J. A. Trskosky, and R. A. Hartwick, J. Chromatogr. 409: 205 (1987).

21. T. V. Raglione, J. A. Trskosky, and R. A. Hartwick, J. Chromatogr. 409: 213 (1987).
22. K. Grob and G. Grob, J. Chromatogr. Sci. 7: 584 (1969).
23. F. J. Yang, A. C. Brown, and S. P. Cram, J. Chromatogr. 158: 91 (1978).
24. D. R. Deans, Anal Chem. 43: 2026 (1971).
25. W. Bertsch, C. S. G. Phillips, and V. Pretorius, J. High Resol. Chromatogr. & Chromatogr. Commun. 6: 234 (1982).
26. K. Grob, Jr., J. Chromatogr. 237: 15 (1982).
27. K. Grob, Jr. and R. Muller, J. Chromatogr. 244: 185 (1982).
28. V. Pretorius, C. S. Phillips, and W. Bertsch, J. High Resol. Chromatogr. & Chromatogr. Commun. 6:232 (1983).
29. G. Takeoka and W. Jennings, J. Chromatogr. Sci. 22: 177 (1984).
30. K. Grob Jr. and B. Schilling, J. Chromatogr. 39: 3 (1987).
31. V. Pretorius and K. Lawson, J. High Resol Chromatogr. & Chromatogr. Commun. 6: 335 (1986).
32. K. Grob Jr. and B. Schilling, Chromatographia 17: 361 (1983).
33. V. Pretorius and W. Bertsch, J. High Resol. Chromatogr. & Chromatogr. Commun. 6: 64 (1983).
34. K. Grob Jr., J. Chromatogr. 279: 225 (1983).
35. K. D. Bartle, B. W. Wright, and M. L. Lee, Chromatographia 7: 387 (1981).
36. B. A. Jones, K. E. Markides, J. S. Bradshaw, and M. L. Lee, Chromatography Forum 1: 38 (1986).
37. H. J. Cortes, C. D. Pfeiffer, G. L. Jewett, and B. E. Richter, J. Microcolumn Separations, 1: 28 (1989).
38. H. J. Cortes, C. D. Pfeiffer, and B. E. Richter, J. High Resol. Chromatogr. & Chromatogr. Commun. 8: 469 (1985).
39. K. Grob, Jr., Ch. Walder, and B. Schilling, J. High Resol. Chromatogr. & Chromatogr. Commun. 9: 95 (1986).
40. I. L. Davies, M. W. Raynor, J. P. Kithinji, K. D. Bartle, P. T. Williams, and G. E. Andrews, Anal. Chem. 60: 683A (1988).
41. D. Duquet, C. Dewaele, and M. Verzele, J. High Resol. Chromatogr. & Chromatogr. Commun. 3: 254 (1988).
42. J. Gluckman and M. Novotny, Microcolumn Separations, Elsevier Scientific Publishers, Amsterdam 1984.
43. P. Kucera, Microcolumn High Performance Liquid Chromatography, Elsevier Scientific Publishers, Amsterdam, 1984.
44. L. Snyder and J. Kirkland, Introduction to Modern Liquid Chromatography, 2nd ed., J. Wiley & Sons, New York, 1975.
45. J. C. Giddings, Dynamics of Chromatography, Marcel Dekker, Inc., New York, 1965.
46. I. Halasz, R. Endele, and J. Asshouer, J. Chromatogr. 112: 37 (1975).

47. J. Gluckman, A. Hirose, V. McGuffin, and M. Novotny, Chromatographia 17: 303 (1983).
48. F. Andreolini, C. Borra, and M. Novotny, Anal. Chem. 59: 2428 (1987).
49. T. Takeuchi and D. Ishii, J. Chromatogr. 213: 25 (1981).
50. F. Yang, Great Britain Patent 2128099A.
51. D. Shelly, J. Gluckman, and M. Novotny, Anal. Chem. 56: 2990 (1984).
52. H. J. Cortes, C. D. Pfeiffer, B. E. Richter, and T. Stevens, J. High Resol. Chromatogr. & Chromatogr. Comm. 10: 446 (1987).
53. D. Ishii, Introduction to Microscale High Performance Liquid Chromatography, VCH Publishers, New York, 1988.
54. Alltech Associates Inc., Deerfield, Michigan.
55. LC Packings, San Francisco, California,
56. K. Grob, Jr., D. Frollich, B. Schilling, H. Neukom, and P. Nageli, J. Chromatogr. 295: 55 (1984).
57. H. J. Cortes, C. D. Pfeiffer, B. E. Richter, and D. E. Jensen, J. Chromatogr. 349: 55 (1985).
58. F. Munari, A. Trisciani, C. Mapelli, S. Trestianu, K. Grob, Jr., and J. Colin, J. High Resol. Chromatogr. & Chromatogr. Commun. 9: 601 (1985).
59. H. J. Cortes and C. D. Pfeiffer, Chromatography Forum 4: 29 (1986).
60. K. Grob, Jr., H. Neukom, and R. Etter, J. Chromatogr. 357: 416 (1986).
61. K. Grob, Jr. and J. Stoll, J. High Resol. Chromatogr. & Chromatogr. Commun. 9: 518 (1986).
62. K. Grob, Jr. and T. Laubli, J. High Resol. Chromatogr & Chromatogr. Commun. 9: 593 (1986).
63. I. Davies, M. Raynor, P. Williams, G. Andrews, and K. Bartle, Anal. Chem. 59: 2579 (1987).
64. K. Grob, Jr., E. Muller, and W. Meier, J. High Resol. Chromatogr. & Chromatogr. Commun. 10: 416 (1987).
65. V. Gianesello, L. Bolzani, E. Brenn, and A. Gazzaniga, J. High Resol Chromatogr. & Chromatogr. Commun. 1: 99 (1988).
66. E. Noroozian, F. Maris, M. Nielen, R. Frei, G. deJong, and U. Th. Brinkman, J. High Resol. Chromatogr. & Chromatogr. Comm. 10: 17 (1987).
67. F. Maris, E. Noroozian, R. Otten, R. van Dijick, G. de Jong, and U. Th. Brinkman, J. High Resol Chromatogr. & Chromatogr. Commun. 11: 197 (1988).
68. Th. Noy, E. Weiss, T. Hrps, H. Van Crutchen, and J. Rijks, J. High Resol. chromatogr. & Chromatogr. Commun. 11: 181 (1988).
69. F. Munari and K. Grob, Jr., J. High Resol. Chromatogr. & Chromatogr. Commun. 2: 172 (1988).

70. I. L. Davies, M. W. Raynor, D. J. Irwin, K. D. Bartle, M. Tlay, E. Ekinci, and H. E. Schwartz, J. High Resol. chromatogr. & Chromatogr. Commun. 11: 792 (1988).
71. V. M. A. Hakkinen, M. M. Virolainen, and M. L. Riekkola, J. High Resol. Chromatogr. & Chromatogr. Commun. 2: 214 (1988).
72. B. Pacciarelli, E. Muller, R. Schneider, K. Grob, W. Steiner, and D. Frohlich, J. High Resol. Chromatogr. & Chromatogr. Commun. 11: 135 (1988).
73. H. J. Cortes, in Microbore Column Chromatography. (F. Yang, ed.), Marcel Dekker, Inc., New York, 1989.
74. A. Pawlese, D. deJong, and J. M. H. Van den Berg, J. High Resol. chromatogr. & Chromatogr. Commun. 8: 607 (1988).
75. D. Duquet, C. Dewaele, M. Verzele, and S. McKinley, J. High Resol Chromatogr. & Chromatogr. Commun. 11: 824 (1988).
76. F. Berthou and Y. Dreano, J. High Resol Chromatogr. & Chromatogr. Commun. 11: 706 (1988).
77. T. V. Raglione and R. A. Harwick, J. Chromatogr. 454: 157 (1988).
78. H. J. Cortes, C. D. Pfeiffer, G. L. Jewett, S. Martin, and C. Smith, Anal. Chem. 61: 961 (1989).

8

Multidimensional Supercritical Fluid Chromatography

ILONA L. DAVIES,* KARIN E. MARKIDES, and MILTON L. LEE
Brigham Young University, Provo, Utah

KEITH D. BARTLE   *The University of Leeds, Leeds, England*

I. SUPERCRITICAL FLUID CHROMATOGRAPHY IN ONE DIMENSION

Supercritical fluid chromatography (SFC) is a chromatographic technique in which the mobile phase is neither a gas nor a liquid, but is a supercritical fluid with physical properties that are intermediate between those of gases and liquids. Supercritical fluids are useful as mobile phases because of their variable solvating power and relatively high diffusivities (1).

A number of substances fulfill the criteria required of supercritical fluid mobile phases. Such fluids should possess suitable critical parameters and good solvating power, while being relatively safe to use. The most common mobile phase for SFC is carbon dioxide (CO_2), because it has practical critical parameters (T_c = 31.3°C; P_c = 72.9 atm), and in addition to being inert, CO_2 is compatible with flame-based detectors. One of the advantages that SFC has over high performance liquid chromatography (HPLC) when using CO_2 as a mobile phase is the ease with which it can be interfaced to a variety of detectors, including UV-absorbance detectors, flame ionization detectors (FIDs), mass spectrometry (MS) and Fourier transform infrared spectrometry (FTIR) (2). SFC also serves as a high resolution chromatographic technique for the analysis of high molecular mass solutes that are beyond the volatility range of gas chromatography (GC).

---

*Current affiliation: Dionex Corporation, Sunnyvale, California

The dependence of CO_2 density on pressure and temperature is illustrated in Figure 1. Near the critical point, small changes in pressure produce large changes in the density (and hence the solvating ability) of the CO_2. Above its critical point, a supercritical fluid has a density and solvating power approaching that of a liquid, but a viscosity similar to that of a gas, and a diffusivity intermediate between the two. The supercritical fluids which possess sufficiently strong intermolecular interactions with solutes have solvating abilities which make their use as chromatographic mobile phases highly favorable.

Four different operating parameters may be varied in SFC to achieve the optimum chromatographic selectivity and retention (Table 1). In common with GC and LC, temperature, mobile phase composition and stationary phase type may be varied in SFC. However, only SFC has the additional characteristic of being able to vary the solvating power of the mobile phase by changing its density during pressure programming. In SFC, more than one of the operating parameters can be varied simultaneously; for example, the temperature may be continuously increased during a density/pressure program.

SFC may be carried out with both packed and open-tubular columns: packed conventional (2.0–4.6 mm i.d.), packed microbore (1–2 mm i.d.), packed fused silica (0.2–0.5 mm i.d.), and open-tubular fused silica (25–100 μm i.d.). There are many advantages associated with the use of microcolumns in SFC. Because the pressure drop along an open-tubular column can be minimized, the length of the column can be increased considerable (3-20m) to yield a higher number of theoretical plates. Consequently, open-tubular SFC is associated with high efficiencies, high sensitivities and low mobile phase flow rates. Packed column SFC has the advantage of speed because relatively short columns (3-50 cm) can be used at high flow rates, and packed columns have high solute loading capacities due to the relatively high mass of stationary phase present.

SFC instrumentation has some similarities to both LC and GC. A syringe or reciprocating pump delivers the SFC mobile phase as a liquid to the injection valve. Microcolumn SFC usually requires a syringe pump to deliver low volumetric flow rates, while reciprocating pumps with cooled heads may be used with packed column SFC. The chromatographic column is usually held in a heated oven, and a pressure restrictor at the end of the column maintains supercritical conditions. SFC can utilize both GC and LC detectors, usually flame ionization detection for microcolumn SFC and UV absorption for conventional column SFC.

SFC with an FID is a good combination for the initial screening of unknown samples. Furthermore, SFC is the most suitable chromatographic technique for a wide range of different applications, includ-

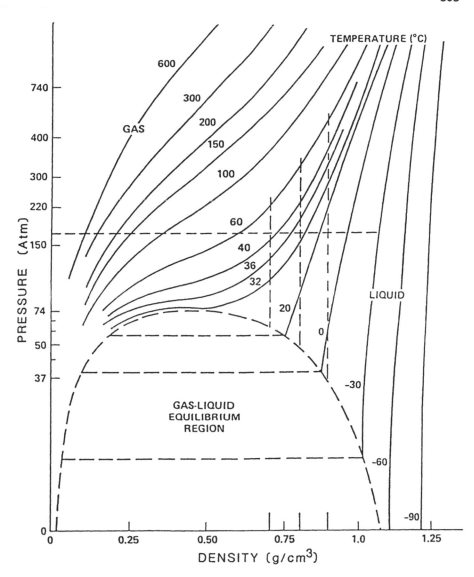

FIGURE 1   Carbon dioxide phase diagram.

ing those compounds which are thermally labile, nonvolatile, or
reactive (3). Thermally labile compounds, such as some carbamate
pesticides, azo dyes, polymer additives, explosives, and drugs,
cannot be analyzed by GC. The relatively low operating temperatures of SFC are essential for the analysis of these compounds.
For some high molecular mass species, such as polymers, biomole-
cules, and fossil fuels, SFC is an attractive alternative to size-
exclusion chromatography (SEC), which suffers from poor efficiency,
and LC, which lacks universal detection. SFC-FID is an excellent
alternative for simulated distillation of high-boiling heavy petroleum
fractions, and for group-type analysis of saturates, olefins, and
aromatics in petroleum fuels (4,5).

SFC is often applied to the analysis of high molecular mass
samples such as fossil fuels, but such samples generally become
more complex with increasing molecular mass because the number
of possible isomers increases. Many samples are so complex that
they cannot be resolved by a single chromatographic method alone.
Multidimensional SFC offers advantages similar to other coupled
chromatographic systems for the analysis of complex samples. Specifically, component resolution and identification are enhanced by
another chromatographic dimension, or by selective introduction or
detection techniques. The selectivity obtained by supercritical fluid
extraction (SFE) prior to on-line chromatography can sometimes
rival that obtained by coupled column methods. In both cases, the
first separation provides a cleanup, fractionation, or concentration
step prior to the final high resolution chromatographic analysis.

II. MULTIDIMENSIONAL CHROMATOGRAPHY WITH SUPERCRITICAL FLUIDS

A. General Concepts

There are many practical reasons why SFC may be desirable in a
multidimensional chromatographic system. Although GC provides a
highly efficient separation method, only the most volatile samples
can be analyzed by GC. Multidimensional GC-GC and LC-GC are
only applicable for the determination of volatile, thermally stable
solutes. Substituting open-tubular SFC for the capillary GC dimension of a coupled column system (that is SFC-SFC or LC-SFC) allows
a wider range of samples to be analyzed, with only a moderate
sacrifice in chromatographic efficiency. Alternatively, the LC dimension of LC-GC may be replaced by packed column SFC in order to
improve the compatibility between two mobile phases and to allow
the FID to be used for both separations. The techniques developed
for solvent evaporation at the LC-GC interface are often not required
in SFC-GC, because the solutes are deposited at the front of the
GC column when the CO_2 decompresses into a gas at the end of the

SFC column. Because of the relatively nonpolar nature of supercritical CO_2, SFC-GC is particularly recommended as a substitute
for many normal-phase LC-GC analyses.

Multidimensional LC techniques are frequently used in the biomedical field for the analysis of samples which are not amenable to
GC. However, on-line LC-LC can involve interfacing normal-phase
and reversed-phase LC eluents, an approach which requires an
intermediate solvent exchange step. The application of CO_2 as a
common mobile phase for both normal-phase and reversed-phase
chromatography is a more simple alternative to many LC-LC methods.
Often, the use of SFC in a multidimensional system (SFC-SFC, SFC-
GC, and LC-SFC) avoids many of the problems associated with the
interfacing of two dissimilar mobile phases.

Simple changes in the column-types and operating conditions of
multidimensional SFC separations can provide either GC-like or LC-
like conditions. SFC-SFC interfaces can be either rotary switching
valves (as for LC-GC) or flow-switching arrangements (as for GC-
GC). Intermediate solute focusing within a cold trap is optional
for multidimensional SFC. Although any type of chromatographic
column can be employed in multidimensional SFC systems, perhaps
the most desirable column combination is a packed primary column
(fast, high loading capacity) and an open-tubular secondary column
(high efficiency).

B. SFC-SFC

Although multidimensional SFC could be applied to a wide range of
sample types, most of the development work on this subject has involved the analysis of fossil fuels. Historically, the petroleum
industry has always provided a major thrust to the development of
new chromatographic techniques. Few other industries have such a
wide range of highly complex products that require detailed characterization by GC, SFC, LC, or multidimensional combinations of
these methods.

Chemical analysis of petroleum products frequently involves the
quantitation of the relative proportions of chemical classes (group-
type determinations) of compounds present, or the determination of
specific components within the fuel. Coupled LC-LC can separate
high boiling petroleum residues into groups of saturates, olefins,
aromatics, and polar compounds. However, the lack of a suitable
mass-sensitive, universal detector in LC makes quantitation difficult.
It was this need to quantitate hydrocarbon groups in fuels that
provided the initial impetus to convert such methods to SFC-SFC,
where the FID could be utilized. In 1985, two research groups
published details of normal-phase LC-reversed-phase LC column
switching methods which had been converted to multidimensional

SFC by simply replacing the two mobile phases with supercritical δ CO$_2$.

The SFC-SFC system of Lundanes and Greibrokk (6) used two 6-port valves and three microbore packed columns in a configuration shown in Figure 2. The columns were (a) a 3 cm × 2.1 mm i.d. cyano-silica packed column, (b) a 25 cm × 1.3 mm i.d. silica packed column, and (c) a 25 cm × 1.3 mm i.d. silver-loaded silica packed column (AG$^+$). With supercritical CO$_2$ mobile phase, both UV and FID could be used in series. The combination of silica and Si-AgNO$_3$ columns gave separation of the saturates from the aromatics, and the trapped polar compounds were backflushed from the cyano column. Figure 3 shows a group separation of a high-boiling petroleum residue (bp > 350°C) using this SFC-SFC system with an FID. Quantitative results were directly obtained from the chromatograms. A later publication (7) reported an investigation of the use of nitrous oxide (N$_2$O) as mobile phase for multidimensional SFC. However, N$_2$O gave a higher background signal in the FID than did CO$_2$.

In 1988, Campbell et al. (8) described the use of SFC-SFC for the quantitative determination of saturates, olefins, and aromatics in gasolines and middle distillate fuels in a single chromatographic run.

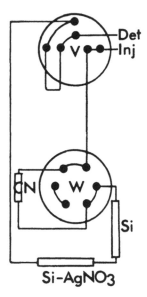

Si-AgNO3

FIGURE 2   Column-switching SFC-SFC system for group-type separation of fossil fuels into saturates, aromatics and polars. CN = 3 cm × 2.1 mm i.d. cyano-silica column; Si = 25 cm × 1.3 mm i.d. silica column; Si-AgNO$_3$ = 25 cm × 1.3 mm i.d. silver nitrate impregnated silica column; V = backflush valve; W = column selection valve. (reprinted with permission from Ref. 6.)

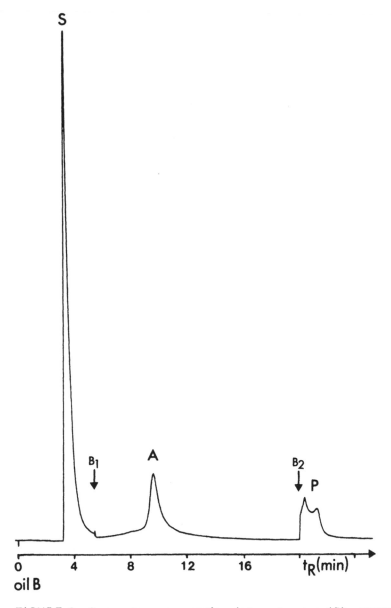

oil B

FIGURE 3   Group-type separation into saturates (S), aromatics (A), and polars (P) of a deasphalted high-boiling residue from a North Sea oil, using the column-switching system of Figure 2 with super-critical CO_2 mobile phase (0.34 mL min^{-1}) at 50°C and 270 bar. A 0.5 μL aliquot of 50% solution was injected, and an FID was used. Backflushing of the aromatics started at B_1, and backflushing of the polars started at B_2. (Reprinted with permission from Ref. 6.)

In order to adequately separate the saturates group from the olefinic group, two 1 mm i.d. microbore columns were used in a column-switching configuration with a mixed mobile phase of 10% CO_2 in sulfur hexafluoride (SF_6), as shown in Figure 4. The aromatic fraction was isolated on a silica gel column at 50°C, while a silver-loaded cation exchange column at 40°C was used to separate the saturates from the olefins. Group quantitation was achieved with an FID, which provided near uniform linear response for the hydrocarbons. Figure 5 shows the hydrocarbon group separation of a standard gasoline using this SFC-SFC system.

Unlike petroleum products, coal tars contain a large proportion of polycyclic aromatic compounds (PAC) with lower concentrations of aliphatic hydrocarbons. Multidimensional chromatography using supercritical carbon dioxide for the separation of PAC in a coal tar was first reported in 1985 by Christensen (9). Two 4.6 mm i.d. packed columns (aminosilane-bonded silica and octadecylsilane-bonded

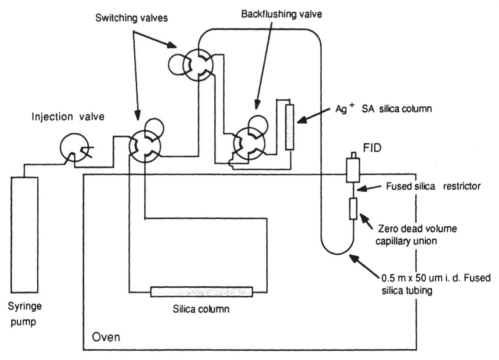

FIGURE 4 Schematic diagram of a column-switching SFC-SFC system for fossil fuel hydrocarbon group-type analysis. Columns: 25 cm × 1 mm i.d. packed silica microbore column; 25 cm × 1 mm i.d. packed silver-loaded sulphonic acid silica microbore column. (Reprinted with permission from Ref. 8. Copyright (1988) American Chemical Society.)

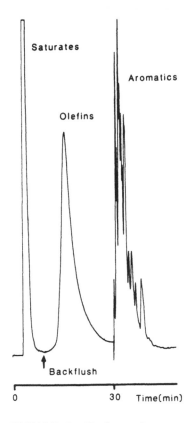

Saturates

Aromatics

Olefins

↑
Backflush

0   30   Time(min)

FIGURE 5 Hydrocarbon group-type separation of a gasoline sample
using the column-switching SFC-SFC system of Figure 4. Mobile
phase: 10% CO_2 in SF_6 (mole percent); detection by FID. (Reprinted with permission from Ref. 8. Copyright 1988, American Chemical
Society.)

silica) were arranged in series with two 6-port switching valves, so
that they could be switched in or out of the mobile phase flow
(Figure 6). This system was adapted from the on-line normal-phase
LC-reversed-phase LC method described by Sonnefeld et al. (10).
(Carrying out both separations using the same mobile phase avoided
the difficulties previously experienced with the intermediate solvent
exchange procedure). When the secondary column was switched out
of the flow, the aminosilane primary column (NH_2) eluted coal tar
PAC at 250 atm and 50°C in order of increasing aromatic ring number,
as shown in Figure 7(a). Shortly before the elution of the "chrysene"
fraction of interest, the secondary octadecylsilane column (ODS) was
switched into the flow path and the NH_2 column was switched out.

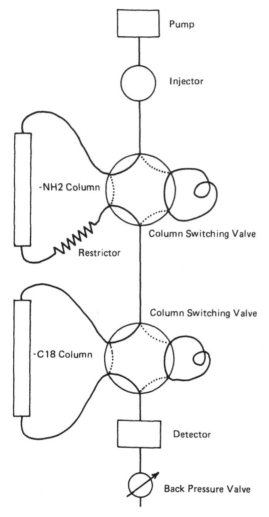

FIGURE 6   Schematic diagram of a multidimensional SFC-SFC system for coal tar PAC fractionation.   Columns: 25 cm × 4.6 mm i.d. packed NH_2-silica column held at 250 atm; 25 cm × 4.6 mm i.d. packed C_{18}-silica column held at 100 atm.   (Reprinted with permission from Ref. 9.)

PAC within this fraction were eluted at 100 atm using a negative temperature ramp (50°C to 35°C at 4°C min^{-1}). The "chrysene" fraction was resolved into triphenylene, benz[a]anthracene and chrysene on the ODS column (Figure 7(b)). Although the use of conventional packed columns in this system resulted in low efficiencies, the high mobile phase flow rates provided fast analysis times.

Davies et al. (11) first reported open-tubular column multidimensional SFC-SFC using 50-µm i.d. fused silica columns, CO_2 mobile phase, and a flow-switching interface. The capillary SFC-SFC system was comprised of a syringe pump, two ovens equipped with FIDs, and two pump/oven controllers (Figure 8). A solvent-vent injection method was incorporated in this system which allowed sample injections of microliter volumes (12). By means of a bypass line and either a T-piece interface or an offset-cross interface (Figure 9), solutes could be directed from the primary column into the secondary column. The solutes in the primary column when the mobile phase flow was diverted to the secondary column remained "parked" there until flow was once again resumed through both columns.

An unknown sample is generally first chromatographed by running it straight through both columns in series. Components present in the sample are identified by comparing their relative retention times with those of known standards. Simultaneous elution of compounds on either column is noted and a fractionation scheme is planned in order to separate the components of interest by taking advantage of the different selectivities of the two columns. The sample is then reinjected and, at a certain predetermined point during the run (monitored by the first detector), the flow is switched to the second column. After all of the desired compounds have eluted through the secondary column, the pressure/density is lowered and the bypass valve is returned to the forward flow position. This fractionating procedure can be repeated as many times as required during a run, with only one sample injection. Standard mixtures can be subjected to the same fractionation method to assist in the identification of unknowns.

This multidimensional microcolumn SFC-SFC method was used to resolve isomeric PAC in a coal tar extract, to isolate a labile pesticide (carbofuran) from a bird extract, and to separate aliphatic/ aromatic hydrocarbons in a high-boiling petroleum distillate. The coal tar chroamtographic results are shown in Figure 10. A standard coal tar extract was subjected to capillary SFC-SFC in an attempt to resolve all of the major PAC isomers in one analysis. Previous work on this sample had required several subsequent LC and GC separations (13). For this application, a 10 m SB-Biphenyl-30 primary column was interfaced to a 6 m SB-Smectic liquid crystal secondary column (both columns obtained from Lee Scientific, Salt Lake City, Utah).

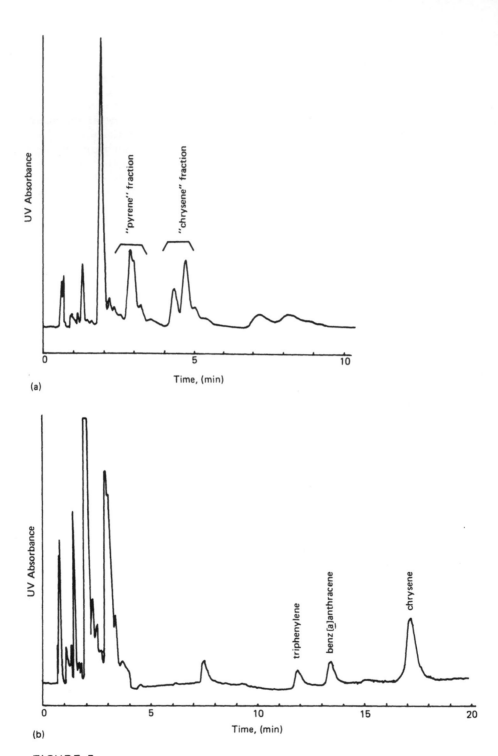

(a)

(b)

FIGURE 7

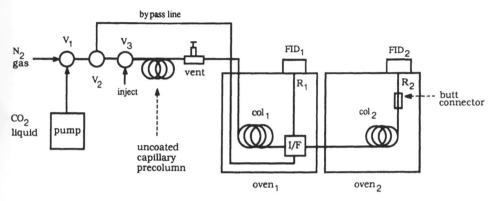

FIGURE 8 Schematic diagram of a multidimensional microcolumn
SFC-SFC system using a flow-switching interface. (Reprinted with
permission from Ref. 11.)

On the first column, the PAC in the sample eluted principally
in groups of increasing molecular mass and polarizability: 3-ring
phenanthrene/anthracene, followed by 4-ring fluoranthene/pyrene
and chrysene/triphenylene/benz[a]anthracene, 5-ring benzofluoran-
thenes and benzopyrenes, 6-ring indeno[1,2,3-cd]pyrene/benzo[ghi]
perylene/anthanthrene, and 7-ring coronene. Many of these isomeric
groups were separated on the secondary column. As can be seen in
Figure 10, the shape-selective liquid crystalline stationary phase in
the second dimension resolved a number of isomers including anthracene and phenanthrene; triphenylene, benz[a]anthracene, and chrysene; and the benzofluoranthene group into the a, j, b, and k isomers. Preliminary quantitative work using this method indicated
that SFC-SFC is reproducible and accurate when using an internal
standard. This coal tar SFC-SFC application shows some similarities
to both LC and GC, depending on the conditions used. The primary
separation can be compared to that of a normal-phase LC aromatic

FIGURE 7 (a) SFC chromatogram of a coal tar extract run on an
NH_2-silica primary column, using the system shown in Figure 6.
The "chrysene" fraction was selected for further resolution on the
C_{18}-silica column. Conditions: 250 atm; 50°C; CO_2 mobile phase at
4 mL min^{-1}; UV detection. (b) Separation of the "chrysene" fraction,
cut from the NH_2-silica primary column, on the C_{18}-silica secondary
column using the system shown in Figure 6. Conditions: 100 atm;
CO_2 mobile phase at 4 mL min^{-1}; UV detection; 50°C to 35°C at 4°C
min^{-1}. (Reprinted with permission from Ref. 9.)

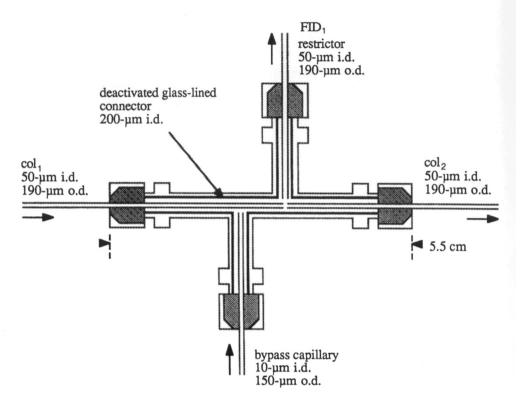

FIGURE 9 Details of the offset-cross flow-switching interface between the primary and secondary columns of the SFC-SFC system of Figure 8. (Reprinted with permission from Ref. 11.)

FIGURE 10 Open-tubular column SFC-SFC chromatograms of a standard coal tar extract (SRM 1597), using the system of Figure 8. Conditions: CO_2 mobile phase, 2 cm s^{-1}; density program from 0.20 g mL^{-1} (hold for 5 min) to 0.76 g mL^{-1} at a ramp rate of 0.005 g mL^{-1} min^{-1}; Column 1: 10 m × 50 μm i.d. SB-Biphenyl-30 at 100°C; Column 2: 6 m × 50 μm i.d. SB-Smectic liquid crystal at 120°C; Interface: T-piece. 0.5-μL aliquot of sample injected using the solvent-vent technique. Peak identifications as follows: 1. phenanthrene, 2. anthracene, 3. fluoranthene, 4. pyrene, 5. triphenylene, 6. benz[a]anthracene, 7. chrysene, 8. benzo[a]fluoranthene, 9. benzo[j]fluoranthene, 10. benzo[b]fluoranthene, 11. benzo[e]pyrene, 12. perylene, 13. benzo[k]fluoranthene, 14. benzo[a]pyrene, 15. indeno[1,2,3-cd]pyrene, 16. benzo[ghi]perylene, 17. anthanthrene, and 18. coronene. (Reprinted with permission from Ref. 11.)

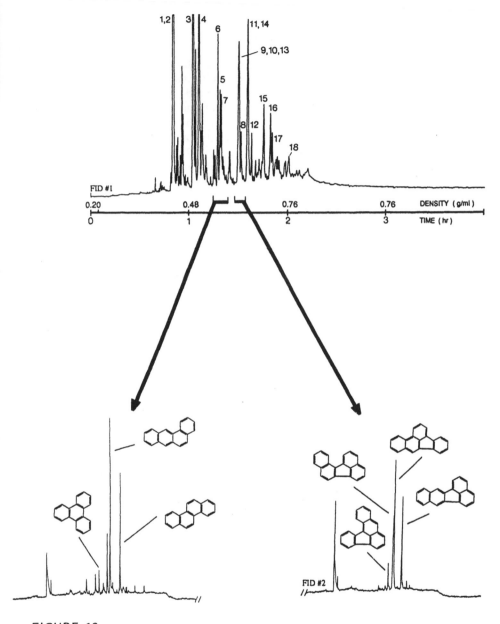

FIGURE 10

ring-number separation, whereas the second separation is similar to that obtained by reversed-phase LC, or by GC using a liquid crystalline phase.

Recently, Payne et al. (14) analyzed the same coal tar standard sample by multidimensional SFC-SFC using two packed-capillary (250-μm i.d.) fused silica columns. Packed microcolumns were chosen in an attempt to shorten the analysis time. A 25 cm packed NH_2-silica column was connected to a 55 cm packed C_8-silica column using the offset-cross interface. Results of this SFC-SFC separation are shown in Figure 11; the primary PAC separation according to the number of aromatic rings is adequate, and the analysis time is considerably shorter than that obtained previously using open-tubular columns (11). However, the secondary separation lacks the selectivity and efficiency required for this particular sample. Obviously, the ideal SFC-SFC system would consist of a short (10–30 cm) packed-capillary primary column interfaced to a long (5–10 m) open-tubular column. Attempts to use the flow-switching interface with such a column combination were unsuccessful due to the different flowrates required for each column type. A rotary valve interface with intermediate solute trapping would be more suitable.

The concept of using a rotary switching valve interface for microcolumn SFC-SFC was discussed by David et al. (15). Systems were proposed for either 100 μm i.d. open-tubular columns or 250 μm i.d. packed-capillary columns. Figure 12 shows one configuration proposed for the direct transfer of selected solutes from the primary to the secondary column by way of a 4-port valve interface. Pressure restrictors leading to FIDs would allow components eluting from each column to be detected. The authors suggested refocusing the transferred solutes by operating the secondary column at a lower density than the primary column. Because two pumps were proposed, it would be possible to run the elution program in the secondary column independent of that in the primary column. Low

---

FIGURE 11 Packed-capillary column SFC-SFC chromatograms of a standard coal tar extract (SRM 1597) using the system of Figure 8. Conditions: column 1: 25 cm × 250 μm i.d. NH_2-silica packed-capillary; CO_2 mobile phase at 0.9 cm s^{-1}; 100°C; column 2: 55 cm × 250 μm i.d. C_8-silica packed-capillary; CO_2 mobile phase at 0.6 cm s^{-1}; 120°C; density program 0.35 (2 min) to 0.755 (hold) at 0.008 g mL^{-1} min^{-1}; 1.2-μL aliquot sample injected using solvent-vent technique. Peak identifications as follows: 1. phenanthrene, 2. fluoranthene, 3. pyrene, 4. benz[a]anthracene, 5. chrysene, 6. benzofluoranthene isomers, 7. benzopyrene isomers, 8. anthanthrene, and 9. coronene. (Reprinted with permission from Ref. 14.)

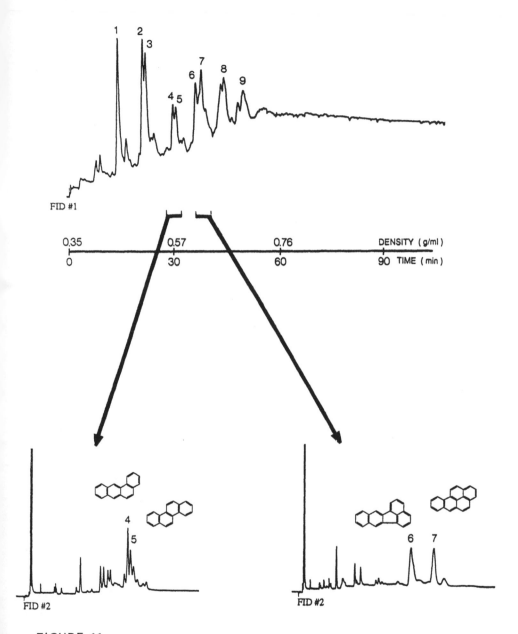

FIGURE 11

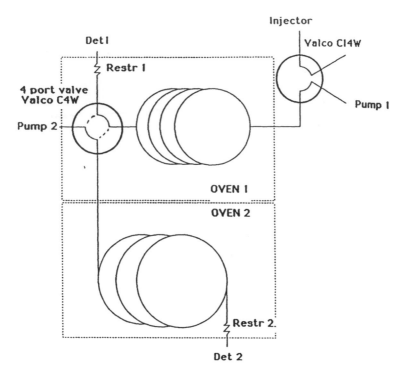

FIGURE 12 Proposed microcolumn SFC-SFC configuration using a
valve interface for direct heartcut analysis. (Reprinted with permission from Ref. 15.)

initial temperatures and pressures were suggested for the secondary
separation; it was thought that solute volatility would hinder solute
focusing if the temperature in the second oven was increased (to
decrease the mobile phase density). The major problem with this
design was thought to be the experience of some difficulty in balancing the pressure at the interface during switching.

    A second proposal for valve-switching in SFC-SFC (Figure 13)
incorporates a cold trap at the interface. Solutes from the first
column can be transferred into the trap by switching a 3-port valve,
while the end of the second column is open to the atmosphere through
a 3-port outlet valve. After transfer, the two 3-port valves are
turned simultaneously back to their respective FID restrictor positions. Unfortunately, this configuration does not provide for monitoring of the FID signal from the primary column during solute transfer into the cold trap.

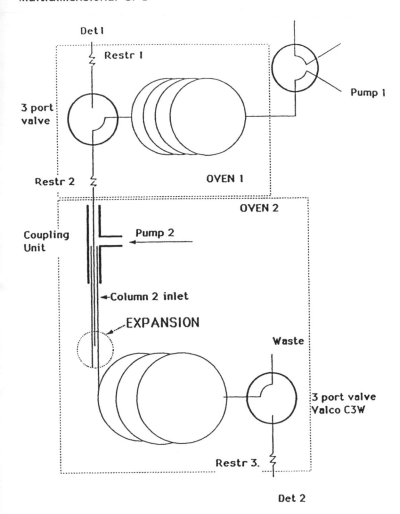

FIGURE 13 Proposed microcolumn SFC-SFC configuration using a valve interface with intermediate solute focusing. (Reprinted with permission from Ref. 15.)

No results were shown to support these SFC-SFC proposals; however, the authors reported that while packed-capillary columns worked well with the latter configuration, several technological (dead-volume) difficulties were experienced when working with 100 µm open-tubular columns. Research in this area will obviously continue, and results of microcolumn (packed-capillary and open-tubular column) SFC-SFC using rotary valve interfaces can be expected within the next few years.

C. SFC-GC

Levy et al. (16) were the first to report coupled SFC-GC using conventional packed column SFC on-line with capillary GC for fossil fuel analysis. A heated capillary restrictor was passed directly from the end of the SFC column into a GC split vaporization injector (Figure 14). Examples of SFC-GC applications reported by this group included the use of combined packed cyano-silica and silica SFC columns for the separation of an aviation fuel into saturates, 1-ring and 2-ring aromatics; and a packed ODS SFC column for the isolation of PAC from complex hydrocarbon mixtures.

SFC-GC also was proven useful for group-type separations of high-olefin gasoline fuels (saturates, olefins and aromatics) (5,17). The SFC separations were performed with CO_2 on four conventional size packed columns in series: (a) silica, (b) Ag^+-loaded silica, (c) cation-exchange silica, and (d) NH_2-silica. Heartcut fractions were transferred into a capillary GC column coated with a methyl polysiloxane stationary phase (Figure 15). Cryofocusing at 50°C was used for focusing during transfer into the GC.

D. LC-SFC

Combining LC with SFC was recently predicted (4,18), but research in this area is so new that only one application has been published. SEC coupled on-line to capillary SFC was reported by Lurie (19) for the determination of high molecular mass trace impurities in cocaine. Both SEC and SFC are used to analyze high molecular mass species, so their combination may be advantageous (20). The two techniques were coupled via a 10-port switching valve, that directed the SEC eluent through a 200 nL injection valve on the SFC instrument (Figure 16). Conventional SEC columns were used with dichloromethane mobile phase, and the SFC separation was performed on a 10 m SB-Biphenyl-30 open-tubular SFC column. In this case, SFC provided the high resolution analytical step after SEC fractionation; the results are shown in Figure 17. Unfortunately, due to the nature of the interface and the use of conventional-size SEC columns, only small portions of the SEC fractions could be analyzed by SFC; packed-capillary SEC would be an obvious improvement to reduce the primary column peak volumes.

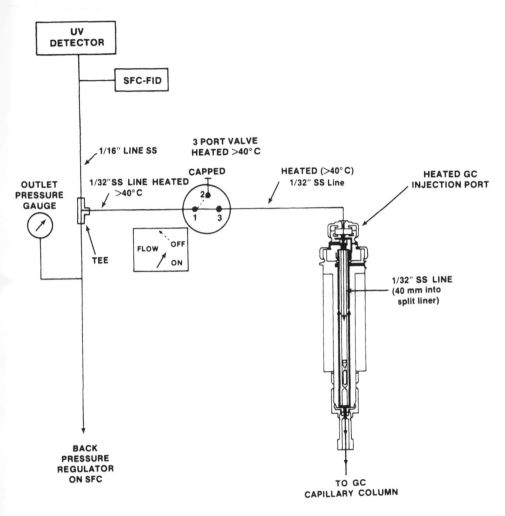

FIGURE 14   Schematic diagram of an on-line packed column SFC-capillary GC interface used for fuel analysis. (Reprinted with permission from Ref. 16.)

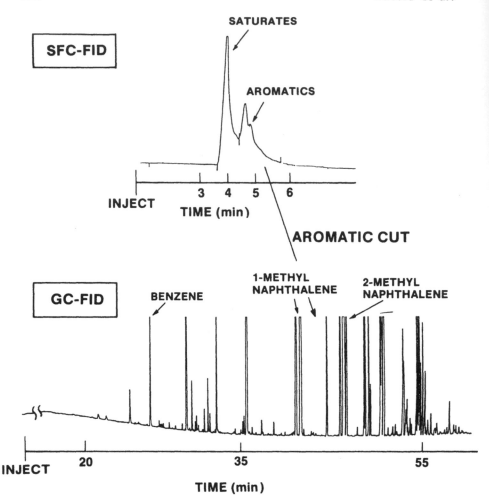

FIGURE 15 SFC-GC characterization of the aromatic fraction of a gasoline fuel. SFC conditions: Four 4.6 mm i.d. columns in series—silica, silver-loaded silica, cation exchange silica, and NH_2-silica; 50°C; 2850 psi; CO_2 mobile phase at 2.5 mL min^{-1}; FID. GC conditions: 50 m × 0.2 mm i.d. methyl silicone column; injector split ratio 80:1; injector temperature 250°C; helium; -50°C (8 min) to 320°C at 5°C min^{-1}; FID. (Reprinted with permission from Ref. 17.)

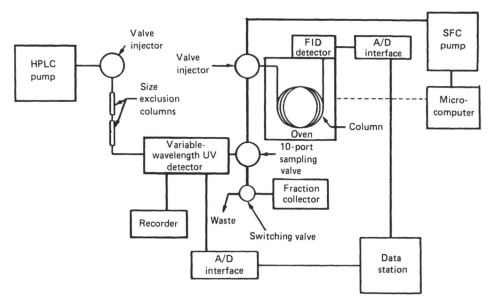

FIGURE 16 Schematic diagram of a coupled LC-SFC system using packed column SEC and open-tubular column SFC. (Reprinted with permission from Ref. 19.)

E. SFE-Chromatography

The unique properties of supercritical fluids, which make them so attractive as chromatographic mobile phases, have also provided the impetus for the development of SFE techniques (21).   Commonly-used supercritical fluids such as CO_2 possess many characteristics that are also favorable for extraction purposes: supercritical fluid CO_2 exhibits rapid mass transfer properties because of its low viscosity and high solute diffusivity, resulting in faster and more efficient extractions than is possible with traditional methods such as Soxhlet extraction; CO_2 also has useful critical parameters which provide mild extraction conditions, avoiding the decomposition of labile components; and CO_2 is a nontoxic solvent which decompresses into a gas at room temperature and pressure, facilitating fraction collection and interfacing to chromatography.

One of the most important factors to be considered in SFE is that the solvent strength of supercritical fluid CO_2 is directly related to its density and dielectric constant, both of which increase as a function of pressure at constant temperature (Figure 18).   The greatest variation in CO_2 solvent strength at 50°C occurs between operating pressures of 70 atm and 150 atm, and this phenomenon

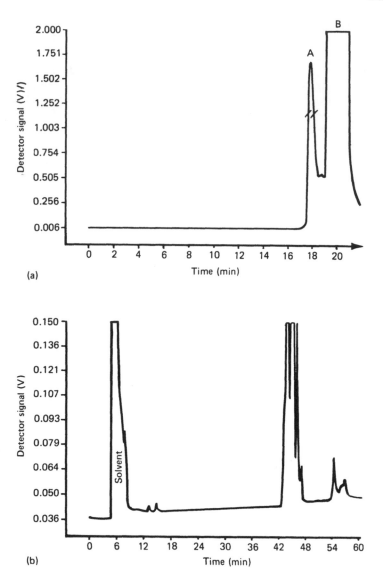

FIGURE 17 (a) SEC chromatogram of illicit cocaine, using the LC-SFC system of Figure 16. Conditions: Two 4.6 mm i.d. SEC columns in series; dichoromethane mobile phase at 1.0 mL min^{-1}; UV detection. A = impurities, B = cocaine fraction. Apex portion of peak A was diverted on-line to capillary SFC for further analysis. (b) On-line open-tubular column SFC chromatogram of SEC peak A. Conditions: 10 m × 50 μm i.d. SB-Biphenyl-30 capillary column; 140°C; CO_2 mobile phase; density program: 0.25 g mL^{-1} (5 min) to 0.35 g mL^{-1} at 0.1 g mL^{-1} min^{-1}, then to 0.65 mL^{-1} at 0.006 g mL^{-1} min^{-1}; FID. (Reprinted with permission from Ref. 19.)

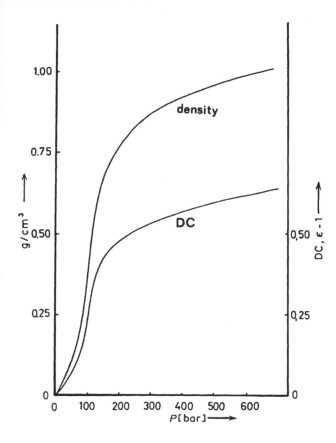

FIGURE 18   Density and dielectric constant (DC) of CO_2 as a function of pressure (50°C).

can be used to selectively extract specific fractions containing components of similar solubilities from a complex mixture (rather than exhaustively extracting all the components at once at high densities). Fractionation schemes can be planned by predicting extraction efficiencies under different practical conditions by calculating solute solubilities in the supercritical phase (22). However, more polar and higher molecular-mass solutes need stronger extraction fluids such as methanol-modified CO_2 to achieve good efficiencies.

Researchers are currently exploring the potential of coupled SFE-chromatographic (LC, SFC, or GC) analytical techniques. Of

TABLE 1 Operating Variables of Some Separation Techniques

Operating variable	GC	LC	SFC	SFE
Temperature	X		X	X
Pressure/density			X	X
Mobile phase		X	X	X
Stationary phase	X	X	X	

particular interest, is the extent to which on-line SFE chromatography may be used instead of coupled-column multidimensional chromatography for the analysis of complex samples (18). Certainly, aliphatic/aromatic selective SFE-GC of diesel exhaust particulates (23) can achieve similar separations to those obtained by Soxhlet extraction followed by LC-GC (24) or SFC-GC analysis, but with far less operator time. Similarly, PAC molecular-mass fractionation with on-line analysis can be provided by either SFE-GC (25) or SFE-capillary SFC (26,27), as shown in Figure 19, instead of by multidimensional SFC-SFC (11).

Limitations to the types of separations that can be achieved using SFE chromatography must exist, however. Because SFE selectivity is determined mainly by solute solubility in the supercritical extraction fluid, the special selectivities provided by certain (chiral, liquid crystal) chromatographic stationary phases are not available for isomeric separations by SFE alone (Table 1). In addition, trade-offs between selectivity and extraction efficiency must be considered, but an an on-line extraction method which can also provide crude primary fractions prior to on-line high resolution chromatography, SFE may be the answer to many analytical problems.

FIGURE 19 On-line SFE-capillary SFC of two fractions of PAC from a coal tar pitch. Conditions: CO_2 mobile phase at 110°C; 0.25 g mL^{-1} (10 min) to 0.74 g mL^{-1} at 0.006 g mL^{-1} min^{-1}; 10 m × 50 μm i.d. open-tubular SFC column (SB-Biphenyl-30); FID. Peak identifications as follows: 1. phenanthrene, 2. fluoranthene, 3. pyrene, 4. benz[a]anthracene, 5. chrysene, 6. benzofluoranthene isomers, and 7. benzopyrene isomers. (a) Extraction for 1 hr at 100 atm and 43°C; injection valve at 5°C (b) Extraction for 1 hr at 200 atm and 43°C; injection valve at 5°C for collection, -40°C for injection. (Reprinted with permission from Ref. 26.)

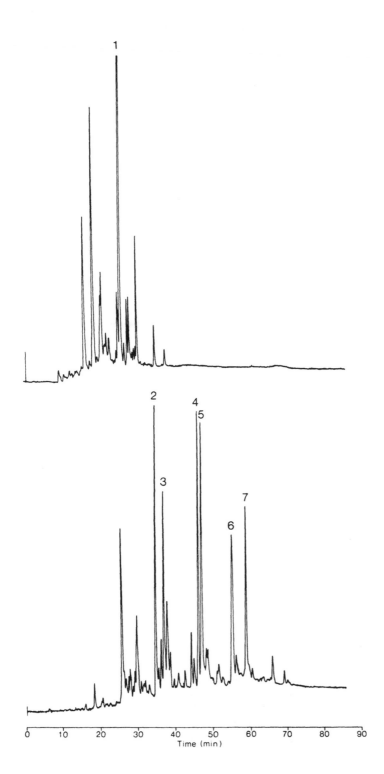

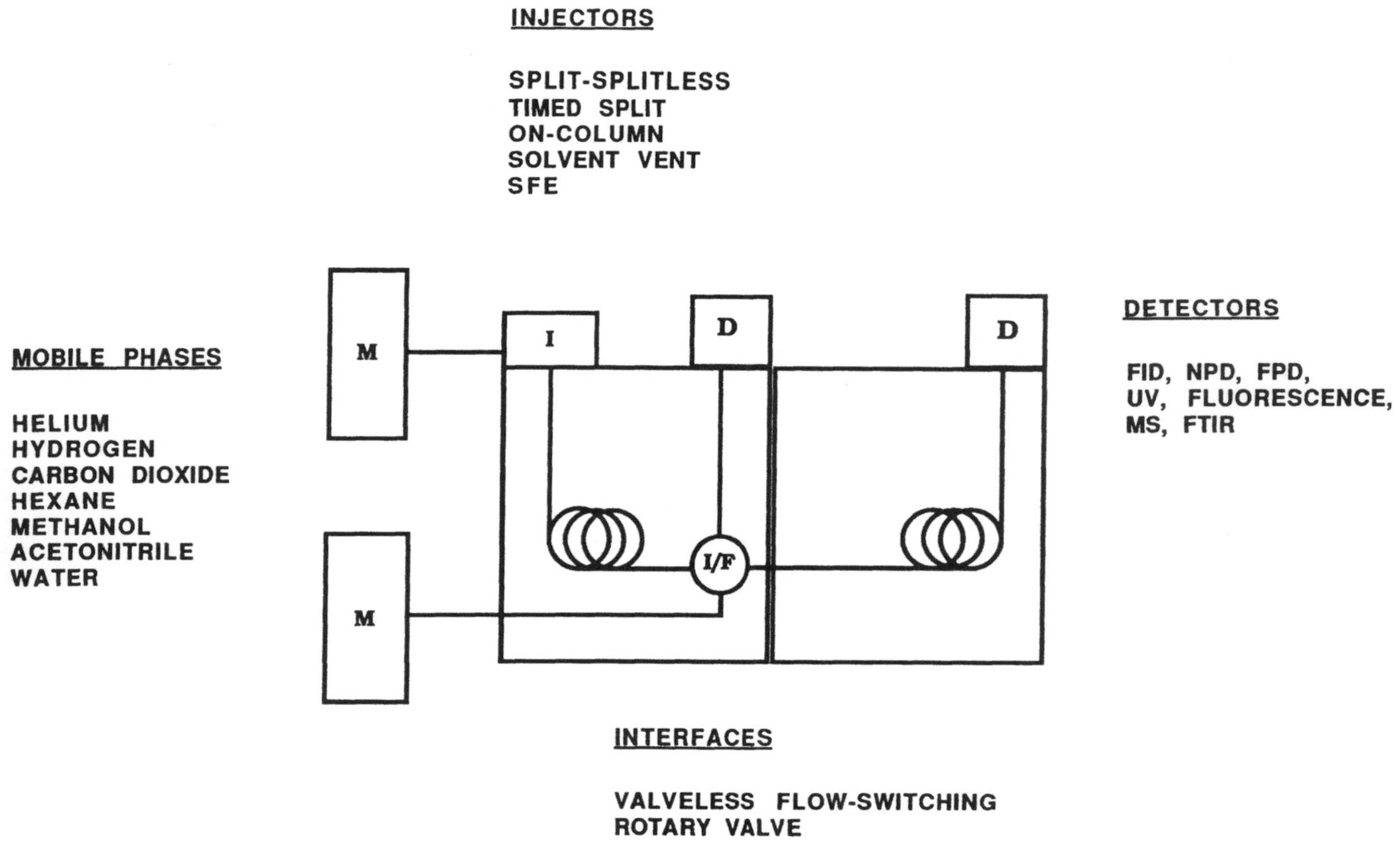

FIGURE 20 Proposed schematic diagram for unified multidimensional microcolumn chromatography. (Reprinted with permission from Ref. 28.)

III. FUTURE TRENDS AND POTENTIAL

Multidimensional SFC is a relatively new technique which is growing at a rapid pace. The major areas of growth in this area over the next few years can be expected to be in the areas of sample introduction (particularly SFE), informative and selective detection (such as FTIR, MS, and element-specific detectors), and interface configurations. A general trend toward the widespread use of microcolumns (packed-capillary and open-tubular columns) is expected, that should ease interfacing difficulties and improve efficiencies. SFE-GC and SFC-GC may replace normal-phase LC-GC, and SFC-SFC may be adopted for a wide range of applications, particularly for solutes that are not amenable to GC and as a replacement for some coupled LC-LC separations. Automated multidimensional microcolumn SFC, using a valve-switching interface and SFE sample introduction, will probably be the next major development in this field (28). In the long term, a unified approach to multidimensional microcolumn chromatography is possible (Figure 20), in which LC, SFC, and GC methods are used interchangeably in both dimensions (29).

REFERENCES

1. M. L. Lee and K. E. Markides, Science 235: 1342 (1987).
2. M. H. Hill, Jr. and C. R. Shumate, in Microbore Column Chromatography: A Unified Approach to Chromatography, (F. J. Yang, ed.), Marcel Dekker, Inc., New York, pp. 267—306, 1989.
3. K. E. Markides, and M. L. Lee, SFC Applications, Brigham Young University Press, Provo, Utah, 1988 and 1989.
4. F. J. Yang, in Microbore Column Chromatography: A Unified Approach to Chromatography, (F. J. Yang, ed.), Marcel Dekker, Inc., New York, 1—36, 1989.
5. H. E. Schwartz, J. M. Levy, and J. P. Guzowski, in Modern Supercritical Fluid Chromatography, (C. White, ed.), Hüthig Verlag, Heidelberg, 135—154, 1988.
6. E. Lundanes and T. Greibrokk, J. Chromatogr. 349: 439 (1985).
7. E. Lundanes, B. Iversen, and T. Greibrokk, J. Chromatogr. 366: 391 (1986).
8. R. M. Campbell, N. M. Djordjevic, K. E. Markides, and M. L. Lee, Anal. Chem. 60: 356 (1988).
9. R. G. Christensen, J. High Resolut. Chromatogr. Chromatogr. Commun. 8: 824 (1985).
10. W. J. Sonnefeld, W. H. Zoller, W. E. May, and S. A. Wise, Anal. Chem. 54: 723 (1982).
11. I. L. Davies, B. Xu, K. E. Markides, K. D. Bartle, and M. L. Lee, J. Microcolumn Separations 1:71 (1989).

12. M. L. Lee, B. Xu, E. C. Huang, N. M. Djordjevic, H-C. K. Chang, and K. E. Markides, J. Microcolumn Separations 1: 7 (1988).

13. S. A. Wise, B. A. Benner, G. D. Byrd, S. N. Chesler, R. E. Rebbert, and M. M. Schantz, Anal. Chem. 60: 887 (1988).

14. K. M. Payne, I. L. Davies, K. D. Bartle, K. E. Markides, M. L. Lee, J. Chromatogr., in press.

15. F. David, P. Sandra, C. Van Tilburg, Ninth International Conference on Capillary Chromatography, California, 1988, Book of Abstracts, (P. Sandra, ed.), Huethig, Heidelberg, p. 288–294, 1988.

16. J. M. Levy, J. P. Guzowski, and W. E. Huhak, J. High Resolut. Chromatogr. Chromatogr. Commun. 10: 337–341 (1987).

17. J. M. Levy and J. P. Guzowski, Fresenius Z. Anal. Chem. 330: 207 (1988).

18. I. L. Davies, M. W. Raynor, J. P. Kithinji, K. D. Bartle, P. T. Williams, and G. E. Andrews, Anal. Chem. 60: 683A (1988).

19. I. S. Lurie, LC.GC 6: 1066 (1988).

20. K. D. Bartle, I. L. Davies, and M. W. Raynor, in Size Exclusion Chromatography (B. J. Hunt and S. R. Holding, eds.), Blackie & Son Ltd., Glasgow, 1988.

21. M. McHugh and V. Krukonis, Supercritical Fluid Extraction. Principles and Practice, Butterworths, London, 1986.

22. G. F. Shilstone, M. W. Raynor, K. D. Bartle, A. A. Clifford, I. L. Davies, and S. A. Jafar. Polycyclic Aromatic Hydrocarbons, in press.

23. S. B. Hawthorne and D. J. Miller, J. Chromatogr. Sci. 24: 258 (1986).

24. I. L. Davies, M. W. Raynor, P. T. Williams, G. E. Andrews, and K. D. Bartle, Anal. Chem. 59: 2579 (1987).

25. B. W. Wright, S. R. Frye, D. G. McMinn, and R. D. Smith, Anal. Chem. 59: 640 (1987).

26. K. D. Bartle, M. P. Burke, A. A. Clifford, I. L. Davies, J. P. Kithinji, M. W. Raynor, G. F. Shilstone, and A. Williams, European Chromatography News 2: 12 (1988).

27. M. W. Raynor, I. L. Davies, K. D. Bartle, A. A. Clifford, P. T. Williams, J. M. Chalmers, and B. W. Cook, J. High Resolut. Chromatogr. Chromatogr. Commun. 11:766 (1988).

28. I. L. Davies, Ph.D. Thesis, The University of Leeds, Leeds, July 1989.

29. J. P. Kithinji, J. Microcolumn Separations 1: 63 (1989).

9

Coupled Supercritical Fluid Extraction – Supercritical Fluid Chromatography

MUNEO SAITO, TOSHINOBU HONDO, and MASAAKI SENDA *JASCO, Japan Spectroscopic Co., Ltd., Tokyo, Japan*

I. INTRODUCTION

Hannay and Hogarth [1] first found that supercritical fluid or dense gas had solvating power more than a century ago. However, in practice, two major applications of supercritical fluids were started only in the 1960s. These are supercritical fluid extraction (abbreviated to SFE in this chapter in consistency with SFC, though many different abbreviations are now used) and supercritical fluid chromatography (abbreviated to SFC). There is one more event in the 1960s that must not be neglected: the advent of high-performance liquid chromatography (HPLC).

HPLC is a separation analysis method not only contemporary with SFE and SFC, but also with a similar history of development. Chromatography originated with Tswett [2] in 1903 as liquid chromatography. Rapid development took place in the late 1960s [3]. These three techniques use many of the same instruments: high-pressure pumps, sample introduction devices, and packed or hollow separation columns. Since the late 1960s, numerous reports on HPLC, SFE, and SFC have been published. However, they seem to have been developing independently, and have so far little to do with each other from the viewpoint of practical instrumentation.

Supercritical fluid chromatography (SFC) that uses supercritical fluid as the mobile phase originated with Klesper and co-workers [4, 5] from high-pressure gas chromatography, and it was developed by several research groups. Myers and Giddings [6–8], Jentoft and Gouw [9–11], Novotny and co-workers [12], Bartmann and Schneider [13], and others in the 60s and 70s. In the early 1980s, advances

in micro-HPLC renewed the interest in SFC. Rapid mass transfer
in supercritical mobile phase attracted researchers as it offers high
speed separation with high resolution on an open tubular capillary
column, which was not very successful in liquid chromatography because of slow mass transfer, and also on a packed capillary column.
The small consumption of a fluid encouraged chromatographers to use
inflammable and even toxic fluids under high pressure and high temperature. Thus, extensive research works have been carried out by
numbers of groups [14—25].

In addition to the use of a simple UV detector with a high-pressure cell, combinations of SFC with new detectors have also been
investigated: SFC-UV spectrometry by using a photodiode-array
multiwavelength UV detector by the authors [26,27] and Jinno and
the authors [28,29], SFC-mass spectrometry (SFC-MS) by Smith and
co-workers [16,21], and Crowther and Henion [30], SFC-Fourier
transform infrared spectrometry (SFC-FTIR) by Shafer and Griffiths
[31], Olesik and co-workers [32], Johnson and co-workers [33],
SFC-infrared spectrometry based on buffer memory technique by Fujimoto and co-workers [34], SFC-flame ionization detection (SFC-FID)
by Rawdon and Norris [35,36], and Chester [37]. SFC systems based
on HPLC instrumentation were also reported by Gere and co-workers
[38], and Greibrokk and co-workers [39].

SFE that uses supercritical fluid as extraction medium was introduced by Zosel [40,41]. Since then the method has developed into
an industrial-scale extraction technique as reported by many research
groups such as Hubert and Vitzthum [42], Peter and Brunner [43,44] ,
Eggers and Tschierch [45], Stahl and co-workers [46,47], Coenen
and Rinza [48], Brogle [49], Filippi [50], Bott [51], Vollbrecht [52],
Calame and Steiner [53], Gardner [54], and others. For obtaining
better performance in SFE, the extract must be subjected to separation analysis and checked if the target component is efficiently extracted in order to optimize extraction conditions, i.e., the suitable
pressure and the temperature of the fluid. For such a purpose,
chromatography is the most essential technique and various types of
chromatography such as thin-layer chromatography (TLC), gas chromatography (GC) and high-performance liquid chromatography (HPLC)
have been employed. The extracts have traditionally been analyzed
by the off-line technique after collecting the extract because the SFE
system generally used was a pilot plant and the primary function of
the system was to collect the extract and not to analyze it. Thus,
the analysis is regarded as a different process. The number of
reports on the on-line analysis of extract is quite limited; Stahl and
Schilts [55,56] developed an extraction system which was combined
with TLC. Nieass and co-workers [57,58] examined the solubility of
organic substances in liquefied carbon dioxide by using a high-pressure cylinder which was connected to an HPLC system. Unger and
Roumeliotis [59] reported a coupling device that allows on-line HPLC

analysis of extracts. Although SFC seems to be closer to SFE than
any other type of chromatography, direct coupling of SFE with SFC
had not been attempted before the authors reported [26,27,60,61]
the direct coupling technique in early 1985.

II. DIRECT COUPLING OF MICROSUPERCRITICAL FLUID EXTRACTION WITH SUPERCRITICAL FLUID CHROMATOGRAPHY

A. Introduction

Supercritical fluid extraction has so far been investigated and utilized
for mainly industrial scale extraction [40–54], though SFE can be
performed in microscale, and such SFE has many advantages over
pilot plant SFE from several viewpoints. The advantages are:

1. ease of building
2. ease of operation
3. small sample quantity
4. low running cost
5. on-line and/or off-line monitoring of extract with analytical
   instruments such as UV, chromatograph, IR, NMR, MS, etc.
6. potential sample pretreatment method for GC, HPLC and other
   chromatographic analyses

In addition, if toxic and environmentally hazardous samples such as
Polynuclear Aromatic Hydrocarbons (PAHs) are to be extracted and
analyzed, a pilot plant SFE is undesirable because serious environmental pollution and damage to operators' health could result from a
large amount of sample required for operation.

Petroleum pitch is known to contain a considerable amount of PAHs
including environmental pollutants [62]. In order to utilize the pitch
as an original material for highly valuable products by utilizing SFE,
reaction in supercritical fluid, etc., it is quite essential to investigate behavior in the extraction of those compounds with a supercritical fluid. For such a purpose, micro-SFE is a very suitable
method, especially when it is directly coupled with a separation analysis method. Because such a system allows the analysis without expos-
ing the extract to oxygen, light and high temperature, the analysis
can be carried out in totally inert ambient, when a fluid such as
carbon dioxide is used. Carbon dioxide is the most preferred fluid
both in SFE and SFC as an extraction medium [26,27,40–59,63] and
as a mobile phase [8,10,11,17,18,20,21,23,26–33,35–39], because it
has a relatively low critical pressure, 72.9 atm, and a low critical
temperature, 31.3°C. In addition, it is nontoxic, nonflammable
nonpolluting and inexpensive. We also used carbon dioxide as both
the extraction medium and the chromatographic mobile phase.

Traditionally, analysis of SFE extracts has been performed off-
line with GC, HPLC, and other types of chromatography. There are
only a few reports on direct coupling of SFE with chromatographic
techniques [55,56,59] as mentioned in the previous section. In 1985,
the authors reported [26,27] the directly coupled micro-SFE with
SFC system in which a photodiode array UV detector was incorporated
as a monitor. Skelton and co-workers [63] reported a sampling system based on micro-SFE for SFC with FTIR as a detector.

B. Instrumentation

In principle, an SFE system consists of a high-pressure pump, an
extraction vessel, a back-pressure regulator and a separation vessel.
Figure 1 shows schematic diagrams of typical SFE systems. There
are two basic types of SFE systems categorized by different separation methods of extracts from extraction media; system (A) is based
on pressure reduction that causes solubility decrease, and system (B)
is based on temperature change that causes solubility change. It
should be noted that, in supercritical fluid, solubility is dependent
not only on temperature but also on pressure. Therefore, temperature reduction does not always cause a decrease in the solubility of
the solute. In practice, both pressure and temperature change techniques are employed in combination for separation of the extract from
the fluid.

Figure 2 shows schematic diagrams of HPLC and SFC systems.
They are generally very similar to each other. However, there is a
very important device in an SFC system that distinguishes the SFC
system completely from the HPLC system. That is a back-pressure
regulator, which pressurizes the system, from a delivery pump
through a detector including an injector and a separation column,
above the critical pressure of the fluid used. Therefore, the main
pressure drop takes place abruptly in the back-pressure regulator.
On the other hand in the HPLC system, the pressure drops gradually
as the mobile phase solvent passes through the clumn to atmospheric
pressure. In addition to the back-pressure regulator, column heating is also necessary to keep the mobile phase temperature above the
critical temperature of the fluid used.

There are two possible methods of coupling SFE with SFC:

1. coupling in parallel
2. coupling in series

The first method of coupling SFE with SFC is parallel connection.
This method requires independent SFE and SFC systems. In this
method, the SFE system can be a pilot plant or even an industrial
scale paint. The SFC system can be an ordinary one comprising
an injection valve, a separation column, a detector and a back-

A

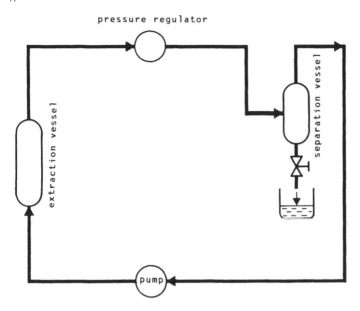

B

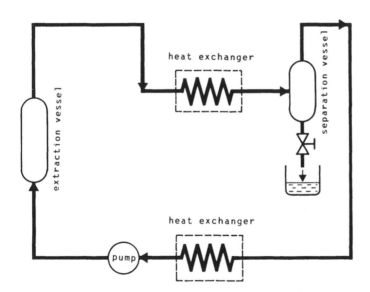

FIGURE 1 Schematic diagram of typical supercritical fluid extraction systems. A type of SFE system based on separation by the pressure reduction (A). Another type of SFE system based on separation by the temperature change (B).

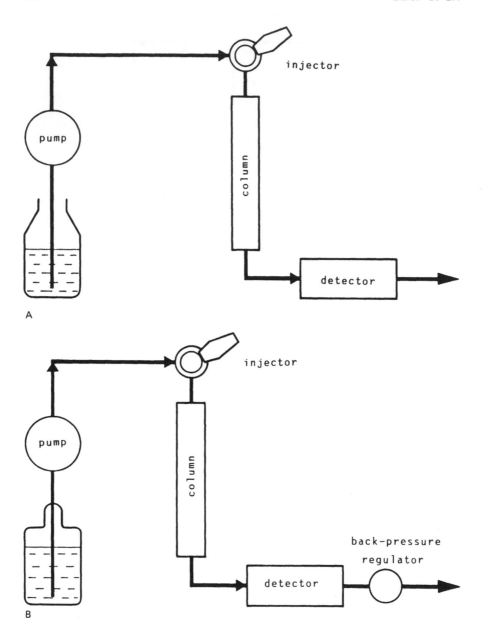

FIGURE 2   Schematic diagram of (A) HPLC and (B) SFC systems.

pressure regulator, as shown in Figure 2. The coupling line is arranged so that the fluid, or a part of the fluid, containing the extract flows through the sample loop of the injection valve in the SFC system. Introduction of the extract can be performed by changing the valve position in a similar way to sample introduction with a loop loading injection in HPLC. This method is more SFE process monitoring than analytical by its nature. We will not discuss this here.

The second method of coupling is series connection of SFE with SFC, which we will discuss in detail in this section. This is simple and recent advances in HPLC instrumentation readily permit us to build a micro-SFE system directly coupled with SFC by this method in a laboratory. As many researchers demonstrated [26,27,38,39], an HPLC pump with cooling jacket for the pump head can be used for delivery of liquefied carbon dioxide. A short empty column can be used as an extraction vessel, and a commercially available back-pressure regulator can conveniently be used for applying the suitable back-pressure without changing the mass flow rate of the fluid.

Figure 3 shows the hydraulic diagram of the system we constructed for direct coupling of micro-SFE with SFC. Liquefied carbon dioxide from cylinder (1) is fed to pump (2) whose pump heads are cooled with a coolant jacket at -5°C (BIP-1 modified for liquefied carbon dioxide delivery, JASCO, Tokyo, Japan). Coolant is supplied by a coolant circulator (3) (CoolPump, Taiyo Scientific Instruments, Tokyo, Japan). An entrainer or modifier solvent is delivered by pump (4) (JASCO BIP-1) and mixed with liquefied carbon dioxide. Liquefied carbon dioxide mixed with a modifier solvent enters an air-circulating oven (21) and (JASCO TU-100) is heated as the fluid flows through a heat exchanger coil (6). Six-way switching valves (7), (9), and (11) (Model 7000, Rheodyne, CA, U.S.A.) are switched in accordance with the desired mode. Effluent from the valve (11) is then led to a photodiode-array multiwavelength UV detector (14) with a high-pressure cell that withstands 300-kg/cm^2 pressure. An adsorbent trap column (16) is placed downstream of the detector and traps the extract dissolved in the fluid. A back-pressure regulator (18) controls the system pressure to desired value. A personal computer-based (NEC PC-9801, Tokyo, Japan) data processor acquires spectral data of the extract or chromatographic components.

The system allows several modes of operation:

1. simple SFC with syringe injection
2. real-time UV-absorption monitoring of extraction process
3. micro-SFE with an adsorbent trap column, that can be followed by off-line GC, HPLC, etc. analyses
4. directly coupled micro-SFE-SFC, i.e., stop-flow SFE with a trap loop, that is followed by SFC analysis with direct sample introduction

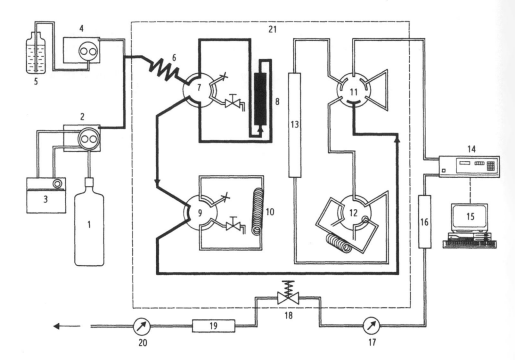

FIGURE 3 Hydraulic diagram of directly coupled micro-SFE-SFC and
flow line for supercritical fluid extraction. Components: 1 = carbon
dioxide cylinder; 2 = liquefied carbon dioxide delivery pump; 3 =
coolant circulating bath; 4 = modifier delivery pump; 5 = modifier
solvent reservoir; 6 = heat exchanger coil; 7 = six-way valve for
bypassing extraction vessel; 8 = extraction vessel; 9 = six-way valve
for bypassing trap loop; 10 = extract trap loop; 11 = six-way valve
for bypassing injector and separation column; 12 = injector; 13 =
separation column; 14 = multiwavelength detector; 15 = detector data
processor; 16 = extract trap column; 17 = back-pressure gauge;
18 = back-pressure regulator; 19 = trap for gas flow meter; 20 = gas
flow meter; 21 = oven; 22 = vent valve for extraction vessel used for
collecting extract; 23 = vent valve for trap loop. After SFE, the
valve (9) is switched to load the extract trap loop (10) with the
extract. The valve is then switched back to bypass the loop for pre-
pressurization and equilibration of the separation column (13), while
the loop holds the extract.

In order to operate the micro-SFE-SFC system successfully:

1. The volume of the extraction vessel should be appropriate for
    the sample size for SFC.
2. The pressure decrease of the supercritical carbon dioxide should
    be kept to a minimum during the transfer of the extract from
    the extraction vessel to the sample loop of the SFC system.
3. The SFC system should be prepressurized and equilibrated at
    the SFC analysis pressure before introducing the extract.

This system can be regarded as a new double-stage separation
analysis method that includes micro-SFE directly coupled with SFC.
In this new method, micro-SFE is utilized as the first separation step
in a similar way to a sample pretreatment in HPLC, and SFC is used
as the second separation step. This configuration allows the analyst
to place a raw and/or solid sample in the system in order to obtain
a chromatogram of the sample extract. Three-dimensional spectro-
metric data, namely absorbance, wavelength and time, graphically
presented in various fashions by the data processor, are very effective in the close examination of compounds in the SFE extract. Furthermore, application of built-in spectral correlation and peak deconvolution software, which was developed by Hoshino, Jinno, and the
authors [28,29], allows further investigation of chromatographic peak
components of the extract.

1. Flow line for simple SFC

In the simple SFC mode, carbon dioxide containing modifier solvent
if necessary directly flows into the separation column (13) via an
injector (12), bypassing the extraction vessel (8) and extract trap
loop (10) by means of valves (7) and (9). The column effluent is
then introduced into a UV detector (14) and all the above components
including the detector cell are pressurized by the back-pressure
regulator (18).

2. Flow line for real-time UV-absorption monitoring of extraction
    process

In the real-time extraction monitoring mode, carbon dioxide flows
through an extraction vessel (8) directly into the UV detector, bypassing all the other components in the oven (21) by means of valves
(7), (9), and (11).

3. Flow line for micro-SFE with adsorbent trap column

In the micro-SFE with adsorbent trap column mode, carbon dioxide
flows through the extraction vessel (8) and detector (14), bypassing
the extract trap loop (10), injector, and separation column (13), via

the adsorbent trap column (16), back-pressure gauge (17), and back-pressure regulator (18), and then is vented to the atmosphere. After SFE, the adsorbent trap column (16) is disconnected from the system, and the extract is eluted with an approprate solvent. Then, the extract is applied to other analytical instruments, such as GC, HPLC, etc.

4.   Flow line for directly coupled micro-SFE-SFC

The general flow line for the directly coupled micro-SFE-SFC mode is as indicated by the heavy line in Figure 4. Carbon dioxide is delivered to an extraction vessel (8), where extraction takes place, then to a six-way valve (9) with an extract trap loop (10), that is purged with carbon dioxide gas at atmospheric pressure prior to the extraction. Valve (11) is set in the nonconnecting position to make

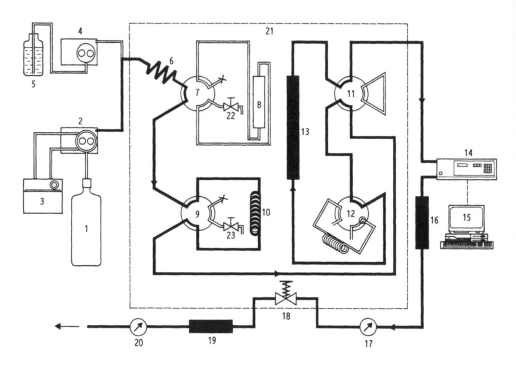

FIGURE 4 Flow line for supercritical fluid chromatography. After equilibration of the separation column under the chromatographic conditions, valve (10) is switched to inject the extract held in the trap loop into the column. Valve (10) is shown in the position for injection in this figure. Note that injector (13) is used only for ordinary syringe injection in the simple SFC mode.

a blocked end for the extraction line as shown in Figure 3 and, at
the same time, valve (11) maintains the pressure of the column that
has been prepressurized and is to be equilibrated at the SFC pressure. The extraction vessel (8), separation column (13), and adsorbent trap column (16) are thermostated in an air-circulating oven
(21).

At the beginning of the extraction, a pump (2) operated in the
constant pressure mode quickly delivers liquefied carbon dioxide at
its maximum flowrate to pressurize the extraction vessel (8). As the
pressure approaches the preset extraction pressure, the flowrate
gradually decreases and, finally, the flow will automatically stop when
the pressure reaches the preset value. Then, the pressure will be
maintained throughout the extraction period. On completion of the
extraction, a six-way valve (9) is switched, in the position shown in
Figure 4, to load a trap loop (10) with the extract. The pump (2)
automatically starts delivery to compensate for the pressure decrease
due to the transfer of carbon dioxide and the extract in the extraction vessel (8) to the trap loop (10) that has been purged with carbon dioxide gas at atmospheric pressure. When the transfer is completed and the pressure is restored, the pump (2) stops. Then, the
six-way valve (9) is switched back again, in the position shown in
Figure 3, so that loop (10) is bypassed, and the extract dissolved
in the supercritical carbon dioxide is held in the trap loop (10) until
the injection is made. Then, valves (7) and (11) are switched to
the SFC separation line, as indicated by the heavy line in Figure 4.
The system is now operated in the chromatography mode for equilibration of separation column (13).

Finally, the six-way valve (9) is switched, in the position shown
in Figure 4, to inject the extract into the separation column (13).
The chromatography mode can be easily converted from SFC to ordinary HPLC by using ordinary solvent without any hardware modification. A multiwavelength UV detector (14) (JASCO MULTI-320) together with its personal computer-based (NEC PC-9801) data processor (15) are used as an extraction and/or chromatographic monitor.
The flow cell, whose volume is 4 μL, is modified to withstand 300
kg/cm^2 pressure to meet pressure requirements in micro-SFE and
SFC. The flow line from the pump through the separation column to
the detector cell is kept under the necessary pressure for SFE and
SFC by a back-pressure regulator (18) (Model 26-3200-24, TESCOM,
Minnesota), where the main pressure drop takes place. The back
pressure is monitored by the back-pressure gauge (17) JASCO model
PG-350. Finally the column effluent is vented to the atmosphere
through a trap (19) and a flowmeter (20).

Various modes of operation of the system permit different types
of applications. In the next section, the analysis of PAHs in petroleum pitch extract will be described by utilizing the system operated
in the simple SFC and directly coupled micro-SFE-SFC modes. Appli-

cations of other modes of operation are described in the literature
[27,28,29].

III.   ANALYSIS OF POLYNUCLEAR AROMATIC
       HYDROCARBONS IN EXTRACT FROM
       PETROLEUM PITCH WITH CARBON DIOXIDE
       BY COUPLED MICRO-SUPERCRITICAL FLUID
       EXTRACTION-SUPERCRITICAL FLUID
       CHROMATOGRAPHY

A.   Experimental

1.   Materials and apparatus

Powdered petroleum pitch was gifted from Mr. Koike, Central R & D
Bureau, Nippon Steel Corporation, Tokyo, Japan. Standard samples
of polynuclear aromatic hydrocarbons (PAHs) were purchased from
Wako Pure Chemical Co., Ltd., Osaka, Japan. NBS standard reference material 1647 was also used as the standard mixture for
chromatographic analysis. Liquefied carbon dioxide, standard
grade, was from Toyoko Kagaku, Kawasaki, Japan. Tetrahydrofuran
(THF), reagent grade from Wako Chemicals, Osaka, Japan, was used
as a reference material to carbon dioxide as an extraction medium as
well as a mobile phase modifier in SFC. An SFC separation column
we used was 4.6 mm i.d. × 150 mm length packed with Develosil ODS
5 μm, Nomura Chemicals, Seto, Aichi, Japan. For gel-permeation
chromatography (GPC) analysis, a FinePak GEL 101 column, 7.2 mm
i.d. × 500 mm length, JASCO, Tokyo, Japan, havng the exclusion
limit of 3,000 as polystyrene molecular weight, was used.
       We constructed and used the directly coupled micro-SFE-SFC sys-
tem described in the previous section which is now commercially available from JASCO as MODEL SUPER-100.

2.   Procedure

First, we separated NBS standard reference material 1647 [62] by
simple syringe injection to the SFC section of the system, in order
to establish the chromatographic conditions for the separation of PAHs
and make a library file containing reference UV spectra for the compounds in the material. The library file was used for assignment of
compounds in the extracts from the petroleum pitch with supercritical
carbon dioxide and THF. Then, 10 mg of the petroleum pitch was
subjected to SFE with carbon dioxide at 300 kg/cm^2 pressure and
40°C temperature. The extract has been directly introduced into
the SFC section of the system by using the injector (12) in Fig. 3.
       For comparing extraction efficiency of supercritical carbon dioxide
for each PAH with that of THF, the THF extract of the same petroleum pitch was introduced into SFC by syringe injection.

Finally, the extracts with supercritical carbon dioxide and THF were subjected to GPC analysis, respectively, to compare molecular weight distributions of the extracts.

B.   Results and Discussion

1.   SFC analysis of test mixture

Figure 5 shows a chromatogram of NBS standard reference material 1647 containing 16 priority pollutant PAHs obtained by syringe injection, 20 μL, into the system operated in the simple SFC mode with supercritical carbon dioxide as the mobile phase at 250 kg/cm^2 pressure and 40°C temperature.   The mobile phase flowrate was 2.4 L/min with gas under the normal conditions, i.e., 0°C and 1 atm.   Although the figure shows only the chromatogram monitored at 255 nm, spectral data in the wavelength range from 195 to 350 nm was detected at an interval of 0.2 sec and stored on a floppy disk.   The spectrum of each component was recorded in the spectrum library file for later use for assignment of compounds in the extracts from the petroleum pitch.

Table 1 lists the retention time and the compound name assigned to each peak.   The component peaks eluted later than benzo[a]pyrene

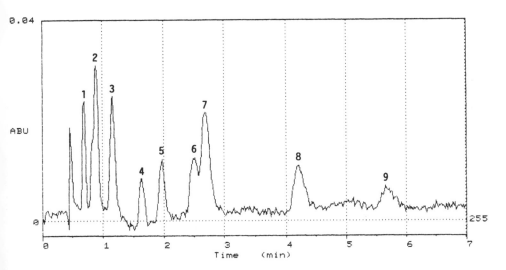

FIGURE 5   SFC chromatogram of standard mixture, NBS standard reference material 1647.   Peak identification is given in Table 1. SFC conditions: pressure, 250 kg/cm^2; temperature, 40°C; flowrate, 2.4 L/min as gas; column, Develosil ODS 5 μm packed in 4.6 mm i.d. × 150 mm long stainless steel tube.

became to broad, under the chromatographic conditions we applied, to obtain high-quality spectra for the library file.

2. Micro-SFE-SFC analysis of petroleum pitch

Figure 6 shows a three-dimensional chromatogram of the extract from the pitch with supercritical carbon dioxide obtained by direct intro- duction into SFC. Ten milligrams of the powdered petroleum pitch was placed in the extraction vessel with 1.0 mL volume and extraction was performed for 5.0 min under the conditions of 300 kg/cm^2 pres- sure and 40°C temperature. Then, the extract was introduced into the SFC section using a 500 μL trap loop by the procedure described in Section II.B.4 The chromatographic conditions were the same as for the test mixture given in the previous section.

The chromatogram looks very complex and suggests that the extract contained a lot of components which were not seen in the chromatogram of the test mixture. It was impossible to assign com- ponents only by their retention data. Therefore, spectral data were examined in detail with reference to the library as well as the reten- tion data for peak assignment.

TABLE 1   Identification of PAHs in Standard Mixture[a]

Peak Number	Retention time (min)	Compound	Injected amount (ng)[b]
1	0.69	naphthalene	450.0
2	0.89	phenanthrene	420.0
3	1.15	anthracene	65.8
4	1.61	fluoranthene	202.0
5	1.95	pyrene	196.8
6	2.49	benz[a]anthracene	100.6
7	2.67	chrysene	93.6
8	4.20	benzo[b]fluoranthene and benzo[k]fluoranthene	102.2 100.4
9	5.66	benzo[a]pyrene	106.0

[a]NBS standard reference material 1647, priority pollutant polynuclear aromatic hyrocarbons (in Acetonitrile).
[b]Calculated from Table 1 in the Data Sheet [62] for the above mate- rial and the injection volume, 20 μL.

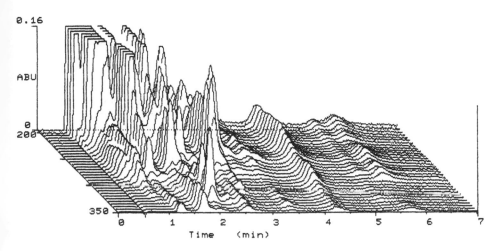

0.16

ABU

0
200

350

0   1   2   3   4   5   6   7
Time   (min)

FIGURE 6   SFE-SFC chromatogram of pitch.   SFE conditions: pressure, 300 kg/cm^2; temperature, 40°C; time, 5 min.   SFC conditions were the same as for the standard mixture separation.

3.   SFC analysis of extract from petroleum pitch with THF

Figure 7 shows a three-dimensional chromatogram of the extract from the pitch with THF.   The powdered pitch was mixed with THF as 10 mg/mL and left for 24 hours.   Then 2 µL of the supernatant was subjected to the SFC analysis by the same procedure as for the analysis of the test mixture.   It was seen that the heights of peaks having relatively shorter retention times are lower than those in the chromatogram of the SFE extract shown in Figure 6.

4.   Comparison of contents of SFE extract with those of THF extract

Figure 8 compares chromatograms of the SFE extract and the THF extract.   Chromatogram (A) is of the SFE extract and chromatogram (B) is of the THF extract.   These chromatograms are presented by the data processor as monitored at 255 nm, using the previous three-dimensional chromatographic data stored on a floppy disk.   Although The degrees of separation are notably different, it is clearly seen that peak heights in chromatogram (A) decreases as the retention time goes longer, while those in chromatogram (B) are generally similar regardless of the retention times.   This suggests that the THF extract contains more amounts of components having relatively higher molecular weights than the SFE extract.   The cause of poorer separation of the peaks of the SFE extract than that of the THF extract may be attributed to the larger volume, 500 µL, of the extract trap loop than the injector loop volume, 20 μL, for the simple SFC mode in which the THF extract was analyzed.

Table 2 lists the retention time and the name of the compound assigned to each peak with reference to the spectral library file and retention times of compounds in the standard mixture. There are 15 chromatographic peaks in the THF extract. However, the spectra of 5 of these peaks were not found in the library. Therefore, assignment of those components was not successful. The correlation between the spectrum of each assigned component with that in the spectral library was higher than 0.95.

Table 3 compares the percent content of each component in the assigned 9 components. Each line of the third column represents the percentage of each compound in the supercritical CO_2 extract (SFE extract). The fourth column shows percent contents in the THF extract, and the fifth column shows the ratio between these values. For instance, naphthalene contains 3.2% in the total sum of the assigned components in the SFE extract, and 0.9% in the THF extract in terms of peak area monitored at 255 nm. Since molar absorbances of these compounds were not taken into calculation, these values do not directly represent the molar contents of the compounds. However, the ratios shown in the fifth column directly compare the extraction efficiencies of PAHs with supercritical carbon dioxide and tetrahydrofuran (THF).

These ratios for the compounds from napthalene through pyrene are more than 1.0, while the values rapidly become less than 1.0 for

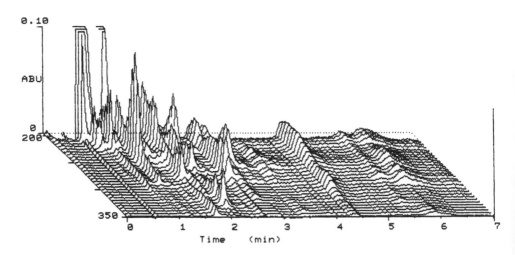

FIGURE 7 SFC chromatogram of pitch extract with THF. SFC conditions were the same as for the standard mixture separation.

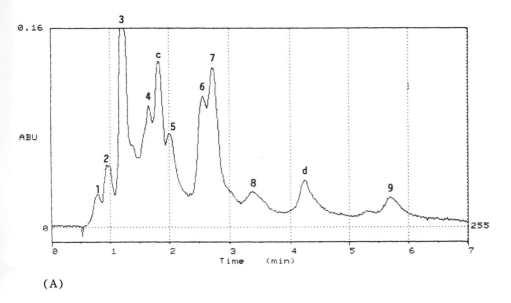

(A)

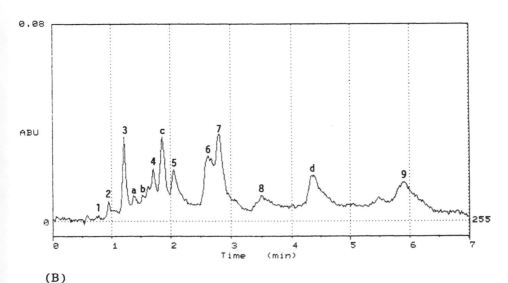

(B)

FIGURE 8 Comparison of chromatograms of pitch extracts with CO_2 and THF. Chromatogram (A) is of the extract with CO_2, chromatogram (B) is of that with THF. Peak identification is given in Table 2. These chromatograms are different representations, single wavelength at 255 nm, of the same three-dimensional chromatographic data shown in Figures 6 and 7.

TABLE 2 Identification of PAHs in Extracts with CO_2^e and THF

Peak Number	Compound	Extracted with CO_2^e	THF	Standard
		Retention time (min)		
1	naphthalene	0.78	0.75	0.69
2	phenanthrene	0.96	0.93	0.89
3	anthracene	1.19	1.21	1.15
a	f	f	1.36	f
b	f	f	1.52	f
4	fluoranthene	1.64	1.61	1.61
c	f	1.81	1.86	f
5	pyrene	1.98	2.04	1.95
6	benz[a]anthracene	2.55	2.61	2.49
7	chrysene	2.70	2.79	2.67
8	benzo[b]fluoranthene and benzo[k]fluoranthene	3.35	3.49	4.20
d	f	4.24	4.35	f
9	benzo[a]pyrene	5.69	5.90	5.66

[e]Supercritical carbon dioxide at 300-kg/cm^2 pressure and 40°C temperature.
[f]Peak components were unidentified.

the compounds from benz[a]anthracene through benzo[a]pyrene. This means that the extraction efficiency of PAHs with supercritical carbon dioxide at 300-kg/cm^2 pressure and 40°C temperature rapidly decreases as the number of benzene rings of the compound increases.

In order to investigate higher molecular weight components in the extracts, chromatographic conditions were changed so that these compounds could elute much earlier. Figure 9 shows the coupled micro SFE-SFC chromatogram of the pitch obtained by adding THF (0.5 mL/min) as a modifier solvent to carbon dioxide. Other conditions were the same for the chromatograms shown in Figure 6. Under these conditions, coronene is known to elute in 1.5−2.0 min by directly injecting the standard compound into the system. However, as it is seen, there is no significant UV absorption after 1.5 min in the chromatogram. Figure 10 shows the SFC chromatogram of the THF extract obtained by the same conditions. There is some UV absorption even after 2 min in the chromatogram. Figure 11 compares these

TABLE 3   Comparison of Extraction Efficiencies of CO_2 and THF
for PAHs

| Peak Number | Compound | Percent content | | Ratio $CO_2{}^a$/THF |
| | | Extracted with | | |
		$CO_2{}^a$	THF	
1	naphthalene	3.2	0.9	3.6
2	phenanthrene	6.5	1.5	4.3
3	anthracene	29.4	10.1	2.9
4	fluoranthene	14.1	6.3	2.2
5	pyrene	12.8	10.8	1.2
6	benz[a]anthracene	10.5	18.8	0.6
7	chrysene	16.6	28.2	0.6
8	benzo[b]fluoranthene and benzo[k]fluoranthene	2.2	8.8	0.3
9	benzo[a]pyrene	4.7	14.6	0.3

[a]Supercritical carbon dioxide at 300-kg/cm^2 pressure and 40°C
temperature.

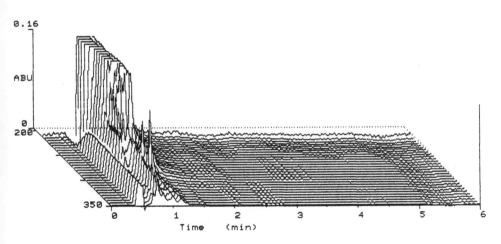

FIGURE 9   SFE-SFC chromatogram of CO_2 extract with modified mobile
phase.   SFE conditions were the same as given in the caption for
Figure 6.   SFC conditions: pressure, 250 kg/cm^2; temperature, 40°C;
CO_2 flowrate, 2.4 L/min as gas; modifier solvent, THF at the flowrate of 0.5 mL/min; column, DEVELOSIL ODS 5 µm packed in 4.6 mm
i.d. × 150 mm long stainless steel tube.

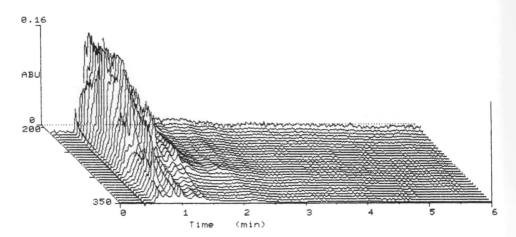

FIGURE 10   SFC chromatogram of THF extract with modified mobile
phase.   SFC conditions were the same as given in the caption for
Figure 9.

chromatograms represented as monitored at 255 nm by the same manner
as for Figure 8.   Chromatogram (A) is of the SFE extract and chromatogram (B) is of the THF extract.   Benzo[a]pyrene was eluted at
approximately 1.2 min by coupled micro-SFE-SFC and that in THF
extract was eluted at approximately 1.0 min.   By close examination
of UV spectra of components eluted after 1.2 min in chromatogram
(A), it is suggested that at least one more component was eluted
after benzo[a]pyrene.   In chromatogram (B), at least three more
peaks can be seen.   However, assignments of these peaks were not
successful.   The difference of retention times of benzo[a]pyrene between two chromatograms is considered to be due to introduction of
the SFE extract with comparatively large volume of carbon dioxide
(500 μL at 300 kg/cm^2 pressure and 40°C temperature), that may have
caused less modifier content, resulting in a longer retention time.

5.   Comparison of chromatograms of SFE and THF extracts by
     gel-permeation chromatography

In order to examine overall profiles of constituents, which exhibit
UV-absorption at 255 nm of both the extracts, they were subjected
to gel-permeation chromatography (GPC).   In advance of the analysis
of the extracts, standard compounds, naphthalene, anthracene,
pyrene, chrysese and coronene were injected respectively into the
GPC system to make a reference table for the relationship between
the retention times (RT) and molecular weights (MW) of these PAHs.
This relationship is presented in Table 4.

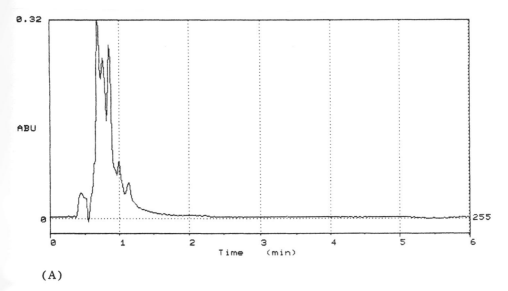

(A)

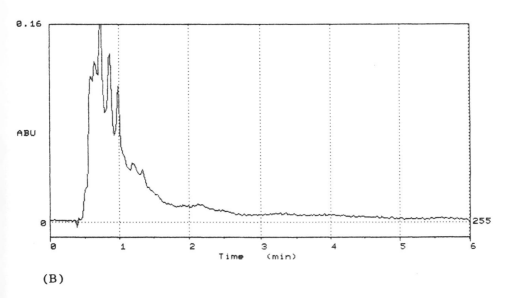

(B)

FIGURE 11   Comparisons of chromatograms of CO_2 and THF extracts
with modified mobile phase.   Chromatogram (A) is of the extract with
CO_2, chromatogram (B) is of that with THF.   These chromatograms
are different representation, single-wavelength at 255 nm, of the same
three-dimensional chromatographic data shown in Figures 9 and 10.

In GPC, generally, the higher the molecular weight, the shorter the retention time becomes. However, although the major mode of chromatography is normally steric exclusion, other modes such as adsorption are inevitably involved. The extent of contribution by other than steric exclusion is dependent on the structures of the compounds to be chromatographed. Polystyrenes, which have a pendant structure with a long alkyl chain, are generally used for the calibration of molecular weight. However, particularly in the analysis of oligomers or comparably low molecular weight compounds, retention times are often affected by the above mentioned reason, resulting in a considerable deviation from the molecular weight calibration curve obtained by polystyrene standards.

PAHs are the typical example of such compounds. As shown in Table 4, there is no systematic relationship between the retention times (RT) and the molecular weights (MW). Coronene (MW = 300) exhibited the largest retention time, though it has the highest molecular weight among all the compounds listed in the table. Similarly, naphthalene (MW = 128), having the lowest molecular weight, is

TABLE 4 Relationship[a] between Retention Times (RT) and Molecular Weights (MW) of PAHs

RT(min)	Compound	MW	Structure
21.9	Chrysene	228	
22.3	Anthracene	178	
22.5	Naphthalene	128	
23.0	Pyrene	202	
23.6	Coronene	300	

[a]Retention time does not relate to molecular weight in the simple manner as in GPC of strain-chain oligomers or polymers, owing to the structures of the compounds.

supposed to have shown the longest retention time. However it had
a shorter retention time than pyrene (MW = 206). Shorter retention
times of chrysene, anthracene, and naphthalene than that of pyrene
and coronene may be explained by the rod-like structures of these
compounds that caused the contribution of steric exclusion to be more
than that of adsorption. On the other hand, the compact structure
of pyrene and coronene caused more contribution of adsorption than
steric exclusion. In this regard, one can never obtain a molecular
weight distribution from the GPC chromatograms of the PAHs. However, PAHs of interest are known by Table 4 to elute in the time
range from 12.1 to 13.0 min, that allows us to overview the profiles
of the constituents of the extracts.

   Figure 12 shows GPC chromatograms of SFE (A) and THF (B)
extracts. The SFE extract was separated from supercritical carbon

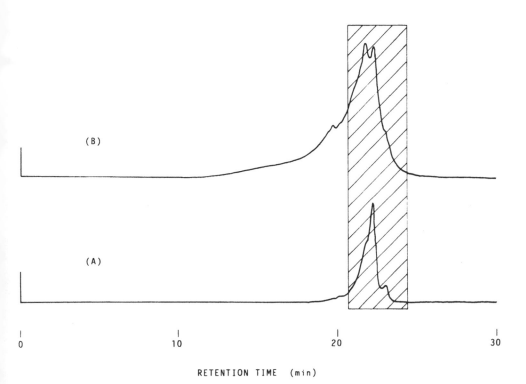

RETENTION TIME (min)

FIGURE 12   GPC chromatograms of CO_2 and THF extracts. (A) THF
extract, (B) SFE extract. Conditions: Column, Finepak GEL 101,
7.2 mm i.d. × 500 mm length having exclusion limit of 3,000 as polystyrene. Mobile phase; THF at 0.6 mL/min. Assigned PAHs are
considered to be eluted in the time range indicated by a hatched box.

dioxide through the vent valve for the extraction vessel, releasing the pressure by opening the valve. Then, the collected extract was dissolved in THF and injected to the GPC system. The time range in which the PAHs were eluted is indicated by the hatched box.

As expected from the data and SFC chromatograms shown previously, chromatogram (A) of the SFE extract shows the narrower distribution, and most of the eluted components are in the time range for the PAHs, showing that major components in the extract are PAHs. On the other hand, the THF extract shown as chromatogram (B) contains a wider range of compounds of higher molecular weight side. The real molecular weights of these compounds which exhibit significant UV-absorption are not known. However, these are considered to be hydrophobic high-molecular weight compounds containing elements other than hydrogen and carbon, such as nitrogen originated from biological products. Visual observation of both the extracts dissolved in THF showed significant difference in color. The SFE extract exhibited fluorescent yellowish green, while the THF extract was dark brown. This also suggests that the PAHs were more selectively extracted from the petroleum pitch with supercritical carbon dioxide than with THF.

IV. CONCLUSION

As we have demonstrated, SFE can be performed successfully in micro-scale by using an apparatus based on HPLC instrumentation technology. Only ten milligrams of the petroleum pitch was enough to investigate extraction efficiency of PAHs with carbon dioxide in reference to that with THF. We have examined the extraction with supercritical carbon dioxide only at 300 kg/cm^2 pressure and 40°C temperature. However, further investigation of extraction with the fluid at different pressures and temperatures, and with different organic solvents such as toluene, xylene, etc. will provide us with more information about the characteristics of supercritical carbon dioxide as extraction medium for PAHs.

It should be noted that the SFE extract was directly introduced into the SFC section of the system as it is in supercritical carbon dioxide without changing the phase of the fluid in a completely closed system. That widens the applications of this method to the extraction and analysis of toxic and polluting substances, such as environmental analysis, utilizing SFE as sample pretreatment that is done separately with solvent in a conventional analysis method.

In conclusion, let us discuss the future prospects of coupled SFE-SFC. The method, which we discussed in this chapter, is simple and easy to build. However, it has a constraint in the quantitative analysis of components in the sample. In the SFE section, the extraction medium flows into the extraction vessel which has a blocked-

end tubing, i.e., the extraction is performed in stop flow. That
does not allow the components to be thoroughly extracted. In order
to extract as much fluid as possible from the compounds, the fluid
flow should be continuous to keep the fresh fluid supplying to the
sample, and the extract should be trapped by using an adsorbent.
That is no longer a direct coupling. However, this method allows
analysis of the extract by using other chromatographic methods, such
GC, HPLC, or other analytical methods, IR spectrometry, mass spectrometry, etc..

   An alternative method is the parallel coupling which was briefly
described in Section II. By this method, the extract itself, as well
as the quantitative information, can be obtained by chromatographic
analysis. At present, however, a commercially available back-pressure
regulator has too large a dead volume, tens of milliliters, to perform micro-SFE that allows fractionation of the extract in continuous
flow mode, i.e., fractionation at the outlet of the back-pressure
regulator.

   In order to improve these situations, a back-pressure regulator
having a low-dead volume, preferably less than 10 µL, is very much
necessary. That may permit extraction of components thoroughly
from the sample by continuous flow SFE and enrich them in a small
column, then introduced into a separation column. SFC peak components can also be fractionated, i.e., preparative scale SFC, by
using such a valve. Now, the development of a back-pressure regulator that satisfies the above requirements is currently under way in
our laboratory.

ACKNOWLEDGMENTS

The authors would like to thank Prof. K. Jinno of Toyohashi University of Technology nd Dr. Y. Yamauchi of JASCO for their instructive advice on GPC analysis.

REFERENCES

1. J. B. Hannay and J. Hogarth, Proc. R. Soc. London 29: 324
    (1879).
2. M. Tswett, Tr. Varshav. Obshch. Estestvoispyt., Otd. Biol.
    14: 20 (1903).
3. J. J. Kirkland (ed.), Modern Practice of Liquid Chromatography,
    Wiley-Interscience, New York (1971).
4. E. Klesper, A. H. Corwin, and D. A. Turner, J. Org. Chem.
    27: 700 (1962).
5. N. M. Karayannis, A. H. Corwin, E. W. Baker, E. Klesper, and
    J. A. Walter, Anal. Chem. 40: 1736 (1968).

6. M. N. Myers and J. C. Giddings, Anal. Chem. 38: 294 (1966).
7. M. N. Myers and J. C. Giddings, Anal. Chem. 37: 1453 (1965).
8. J. C. Giddings, M. N. Myers, and J. W. King, J. Chromatogr. Sci. 7: 276 (1969).
9. R. E. Jentoft and T. H. Gouw, J. Chromatogr. Sci. 8: 138 (1970).
10. R. E. Jentoft and T. H. Gouw, Anal. Chem. 44: 681 (1972).
11. T. H. Gouw and R. E. Jentoft, J. Chromatogr. 68: 303 (1972).
12. M. Novotny, W. Bertsch, and A. Zlatkis, J. Chromatogr. 61: 17 (1971).
13. D. Bartmann and G. M. Schneider, J. Chromatogr. 83: 135 (1973).
14. M. Novotny, S. R. Springston, P. A. Peaden, J. C. Fjeldsted, and M. L. Lee, Anal. Chem. 53: 407A (1981).
15. P. A. Peaden, J. C. Fjeldsted, M. L. Lee, S. R. Springston, and M. Novotny, Anal. Chem. 54: 1090 (1982).
16. R. D. Smith, W. D. Felix, J. C. Fjeldsted, and M. L. Lee, Anal. Chem. 54: 1883 (1982).
17. P. A. Peaden and M. L. Lee, J. Liq. Chromatogr. 5: 179 (1982).
18. P. A. Peaden and M. L. Lee, J. Chromatogr. 259: 1 (1983).
19. S. R. Springston and M. Novotny, Anal. Chem. 56: 1762 (1984).
20. J. C. Fjelsted and M. L. Lee, Anal. Chem. 56: 619A (1984).
21. R. D. Smith, H. T. Kalinoski, H. R. Udseth, and B. W. Wright, Anal. Chem. 56: 2476 (1984).
22. Y. Hirata and F. Nakata, J. Chromatogr. 295: 315 (1984).
23. T. Takeuchi, D. Ishii, M. Saito, and K. Hibi, J. Chromatogr. 295: 323 (1984).
24. Y. Hirata, J. Chromatogr. 315: 31 (1984).
25. Y. Hirata, J. Chromatogr. 315: 39 (1984).
26. M. Saito, K. Sugiyama, T. Hondo, M. Senda, and S. Tohei, International Symposium Kyoto, Jan., 1985, Abstract pp. 84—86.
27. K. Sugiyama, M. Saito, T. Hondo, and M. Senda, J. Chromatogr. 332: 107 (1985).
28. K. Jinno, M. Saito, T. Hondo, and M. Senda, Chromatographia 21: 219 (1986).
29. K. Jinno, T. Hoshino, T. Hondo, M. Saito, and M. Senda, Analytical Letters 19: 1001 (1986).
30. J. B. Crowther and J. D. Henion, Pitts. Conf. Abs., 1985, No. 539.
31. K. H. Shafer and P. R. Griffiths, Anal. Chem. 55: 1939 (1983).
32. S. V. Olesik, S. B. French, and M. Novotny, Chromatographia 18: 489 (1984).
33. C. C. Johnson, J. W. Jordan, R. J. Skelton, and L. T. Taylor, Pitts. Conf. Abs., 1985, No. 538.
34. C. Fujimoto, Y. Hirata, and K. Jinno, J. Chromatogr. 332: 47 (1985).
35. M. G. Rawdon, Anal. Chem. 56: 831 (1984).

36. T. A. Norris and M. G. Rawdon, Anal. Chem. 56: 1767 (1984).

37. T. L. Chester, J. Chromatogr. 299: 424 (1984).

38. D. R. Gere, R. Board, and D. McManigill, Anal. Chem. 54: 736 (1986).

39. T. Greibrokk, A. L. Blilie, E. J. Johansen, and E. Lundanes, Anal. Chem. 56: 2681 (1984).

40. K. Zosel, Austrian Patent Appl. 16.4. 1963.

41. K. Zosel, in Extraction with Supercritical Gases (G. M. Schneider, E. Stahl, and G. Wilke, eds.), Verlag Chemie, Weinheim, p. 1, 1980.

42. P. Hubert and O. G. Vitzthum, in Extraction with Supercritical Gases (G. M. Schneider, E. Stahl, and G. Wilke, eds.), Verlag Chemie, Weinheim, p. 25, 1980.

43. S. Peter and G. Brunner, in Extraction with Supercritical Gases (G. M. Schneider, E. Stahl, and G. Wilke, eds.), Verlag Chemie, Weinheim, p. 141, 1980.

44. G. Brunner and S. Peter, Ger. Chem. Eng. 5: 181 (1982).

45. R. Eggers, in Extraction with Supercritical Gases (G. M. Schneider, E. Stahl, and G. Wilke, eds.), Verlag Chemie, Weinheim, p. 155, 1980.

46. E. Stahl, E. Schutz, and H. Mangold, J. Agr. Food Chem. 28: 1153 (1980).

47. E. Stahl, E. Schutz, Planta Med. 40: 262 (1980).

48. H. Coenen and P. Rinza, Tech. Mitt. Krupp-Werksberichte 39(1981)H1, Z1.

49. H. Brogle, Chem. Ind. (London), 19 June 1982.

50. R. P. de Filippi, Chem. Ind. (London), 19 June 1982.

51. T. R. Bott, Chem. Ind. (London), 19 June 1982.

52. R. Vollbrecht, Chem. Ind. (London), 19 June 1982.

53. J. P. Calame and R. Steiner, Chem. Ind. (London), 19 June 1982.

54. D. S. Gardner, Chem. Ind. (london), 19 June 1982.

55. E. Stahl and W. Schiltz, Z. Anal. Chem. 280: 99 (1976).

56. E. Stahl, J. Chromatogr. 142: 15 (1977).

57. C. S. Nieass, M. S. Wainwright, and R. P. Chaplin, J. Chromatogr. 195: 335 (1980).

58. C. S. Nieass, R. P. Chaplin, and M. S. Wainwright, J. Liq. Chromatogr. 5: 2193 (1980).

59. K. K. Unger and P. Roumeliotis, J. Chromatogr. 282: 519 (1983).

60. K. Sugiyama, M. Saito, and A. Wada, Japanese Pat. Appl. No. 58-117773 (1984).

61. K. Sugiyama, M. Saito, and A. Wada, U.S. Pat. No. 4,597,943 (1986).

62. NBS Standard Reference Material No. 1647, National Bureau of Standards, Washington, D.C. 20234, Dec. 7, 1981.

63. R. J. Skelton, Jr., C. C. Johnson, and L. T. Taylor, Chromatographia 21: 4 (1986).

10

Hardware Considerations for the Automation of Multidimensional Chromatographs

L. DAVID ROTHMAN *The Dow Chemical Company, Midland, Michigan*

I. INTRODUCTION

There are few multidimensional column chromatography (MDCC) schemes which would not benefit from automation of the equipment. Most of the literature references to MDCC applications describe the use of automated instrumentation and those which do not, often refer to the ease with which the experiment could be automated. The value of automation in such applications is clear: few chromatographers want to hover over their instruments during an entire separation so that they can perform simple functions at predetermined times following sample injection. Instead, these functions, which include actuation of valves, controlling the flow of fluids, heating or cooling columns or changing detector operating parameters, are often left to mechanical devices and timing equipment which will perform these tasks reliably and reproducibly. Many articles describing multidimensional chromatography applications seem to assume that readers are familiar with both the practice and advantages of instrument automation and infrequently discuss either in detail. Consequently, readers inexperienced in electronics and other aspects of instrument automation may wish to begin using this technology, but find it difficult to successfully assemble an instrument for automated MDCC applications. The purpose of this chapter is to discuss automation of these experiments, including the actual hardware available for the purpose, and some practical points on assembling and operating such equipment.

II. THE IMPORTANCE OF AUTOMATION IN MDCC

Anyone familiar with analytical column chromatography of any type
appreciates the importance of run-to-run reproducibility in both the
transfer of sample mass into the column system and the elution times
of the analytes.   Accuracy in qualitative chromatographic analysis
depends on analyte identification by retention time.   In addition,
accuracy in quantitative analysis depends on reproducible or measur-
able transfer of sample mass into the column.   In both cases, high
accuracy is achieved by use of the proper equipment for regulating
flowrates of mobile phases (flow controllers or pumps), the sample
injection operation (loop injectors or autosamplers), column temperature
and mobile phase composition.   In quantitative analysis by MDCC,
accuracy depends on these same variables, plus elution time reproducibility of the analyte(s) from at least one additional column and
the reproducibility of the transfer of analyte mass between columns
during the separation. These additional variables affecting the overall accuracy of the measurement make it desirable to have very repro-
dicible control over all aspects of the instrumentation and this is best
achieved by automation of as many instrument functions as possible.
A partial list of instrument functions which have been automated in
MDCC applications includes:

1.   Valve control for:
        mobile phase diversion
        cryogenic fluids
        fluidic switching
        column selection
2.   Mobile phase flow rate changes by:
        changing pump or flow controller operation
        selection of different pumps or flow controllers
        diversion of a portion of the mobile phase
3.   Mobile phase composition changes by:
        altering phase composition at the flow controller inlet
        altering the mixing of flow controller outputs
        selecting different flow controllers
4.   Column temperature alteration
5.   Analyte "trapping," including:
        flowrate control for cryogenic fluids
        cold trap heating
        mobile phase flow in the trap
6.   Operation of data collection equipment
7.   Monitoring instrument performance

     This chapter will discuss some of the considerations in building
practical multidimensional liquid and gas chromatographs for quantitative analysis applications.

III. CONSIDERATIONS FUNDAMENTAL TO
AUTOMATED COLUMN CHROMATOGRAPHY

The two most fundamental considerations in automating column chromatography instruments are reliable operation of the basic chromatography equipment and reproducibility of instrument control event
timing. Fortunately, both are fairly simple to achieve with properly
used modern chromatography instrumentation. This section will discuss topics related to improving the reliability of gas and liquid
chromatographs and the various means by which timing control may
be achieved.

IV. EQUIPMENT RELIABILITY IN MULTIDIMENSIONAL
CHROMATOGRAPHY

Automation of any experiment is of little value if the experimental
equipment does not operate reliably. Minor variations in system performance which are often tolerated in conventional chromatography,
such as retention timer drift caused by slow degradation of an HPLC
column or changes in peak shape caused by changes in column performance, may spell the difference between success or failure in a
long-term application of automated multidimensional chromatography.
As an example, consider the coupling of size exclusion chromatography
(SEC) to reversed phase (RP) chromatography for the purpose of
copolymer characterization. (1) The goal of the experiment is to first
separate the polymer molecules by "size," then separate various discrete size fractions on the reversed phase column to determine their
composition, leading eventually to a three-dimensional map of copolymer composition vs. molecular weight. Such an experiment could be
easily automated. The experiment could be performed by making a
series of injections of the polymer sample into the SEC column and
diverting different retention time window fractions to the reversed
phase column in successive injections. A series of such experiments
allows the user to generate a family of RP chromatograms which represent the composition of analyte components in the SEC column effluent
at different retention times, thus providing the data needed for the
map. It quickly becomes apparent that the reproductibility of the SEC
experiment is of great importance, because it is the factor which
determnes how well the "size" of the molecules in the fractions diverted
to the RP column is known and thus determines the accuracy and
precision of the data in one of the dimensions of the map. Therefore,
it would be important to perform the SEC separations with highly reliable equipment. By the same token, the reproductibility of the RP
experiment determines the accuracy and precision of the data for the
second dimension of the map. Given the time required to generate
the data for the map (at a minimum, the time for a single SEC experi-

ment multiplied by the number of discrete fractions sampled for further separation on the RP column), it only makes sense to assemble a
system in a manner designed to give utmost confidence in the resulting data. This not only makes best use of laboratory time, but also
allows decisions to be made confidently based on the results.

In general, the precautions to be taken to ensure reproducible
instrument operation for multidimensional experiments are little different than those taken for conventional experiments. There are concerns unique to HPLC and GC and these will be discussed separately.
In either case, proper instrument setup and periodic maintenance
pay substantial dividends in more reliable performance.

V. CONCERNS UNIQUE TO HPLC

Factors which affect the stability and reproducibility of liquid chromatographic experiments include column behavior, mobile phase pump
operation and mobile phase composition. Columns in HPLC are subject to temporary or permanent changes in behavior due to column
temperature fluctuations and three general forms of degradation: loss
of the stationary phase (if any) due to hydrolysis or dissolution,
collapse of the column solid support due to either physical rearrange-
ment of the individual particles or dissolution of the solid support
(silica) and contamination of the packing or frits by either the samples
or mobile phase. Pumps may malfunction due to wear of mechanical
parts or failure of the pump seals or check valves, leading to changes
in flow rate and, possibly, to undesirable variations in mobile phase
composition in multisolvent pump systems. Leaks in fittings may alter
the actual column flow rate. An recent article described a number
of modifications to a basic HPLC system which were intended to improve reliability (2) and these precautions are at least equally desirable for multidimensional liquid chromatography separations.

Precautions in instrument operation may include: prefiltration of
the mobile phase, bubble traps in low pressure solvent supply lines,
presaturation of the mobile phase with silica where possible, pump
pressure pulse dampening, filtration of the mobile phase prior to and
following the injector, thermostatting of the column (the entire instrument, if possible) and monitoring of the pump back-pressure with
alarms to indicate undesirable high or low excursions from normal
operations. During routine operation of a multidimensional instrument,
standards should be periodically injected onto each column and the
effluent monitored to determine if retention time, selectivity and band
shape are within acceptable limits. These limits must be determined
on a case-by-case basis by the system operators. While specific rules
for determining these limits cannot be stated, the behavior of individual components in the chromatograph should at least meet the standards of acceptability one would normally apply to conventional single
column separations.

Column degeneration during use is a major concern with some column types, e.g., phase hydrolysis with some ion exchange or short-chain alkyl phases bonded to silica. Adapting such separations to more stable columns is advisable, as it would be in any analysis where long-term retention time reproducibility is important. A consideration is that retention time reproducibility is of greater importance in the first column in a two-dimensional separation, where the analyte retention time must be well known to perform quantitative transfer to the second column. Automatic adjustment of event timing (such as valve position switching), while possible, is not commonly done, nor is this a feature offered by many data collection systems. Retention time drifts of the analyte on the second column are more easily dealt with by readjustment of the peak identification parameters in data collection equipment, a common feature, as long as adequate separation of the analyte from interferences is maintained. Where the option exists, the more stable column in a two-dimensional separation scheme should be used for the first separation. Improved HPLC column stability is a matter which vendors are addressing. As better products are introduced, they are certain to be desirable for MDCC applications.

Column lifetimes are often greatly increased by some pretreatment of the sample, such as passing it through a filter and/or short low pressure column. Many column-switching HPLC schemes have been developed to minimize the amount of sample handling prior to injection. The expectation in some cases is that performance of the first column will gradually degrade during use, eventually requiring the replacement of that column with another. Often, short "precolumns" are used before the main analytical column as such sacrificial devices. In a fully automated instrument, replacement of a contaminated first column can also be automated with a multicolumn selector valve, switching new columns into the system as old ones are depleted. The decision to change columns may be based on a number of criteria, such as: after a specific number of samples (based on experience), after performance decreases (based on monitoring peak shapes and retention times in the separation) or upon development of excessive column head pressure. In the latter two cases, it is apparent that some program for monitoring the separation process on the first column is essential. This can be done by passing the column effluent through a detector prior to the second column or by monitoring the pressure at the pump used for pumping the eluent through the first column.

VI.   CONCERNS UNIQUE TO GC

Just as with conventional single-column GC separations both packed and capillary columns have been applied to MDCC separations. These have been linked in all the possible combinations (packed/packed, capillary/capillary, and packed/capillary). Some of the considerations discussed here are universal, regardless of column type, while others are dependent on the particular column mix in the actual application.

Columns in GC are subject to three general forms of degradation: loss of stationary phase due to "column bleed," alteration of the phase due to oxidation or other chemical changes and contamination of the column with high-boiling components, either from samples or contaminated carrier gases. Leaks may develop in fittings and valves, particularly in applications which involve temperature programming and affect component retention times and mass transfer between columns.

Precautions which may be taken include the exclusive use of "bonded" stationary phases in capillary GC or packed columns to eliminate "bleed," use of efficient oxygen, water and organic contaminant scrubbers on the carrier gas stream and use of some disposable insert (often a short packed precolumn) at the inlet end of the column to retain high-boiling components. Periodically, the insert can be replaced and the retention time of known components can be measured to ensure that system performance has not changed due to column degradation or leaks. As in the case of liquid chromatography periodic verification of column performance is important. A difficulty with GC is that many of the common detectors are destructive, e.g., the flame ionization and thermionic emission detectors. Column effluent cannot be passed through these prior to the second column as it can with absorbance or fluorescence detectors in HPLC. In GC, this problem is sometimes addressed by splitting off a portion of the effluent from the first column and sending it to a destructive detector. This allows continuous monitoring of the separation on the first column, but prevents true quantitative transfer of analytes between columns. However, since flow-stream splitting in GC can be performed reliably and reproducibly and since the sensitive detectors available allow adequate sensitivity with a small fraction of the total column effluent split off to the detector, this approach has been successfully applied.

Sample pretreatment to eliminate high-boiling or other column-modifying components is more difficult in GC than in HPLC, since there is no simple gas-phase analog to filters and open LC columns used by HPLC practitioners for sample "cleanup." The best sample pretreatment tool for GC is probably a packed GC column, since the effluent from such a column is almost certain to be compatible with any subsequent GC separation, regardless of column type. Packed GC columns are somewhat more rugged than capillary columns, in that the larger volume of stationary phase in packed columns generally results in less change in column behavior than is typically seen in capillary columns when similar amounts of column-modifying components are injected into each. Therefore, the packed-to-capillary column multidimensional GC system may be the best system in terms of reliabiliy and resolution, since it combines the ruggedness of a packed column at the sample introduction end of the system with the resolution of a capillary column for the analytical separation. In situations where minimal pretreatment of complex samples is desirable, the packed/capillary combination may be the best choice for reliable operation with a minimum of instrument mantainance. Where the experiment demands the use of capillary /capillary separations, the use of injector inserts packed with liquid phase-coated solid suports is recommended where possible.

VII. EVENT CONTROL TIMING

A number of events may take place during an MDCC experiment, including valve manipulation, mobile phase control and temperature changes. Control of event timing is therefore important in MDCC if accurate, reproducible results are to be obtained. Timing control involves the use of some clock-driven device controlling one or more switches which are in turn used to control the various automated portions of the experiment. Table 1 contains a representative list of devices which may be used to provide timing control for chromatographic experiments. It is interesting to note that nearly any of the basic components of a liquid or gas chromatograph may offer timing control as a built-in feature. Typically, the beginning of a cycle in a timing device can be triggered by a remote contact closure or manual actuation of a pushbutton. Both manual and automatic sample injection systems can be equipped to provide such a contact closure signal at the moment of injection.

Timing controllers typically provide several independently-actuated switches with precise timing and resolution on the order of 0.01 to 0.001 minutes, but often do not provide electrical switching capabilities directly suitable for the needs of the experiment. Switching needs may range from low current, low voltage DC signals, such as those used to control devices with logic level inputs (e.g., integrator or data system remote start inputs) up to high current, high voltage AC power, such as may be needed to control a heater or pump. Rather than try to provide for all possible switching needs, most timing controller vendors provide low-current contact closures or logic-level signals with limited switching capabilities. A typical timing control switch may permit switching no more than 40 mamp of current at a maximum of 5 volts DC. It is up to the experimenter to make the switch compatible with any more stringent switching requirements the experiment. Readers interested in a further discussion of switching concepts are referred to basic texts on instrumentation electronics (Malmstadt et al. (3)).

The experimenter should consider several things in deciding how to interface the timing control device to the experiment. The current/voltage ratings of the timing controller switches must not be exceeded. When the demands of the controlled equipment are incompatable with the switches, some type of relay must be interposed between them. This relay will be controlled by the timing device and must be chosen to have suitable electrical characteristics for switching the controlled device. In addition, it is wise to provide some electronic buffer between the controller and the equipment which is being controlled, so that noise, voltage surges, etc. occurring in the external circuitry will not feed back through the control circuit connections and interfere with the controller. The best answer is to use some of the many types of optically isolated solid-state relays available on the electronics market. Table 2 lists examples of these devices, which are generally available from a variety of electronic parts supply houses. As may be seen, devices are available for switching AC or DC signals, with modest electrical demands on the controlling device (commonly met by the characteristics of the switches in the controllers in Table 1) and yet allowing control of substantial AC or DC voltages and currents. The optical isolation built into the relay electrically isolates the controller from the experiment and protects the controller from noise,

TABLE 1   Examples of Devices for Event Control Timing in Chromatography

Device/Model	Manufacturer	Controls provided (number)	Approximate cost
Chrontrol		relay contact (2—4)	$250
Time delay relays (various suppliers)		relay contact (1)	$50—100 each
HPLC gradient controllers[*]			
600	Waters Assoc.	relay contact (4)	included
721	Waters Assoc.	relay contact (8)	included
HPLC pumps			
590	Waters Assoc.	relay contact (8)	included
1090	Hewlett-Packard	relay contact (4)	included
Brownlee	Brownless Assoc.	relay contact (4)	included
HPLC detectors			
783	Kratos Analytical	relay contact (4)	included
490	Waters Assoc.	relay contact (8)	included
Chromatographic integrators			
4270/4290	Spectra-Physics	logic-level (12)	option
339x series	Hewlett-Packard	contact and triac (8)	option

[*]For data systems, see discussion later in text. These devices are often based on general purpose mini- or personal computers, in which case the availability of timing control is almost certain.

TABLE 2 Examples of Optically Isolated Switches

Vendor	Input control signal	Output switching capabilities
Gordos	3—32 volts DC	120—480 v AC @ 1.5—45 amps
OPTO-22 system	5 volts DC @ 22 mamps	12—140 v AC @ 3 amps (OAC5) 0—60 v DC @ 3 amps (ODC5)

Devices such as these may be obtained from a number of general electronics supply houses, such as Pioneer, Time, Newark, and Shand.

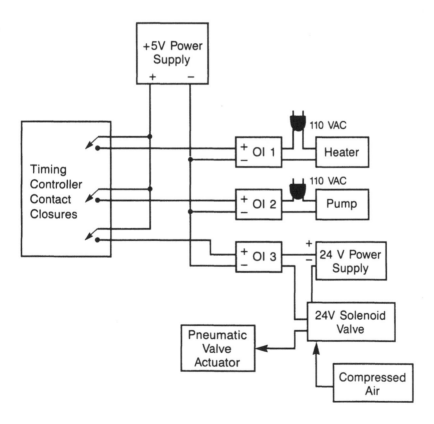

FIGURE 1 Timing device with contact closures.

transients, etc. in the external circuit. Figure 1 shows a typical circuit diagram allowing a timing device with contact closures to control a high voltage heater, an HPLC pump and a valve by means of solid-state switches. In addition to these devices, one may find optical isolation built into many commercial valve actuator packages.

VIII. VALVES FOR MULTIDIMENSIONAL CHROMATOGRAPHY

Valves used for chromatographic applications fall into several different categories, which include low dead volume, high pressure, moderate temperature valves for HPLC eluents, low dead volume, low pressure, high temperature valves for gas chromatography mobile phases, low dead volume, high temperature, high pressure valves for supercritical fluid chromatography eluents, low pressure, low temperature valves for solvent switching at HPLC pump inlets, high or low pressure cryogenic valves for thermal focussing in gas chromatography and low pressure valves for control of compressed air for a variety of applications. Regardless of the valve type, there are two principle means of mechanical valve actuation: pneumatic and electromechanical. Either of these may be easily automated, using a variety of automatic valve actuators available on the market today. These devices are generally available from a number of chromatography supply vendors. Examples of automatic valve actuators are listed in Table 3. Pneumatic valve actuators have the more complicated interface between the controller and the experiment, requiring a relatively high current, possibly high voltage signal to energize solenoids which actuate pneumatic valves, which in turn supply gas under pressure to operate the chromatographic valve actuator. Some vendors have packaged these functions into integrated units or kits, as listed in Table 3. These make the operation of pneumatic valve actuators via logic level signals very simple and may include the optically-isolated relays discussed earlier.

The Valco digital valve interface (DVI) is a particularly useful device for the control of pneumatic valve actuators. It is readily controlled by a low current, low voltage (logic level) external signal and has two built-in solenoid-operated valves for controlling the flow of pressurized gas to valve actuators. The package is relatively small, reasonably priced, easily mounted on a small panel and appears to work very reliably. Tubing and fitting kits are available from Valco to connect the DVI to pneumatic valve actuators. In addition the DVI allows the option of direct manual control of the valve actuator via a switch, so that experiments can be performed manually in addition to being controlled by some programmable timing device. This is often of value in developing or troubleshooting MDCC applications.

A function less often provided by vendors is some sort of monitoring signal which will sense the position of the controlled valve and allow the user to verify, either visually or electronically, that the valve is in the desired position at all times. This is important in automated instruments, where the reliability of the automated valving system is as important a part of the overall instrument as any other. Some devices do provide valve position signals (see, for example, some of the products offered by Autochrom or Rheodyne). It is possible to attach position-sensing switches to valves lacking them, given some mechanical ingenuity. The signals from these switches should be monitored by the instrument control event timer and used to establish that the experiment is proceeding properly. Unfortunately, many of the timing controllers listed in Table 1 are incapable of this sort of experiment status monitoring. It is typically left to the user to either trust the equipment to be reliable or to use some ingenuity to design a system for instrument status monitoring which can also track valve position. Controllers based on mini- or microcomputers, such as data systems and some HPLC gradient controllers are capable of

TABLE 3 Automatic Valve Actuators

Vendor and model	Valve type	Actuator type	Control requirements
Rheodyne			
7001	HPLC	pneumatic	40–125 psi
5700 series	HPLC	pneumatic	80–125 psi
5300	low pressure	pneumatic	20–60 psi air or N2
Valco			
A784X series	HPLC/GC	pneumatic	
H17xx series	HPLC/GC	electromechanical	24 v DC, 110 v AC, 240 v AC
Autochrom			
101	solvent selection	pneumatic with built-in control	1 mamp

Pneumatic valve kits

Vendor and model			
Rheodyne	7163		12 v DC, 420 mamp 120 v AC, 50 mamp 240 v AC, 25 mamp
Valco	DVI		5 v DC, logic level

such monitoring, but this function is typically not provided as one of the system utilities. Figure 2 shows a block diagram of a device which could be constructed to provide the monitoring function. This device uses a digital comparator to compare the state of the control signals from the timing controller to the detected state of the controlled devices. There is a time delay after the comparator to allow the mechanical portions of the equipment to change their position when the control signals are changed (1–2 seconds would normally be ample for valve actuation). If the control signals and the position-sensing switch signals do not match after the delay time, it is assumed that there is a problem in the valve mechanism and the latch is switched,

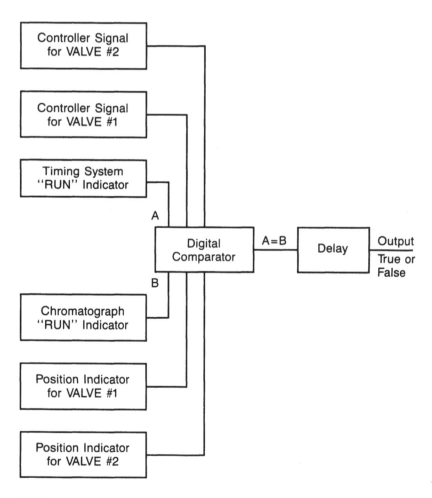

FIGURE 2   Block diagram of device to provide monitoring function.

changing the state of the output. This change can be used to trigger an alarm and/or interrupt the operation of the instrument. A device such as this is not now available commercially, but could be assembled from readily-available electronic digital logic chips and other components. Such a device would be desirable in an instrument which must operate unattended or which has a complex automation scheme.

It should be noted that the flowrate interruptions caused by valve switching in HPLC leads to pressure pulses in the system that are potentially detrimental to column lifetimes. Minimizing switching times and therefore the amplitudes of the pressure pulses, appears to be desirable. This can be accomplished with pneumatically actuated valves by adopting the pneumatic plumbing scheme described by Harvey and Stearns (4).

A number of considerations exist related to choosing the materials of construction, fittings, plumbing schemes and automation of valves. Harvey and Stearns (5,6) have discussed this subject in some detail.

IX. SOURCES OF HARDWARE FOR CHROMATOGRAPHY AUTOMATION

Here is a list of vendors of a broad range of chromatography supplies:

Alltech Associates, Inc.
2051 Waukegan Road
Deerfield, Illinois 60015

American Scientific Products
American Hospital Supply Corporation
1430 Waukegan Road
McGaw Park, Illinois 60085

The Anspec Co., Inc.
50 Enterprise Drive
Ann Arbor, Michigan 48107

Supelco, Inc.
Supleco Park
Bellefonte, Pennsylvania 16823-0048

Autochrom, Inc. P.O. Box 207 Milford, Massachusetts 01757-0207	Provides several products useful for automated chromatography
Valco Instruments Co. P.O. Box 55603 Houston, Texas 77255	Provides valves, actuators, and controllers suitable for gas and liquid chromatography
Rheodyne, Inc. P.O. Box 996 Cotati, California 94928	Provides valves and actuators suitable for liquid chromatography

---

Other vendors include most manufacturers of gas or liquid chromatographic instrument components and of chromatographic data collection
and processing equipment.

X. DATA COLLECTION AND TREATMENT FOR
MULTIDIMENSIONAL CHROMATOGRAPHY

Data collection and treatment are an integral part of chromatography.
No information is gained from any experiment if the data cannot be
properly acquired and examined. The general requirements for data
collection and treatment in MDCC applications are the same as those
for more conventional separations schemes and will be briefly outlined here. In modern instrumentation, data are collected, converted
from an analog to a digital format and then processed by computer
algorithms on a wide variety of computer equipment (e.g., digital
integrators, personal computers and larger systems) to obtain the
retention times, areas, and heights of the peaks in the chromatographic signal. These values are then used for analyte identification and quantitation by different calculational approaches . In conventional separations that is the usual extent to which the data are
used. In MDCC separations, it is possible to obtain additional value
from the processed data generated by some digital equipment. As
an example, consider a two-column MDCC instrument connected to a
chromatography data system. The instrument either performs simple
one-dimensional separations with column 1 always feeding a detector
or else traps an analyte from column 1 for quantitative analysis on
column 2. Some chromatography data systems store, in easily-avail-
able form, the times determined by the data system to be the beginning, maximum, and end of each chromatographic peak. An experiment can be performed in which a solution containing the analyte is
separated in the one-dimensional mode. The peak start and end times
are calculated by the data system. If the experimenter then wishes
to quantitatively trap components from column 1 for analysis on
column 2, the computer-generated values for peak start and end time
make it simple to specify the beginning and end times for trapping
individual components quantitatively and periodic re-injection of the
standard can be used to redetermine correct trapping times based
on computed changes in the start and end times of the peak. Known
delays caused by connecting tubing, etc., can be easily factored into
the determination of trapping times. If the data system also provides
the timing control used for the MDCC experiment, the instrument
system is now capable of self-correction for drifts in column behavior,
mobile phase composition, etc. It is important to note that, while
such a feature is within the capabilities of current data collection
equipment, data system vendors typically do not provide the programs
to perform the self-correction tasks just described.

Table 4 lists a representative sampling of digital data collection equipment available today. A wide variety of products exist, many of which are small data systems based on various personal computers. The scope of this chapter does not include an in-depth treatment of digital data collection devices. Readers interested in more discussion of the subject may wish to read some related articles.(7) Essentially all of the devices in Table 4, and many others, would be suitable for

TABLE 4   Examples of Digital Data Collection Devices for Chromatography

Types of devices	Number of channels[a]	Base computer	Comments
Stand-alone integrators			
HP 3392	1	—	b
SP 4200/4270/4290	1-2	—	b—d
Small data systems			
Nelson TURBOCHROM	1-15	IBM PC	
Varian 6000 series	1-16	proprietary	
Waters 820	1-4	IBM PC	
Waters 840	1-4	DEC PRO 380	
HP CHEMSTATION	1	HP 300 series	
Gilson	1	APPLE Macintosh	
Large data systems			
Nelson ACCESS*CHROM	e	DEC VAX	
Beckman CALS	e	HP 1000 or DEC VAX	
VG MULTICHROM	e	DEC PDP-11 or VAX	
HP LAS	e	HP-1000A series	

[a]The number of different chromatographs which can be monitored simultaneously by the same data collection product. Many of the data systems are capable of monitoring more than one detector per instrument in addition to this.
[b]May be used as data collection device for data system available from same vendor.
[c]Same device available from other instrument vendors.
[d]Has provision for serial data output to personal computer with support software supported by vendor.
[e]The maximum number of instruments which may be interfaced varies widely, depending on processor speed, number of capillary GCs involved, etc. Typically, the number will be in the range of 30 to 60 per processor. Processors may be networked to produce very large systems.

use in MDCC applications for both data collection and timing control.
Fewer would allow the user access to the computed peak start and
end times. Due to the ease with which computer programs can be
changed and the introduction of new equipment, it would be best to
review the current capabilities of any vendor's equipment to find if
the desired features are present. Those which may not be available
today often appear a short time later in revised programs or new
products.

XI. CONCLUSIONS

This chapter has presented an overview of topics of practical importance in the automation of MDCC experiments. In addition to the
understanding of chromatography needed to successfully develop an
automated MDCC system, a few other skills are useful. These would
include a basic understanding of electrical switching concepts (which
may be gained from texts or short courses on electronics) and some
familiarity with digital data collection equipment to allow timing control of the experiment and acquisition and processing of the data.

REFERENCES

1.   G. Glockner, Advances in Polymer Science, 79, Springer-Verlag,
     Berlin Heidelberg, 1986, pp. 159–214. See, in particular, pp.
     204–210.
2.   John W. Dolan and Vern V. Berry, LC, 1(9): 542–544 (1983).
3.   H. V. Malmstadt, C. G. Enke, and S. R. Crouch, Electronics
     and Instrumentation for Scientists, Benjamin/Cummings Publishing,
     1981.
4.   M. C. Harvey and S. D. Stearns, Analytical Chemistry 56: 837
     (1984).
5.   M. C. Harvey and S. D. Stearns, in Liquid Chromatography In
     Environmental Analysis, (J. F. Lawrence, ed.), The Humana
     Press, 1983, pp. 301–340.
6.   M. C. Harvey and S. D. Stearns, LC 1: 3 (1983).
7.   Chromatography Forum 1: 2 (1986), and Chromatography 2: 1
     (1987).

Index